Bionanotechnology to Save the Environment

Pierfrancesco Morganti (Ed.)

Bionanotechnology to Save the Environment

Plant and Fishery's Biomass as Alternative to Petrol

MDPI • Basel • Beijing • Wuhan • Barcelona • Belgrade

EDITOR
Pierfrancesco Morganti
Dermatology Unit, University of Campania "Luigi Vanvitelli", Naples, Italy;
China Medical University, Shenyang, China;
Director of the R&D Nanoscience Centre MAVI, MAVI Sud Srl, Aprilia (Lt), Italy.

Editorial Office
MDPI
St. Alban-Anlage 66
4052 Basel, Switzerland

For citation purposes, cite each article independently as indicated below:

LastName, A.A.; LastName, B.B.; LastName, C.C. Chapter Title. In *Bionanotechnology to Save the Environment. Plant and Fishery's Biomass as Alternative to Petrol*; Pierfrancesco Morganti, Ed.; MDPI: Basel, Switzerland, 2018; Page Range.

ISBN 978-3-03842-692-9 (Hbk)
ISBN 978-3-03842-693-6 (PDF)

doi:10.3390/books978-3-03842-693-6

Cover image courtesy of Pierfrancesco Morganti.

This book is dedicated to my loving wife Paola, my children Gianluca and Adriana, my daughter-in-law Daniela, my beloved nephew Alessandro and to all the professionals who make the world a healthier and happy place and helped me thinking the education as a discovery of our ignorance.

Who neglects learning in his youth, loses the past and is dead for the future.
– Euripides

The authority of those who teach is often an obstacle to those who want to learn.
– Cicero

Learning without thought is a labor, thought without learning is perilous.
– Confucius

Education is the key to unlock the golden door of freedom.
– George Washington Carver

Contents

About the Editor

Pierfrancesco Morganti is a Temporary Professor of Skin Pharmacology and Applied Cosmetic Dermatology in the Department of Physical and Mental Health and Preventive Medicine of the Dermatology Unit, at the University of Campania "Luigi Vanvitelli" in Naples, Italy, as well as a Visiting Professor at China Medical University of Shenyang.

He is also an industrial expert; in fact, since 1980 he is CEO and R&D Director of the Nanoscience Centre of MAVI sud Srl, an Italian company manufacturing and distributing clinically correct cosmetics.

In 1960 he brilliantly graduated from the University of Roma "La Sapienza", and over the years has taught as temporary professor of Applied Cosmetic Dermatology and Skin Pharmacology at the universities of L'Aquila, Milan, Pavia, and Rome, in addition to chairing many scientific seminars as keynote speaker expert in Cosmetic Dermatology at international conferences in Europe, Russia, China, East Asia, Asia-Pacific and the USA.

His current research interests are focused on the development of biomaterials based on the use of chitin nanofibrils and other polymers obtained from agricultural and industrial waste by a nanotechnological biomimetic approach. Due to this expertise, as R&D Director of MAVI, he has been the coordinator of two innovative European research projects: Chitofarma about the production of non-woven tissues to be used in advanced medications, n-Chitopack about production of biodegradable and compostable food grade packaging, participating also in Bio-mimetic and Polybioskin projects about production of bio-based films to be applied in cosmetic emulsions, and innovative and smart beauty masks to be used as innovative cosmeceuticals.

Author of over 400 publications, including peer-reviewed journal papers, conference proceedings, book chapters, authored books, edited books and patents relevant to cosmetic dermatology and bio-nanomaterials, he is a fellow of the Italian Society of Pharmacology, the European Academy of Dermatology, the American Academy of Dermatology, the Italian Society of Chemistry and Cosmetic Sciences, and he is the co-founder and Secretary General of the International Society of Cosmetic Dermatology.

Founding editor-in-chief of the Journal of Applied Cosmetology, and member of the R&D Commission of Federchimica (Italian Federation of Chemical Industries), in the year 2000 he has been recognized as Academician by the Italian Academy of History of Medicine.

List of Contributors

ANNIBOLETTI, TOMMASO
Department of Plastic and Reconstructive Surgery and Burn Center, Sant'Eugenio Hospital, Rome, Italy.

BARONI, ADONE
Department of Mental Health, Physic and Preventive Medicine, Dermatology-Unit, University of Campania "Luigi Vanvitelli", Naples, Italy.

BROŽOVÁ, LIBUŠE
Institute of Macromolecular Chemistry AS CR, v.v.i., Prague, Czech Republic.

BRUNO, AGOSTINO
Department of Plastic and Reconstructive Surgery and Burn Center, Sant'Eugenio Hospital, Rome, Italy.

CHANOVÁ, ELIŠKA
Institute of Macromolecular Chemistry AS CR, v.v.i., Prague, Czech Republic.

CHEN, HONG-DUO
Department of Dermatology, No.1 Hospital of China Medical University, Shenyang, China.

CHIANESE, ANGELO
Department of Chemical Material Environmental Engineering (DICMA), Sapienza University of Rome, Rome, Italy.

CINELLI, PATRIZIA
Department of Civil and Industrial Engineering, University of Pisa (DICI-UNIPI), Pisa, Italy; National Interuniversity Consortium of Materials Science and Technology (INSTM), Firenze, Italy; Institute for the Chemical and Physical Processes (CNR-IPCF), National Research Council of Italy, Pisa, Italy.

COLLINA, AMILCARE
Responsible for relationships with Scientific Community, MAPEI GROUP. MAPEI S.p.A, 20158 Milano, Italy; Member of Technical Committee, CONFINDUSTRIA (Association representing manufacturing and services companies in Italy), Italy; Member of Research & Innovation Committee, CEFIC (European Chemical Industry Council).

COLTELLI, MARIA-BEATRICE

Department of Civil and Industrial Engineering, University of Pisa (DICI-UNIPI), Pisa, Italy; National Interuniversity Consortium of Materials Science and Technology (INSTM), Firenze, Italy; Institute for the Chemical and Physical Processes (CNR-IPCF), National Research Council of Italy, Pisa, Italy.

CRESTINI, CLAUDIA

Department of Chemical Sciences and Technologies, University of Rome 'Tor Vergata', Rome, Italy.

DEL CIOTTO, PAOLA

R&D Nanoscience Centre MAVI, MAVI Sud Srl, Aprilia (Lt), Italy.

DONNARUMMA, GIOVANNA

Department of Experimental Medicine: Microbiology section, University of Campania "Luigi Vanvitelli", Naples, Italy.

FEBO, PIETRO

Atertek S.r.l., Pescara, Italy.

FORTUNATI, ELENA

University of Perugia, Civil and Environmental Engineering Department, Terni, Italy.

GAGLIARDINI, ALESSANDRO

R&D Consultant: Texol Srl, ICO Srl, Progetto Igien.

GAO, XING-HUA

Head of the Department of Dermatology, No.1 Hospital of China Medical University, Shenyang, China.

GERMINARIO, CLAUDIO

European & Italian Patent Attorney, Società Italiana Brevetti (SIB), Rome, Italy.

GIGANTE, VITO

Department of Civil and Industrial Engineering, University of Pisa (DICI-UNIPI), Pisa, Italy.

HAŠEK, JINDŘICH

Institute of Biotechnology AS CR, v.v.i., Prague, Czech Republic.

KAPRÁLKOVÁ, LUDMILA
Institute of Macromolecular Chemistry AS CR, v.v.i., Prague, Czech Republic.

KELNAR, IVAN
Institute of Macromolecular Chemistry AS CR, v.v.i., Prague, Czech Republic.

KENNY, JOSÉ M.
University of Perugia, Civil and Environmental Engineering Department, Terni, Italy.

KOBERA, LIBOR
Institute of Macromolecular Chemistry AS CR, v.v.i., Prague, Czech Republic.

KOVÁŘOVÁ, JANA
Institute of Macromolecular Chemistry AS CR, v.v.i., Prague, Czech Republic.

KUBIES, DANA
Institute of Macromolecular Chemistry AS CR, v.v.i., Prague, Czech Republic.

KUNAVER, MATJAŽ
National Institute of Chemistry, Ljubljana, Slovenia.

KUŽEL, RADOMÍR
Faculty of Mathematics and Physics, Charles University, Prague, Czech Republic.

LANGE, HEIKO
Department of Chemical Sciences and Technologies, University of Rome 'Tor Vergata', Rome, Italy.

LAZZERI, ANDREA
Department of Civil and Industrial Engineering, University of Pisa (DICI-UNIPI), Pisa, Italy; National Interuniversity Consortium of Materials Science and Technology (INSTM), Firenze, Italy; Institute for the Chemical and Physical Processes (CNR-IPCF), National Research Council of Italy, Pisa, Italy.

LI, YUAN-HONG
Department of Dermatology, No.1 Hospital of China Medical University, Shenyang, China.

MALLEGNI, NORMA
Department of Civil and Industrial Engineering, University of Pisa (DICI-UNIPI), Pisa, Italy; National Interuniversity Consortium of Materials Science and Technology (INSTM), Firenze, Italy; Institute for the Chemical and Physical Processes (CNR-IPCF), National Research Council of Italy, Pisa, Italy.

MIKEŠOVÁ, JANA
Institute of Macromolecular Chemistry AS CR, Prague, Czech Republic.

MORGANTI, GIANLUCA
Nanoscience Center MAVI , Rome, Italy.

MORGANTI, PIERFRANCESCO
Professor of Skin Pharmacology, Postgraduate School in Dermatology and Venereology, Dermatological Unit, University of Campania "Luigi Vanvitelli", Naples, Italy; Visiting Professor, Department of Dermatology, No.1 Hospital of China Medical University, Shenyang, China; Director of the R&D Nanoscience Centre MAVI, MAVI Sud Srl, Aprilia (Lt), Italy; Department of Dermatology, No.1 Hospital of China Medical University, Shenyang, China.

MORONI, SIMONE
Department of Plastic and Reconstructive Surgery and Burn Center, Sant'Eugenio Hospital, Rome, Italy.

NETOPILÍK, MILOŠ
Institute of Macromolecular Chemistry AS CR, v.v.i., 162 06 Prague 6, Czech Republic.

NUNZIATA, MARIA LUISA
MAVI Sud, Aprilia (LT), Italy.

PALMISANO, PIETRO
Department of Plastic and Reconstructive Surgery and Burn Center, Sant'Eugenio Hospital, 00144 Rome, Italy; Biochemtex S.p.A., Rivalta Scrivia (AL), Italy.

PALOMBO, MARCO
Department of Plastic and Reconstructive Surgery and Burn Center, Sant'Eugenio Hospital, Rome, Italy.

PALOMBO, PAOLO
Head of the Department of Plastic and Reconstructive Surgery and Burn Center, Sant'Eugenio Hospital, Rome, Italy.

PAOLETTI, IOLE
Department of Experimental Medicine: Microbiology section, University of Campania "Luigi Vanvitelli", Naples, Italy.

PAVLOVÁ, EWA
Institute of Macromolecular Chemistry AS CR, v.v.i., Prague, Czech Republic.

PEKÁREK, MICHAL
Institute of Macromolecular Chemistry AS CR, v.v.i., Prague, Czech Republic.

PERFETTO, BRUNELLA
Department of Experimental Medicine: Microbiology section, University of Campania "Luigi Vanvitelli", Naples, Italy.

PIGLIONE, MARIA CHIARA
Biochemtex S.p.A., Rivalta Scrivia (AL), Italy.

SEDLÁKOVÁ, ZDENKA
Institute of Macromolecular Chemistry AS CR, v.v.i., Prague, Czech Republic.

SELLETTI, SONIA
Studio Legale Astolfi e Associati, Milan, Italy.

ŠPIRKOVÁ, MILENA
Institute of Macromolecular Chemistry AS CR, v.v.i., Prague, Czech Republic.

STOLLER, MARCO
Department of Chemical Materials Environmental Engineering, Sapienza University of Rome, Rome, Italy.

TISHCHENKO, GALINA
Institute of Macromolecular Chemistry AS CR, v.v.i., Prague, Czech Republic.

TUFANO, MARIA ANTONIETTA
Head of the Department of Experimental Medicine: Microbiology section, University of Campania "Luigi Vanvitelli", Naples, Italy.

XU, XUE-GANG
Department of Dermatology, No.1 Hospital of China Medical University, Shenyang, China.

Foreword

Nanotechnology is the science of manipulating atoms and molecules in the nanoscale thousand times smaller than the width of a human hair.

The world market for products that contain nanomaterials is expected to increase enormously in the future. The use of nanotechnology has stretched across various streams of science, from electronics to medicine and has also found applications in the field of cosmetics.

How will this revolution impact our lifestyle and our planet?

Very often the progresses of science, human knowledge and evolution of our lifestyle has been associated with devastating effects on our forests, oceans and more in general on our planet. The real challenge in the years to come is the sustainability of human evolution.

The reader of this interesting book will discover how nanotechnology, and in particular nanomaterials derived from plant biomass and fishery's waste, can improve the quality of our environment by reducing carbon emissions, improving the recycling of materials and even, in the long run, became a profitable business.

Green nanotechnologies can be applied to a huge number of products ranging from intelligent textiles to smart drugs or functional polymers which can have a big impact on our daily lives, but nevertheless help us in saving our biodiversity and our planet.

However, to fully achieve all these benefits, companies and scientists should be supported by National and International Agencies and Institutions in order to facilitate and support scientific development in this field allowing from one side the protection of intellectual property, but on the other giving accessibility of these technologies to emerging countries for improving the quality of life and the environment all over the world equally.

Enzo Berardesca, MD
Director Clinical Dermatology
San Gallicano Dermatological Institute
Rome, Italy

Preface

The main aim of the current policy is to keep the environment in orbit with economics at the center, not considering *nature* or *environment* as simple commodities. Preserving planet Earth's biodiversity is crucial in order to keep its ecosystems in equilibrium.

For this purpose, it is necessary to produce goods and tools using bio and eco-compatible methodologies whilst also increasing knowledge on the concept of industrial sustainability. In fact, sustainability has to be based on the 3P pillars: Planet, People and Profit, i.e., (a) the preservation of the planet's environment; (b) the respect of people's safety and well-being, to meet the social expectations; (c) the maintenance of the industrial profit to manufacture and compete at a global level.

Without entering into any kind of neoliberal education policy, this book aims to present new ideas to encourage those in the manufacturing industry to use both industrial and agricultural biomass in order to produce goods in a greener way. It may also encourage scientists and marketing professionals to educate the consumer on the necessity of maintaining biodiversity without impoverishing our planet of crucial raw materials.

The book is organized in the following five parts:

Part I *INTRODUCTION*, where, in *Chapter 1*, the necessity to recycle waste using bionanotechnology is reported, underlining that our health, together with economical and social progress, is closely linked to the quality of the environment. For this purpose, the strategic necessities to use plant biomass and waste from fisheries as raw materials are reported and discussed in six chapters (Chapters 2 to 7) in **Part II** *BIOMASS AS RENEWABLE RAW MATERIAL FOR A GREEN ECONOMY*.

Chapter 2 deals with the current Chemical Industry Strategy of Innovation, based on the necessity to maintain the ecosystem and biodiversity of our earth by increasing the use of by-products, as basic raw material for producing goods. Sustainability, in fact, must be the key factor dictating the progress of the European Chemical Industry in the next decade.

Chapter 3 describes the quality and quantity of available biomass as a basic source of raw material; therefore, much effort has to be devoted to converting it into useful industrial and commercially viable products. Biomass represents, in fact, an immense and renewable source for the production of bio-fuels and valuable ingredients, despite only a small amount being utilized to make goods. For this reason, *Chapter 6* reports the most fundamental chemicals and fuels that can be produced by a bio-refinery, simply by processing feedstock rich in cellulose and lignin. Therefore, the use of biomass in nanobiotechnological processes is useful to achieve a sizeable bio-green economy as a base for our future progress, indispensable to reduce the social differences as well. In *Chapters 4* and *5*, availability, production and use of Chitin, Cellulose, and Lignin are discussed. These polymers, which represent the most abundant natural and underutilized materials present in the world, are precious raw materials that are useful to the manufacture of many kinds of products.

Chitin Nanofibrils (CN), obtained from crustacean waste, and lignin, from plant biomass, have been shown to be interesting natural polymers useful for producing, for example, non-woven tissues to make innovative beauty masks and/or advanced medications. It is worth noting that the use of nanocomposites made from the CN–Lignin block copolymeric micro/nanoparticles has shown to be effective for producing non-woven tissues and emulsions with interesting anti-ageing and/or anti-inflammatory effectiveness. During the polymerization process—necessary to produce the micro/nanoparticles—it is possible, in fact, to entrap different active ingredients to characterize the activity of the final products.

Chapter 7 explains and reviews the definition of bio-based, recyclable, biodegradable and compostable materials by the current standards and legislation. The general assessment of biodegradability of a plastic material, in fact, is not sufficient for fulfilling the requirements of the consolidated international standard for composting waste. Indeed, if not properly planned and addressed, the production of compostable materials might be less sustainable than that of petrol-derived, non-biodegradable polymers.

Part III *FUNDAMENTALS*, comprises five chapters, reporting and discussing (from Chapters 8 to 12) the physicochemical characteristics of the principal biopolymers obtained by the plant biomass, such as cellulose and lignin, together with the more known chitin from fisheries' waste. Thus, *Chapter 8* introduces the general characteristics and peculiarities of lignin as a biopolymer, presenting its structural issues. For this purpose, the techniques necessary to investigate these interesting macromolecules are reported, highlighting the latest technologies created to isolate lignin from plant biomass, controlling and valorizing its interesting polyphenolic structure.

Chapter 9 is focused on the biomedical applications of cellulose nanocrystals, highlighting their use in drug-delivery systems and in tissue engineering. Cellulose, in fact, as a natural polymer and biomaterial, plays an important role in regenerative medicine to control cellular activities and functions, when used to produce and reinforce non-woven tissues, because of its non-toxic and biodegradable properties.

Chapter 10 reports the structural (morphological and crystalline) and dynamic (rheological) properties of chitosan (CS) and chitin nanofibrils (CNs), used to produce composite films by casting technology. The rheological tests provide useful information about the rheological changes of CS/CN slurries over time, which are necessary to determine the limit of their storage. Based on the analysis of the permeability for gases and water vapor, thermal and mechanical stability of these innovative films it can be concluded that they are suitable as disposable packages for dry products. Replacing cellulose paper with CS/CN-based films may have an effect on reducing deforestation and the associated climate change.

Chapter 11 gives an excursus on the possible use of biodegradable raw materials to make safe baby diapers. It shows the historical use of these diapers for protecting the baby skin from the common rash problem, reporting the different shapes used during the years and the safety and security of the actual based baby diapers.

Chapter 12 reports data and operating conditions of the electrospinning technique used to produce non-woven-tissues by a blend of chitin nanofibrils and other natural polymers.

Part IV *APPICATIONS* contains five chapters (from Chapters 13 to 17), reporting the industrial applications that chitin and its derivatives could have in different and important economical fields, such as cosmetics, advanced medications, and food packaging.

Chapter 13 describes the safeness and effectiveness of the block co-polymeric chitin nanofibril–hyaluronan (CN–HA) as a skin anti-wrinkling agent, underlining its efficacy to neutralize the free radical activity and to regularize the correct cell turnover. Moreover, it has also been shown that the controlled delivery and release of active ingredients throughout the skin layers is of fundamental importance to achieve the effectiveness of the topically applied products.

Chapter 14 highlights the recent researches in tissue engineering, underlining the different aspects of chitin-based nanocomposites to produce skin-friendly scaffolds. *Chapter 15* reports the employment of different polymers obtained from biomass in the production of soft container packaging to reduce waste production and related greenhouse emissions. *Chapter 16* is focused on the possibility to use bio-based polymers for the industrial processing of hard containers for food and cosmetic purposes, reporting the more recent technologies adopted for maximum valorization of bio-based polymers. This application outlines the potentiality and the growing attention, not only from researchers but also from producers and consumers, towards bio-based materials. The modification and processing of bio-based products with additives, polymers, and natural fibers is also discussed.

Part V *BIOLOGICAL ACTIVITY, SAFETY,* PATENT APPLICATIONS AND LEGAL CONSIDERATIONS, consisting of four chapters (from Chapters 17 to 20), reports update studies, in vitro and in vivo, showing the safeness and effectiveness of chitin nanofibrils used for medical purposes, underlining the necessity to patent all the innovations achieved.

In the last few years, many studies have focused their attention on the biomedical application of natural biocompatible polymers used to produce nanocomposites for tissue regeneration.

Chapter 17 is therefore focused on the study of the biological properties of the new polymeric nanoconstructs which, being biodegradable and biocompatible, have the capacity to exploit the body's natural biological response, at the same time respecting the environment equilibrium. This is the reason why chitin nanofibrils and lignocellulosic polymers, possessing these characteristics, are at the center of many therapeutic applications.

Chapter 18 reports the latest in vivo studies on innovative non-woven tissues made by chitin nanofibrils entrapping nanostructured silver, in a very low dose. These particular tissues, typically applied for a period of 6 days on skin affected by first- and second-degree burns, resulted in quickly regenerating the skin tissues, temporarily slowing down the bacterial growth, without causing any toxic side effect. It is worth noting that the non-woven tissues, made prevalently by chitin nanofibrils, have shown *in vitro,* in keratinocyte cultures, an interesting effectiveness to modulate the cell production of defensines, balancing the imetalloproteinases activity also.

The last two chapters are concerned with the EU regulations on c osmetic products and the use of nanomaterials, underlining the necessity to protect the industrial innovations by organized patent applications. *Chapter 19* outlines the main features of the recent European recast of the Cosmetic Regulation, touching upon the use of nanomaterials and the necessity to respect the environment's equilibrium. *Chapter 20* highlights how the protection of nanotechnological inventions implies a new interpretation and application of the general requirements of patent ability.

This book represents a good asset for graduate students, researchers, academicians, and industrial experts working in the field of natural polymers who wish to maintain the biodiversity of our planet, improving our quality of life by the use of green bionanotechnologies.

Pierfrancesco Morganti
Editor

Acknowledgments

I express my sincere gratitude to Dr. Maria Luisa Nunziata for the continuous support given during the different steps of this work.

I thank the Publisher for giving me the opportunity to write this book, and all the professionals who accepted to be part of it, contributing to enrich the content with their great experience. Moreover I have to thank Biobased Industries and Horizon 2020 for the fund received to support part of the reported studies.

Last but not least, I'd like to thank TEXOL Srl, Italy for the support given in the making of this book.

PART I
Introduction

Nanotechnology, Nanobiotechnology, and the Environment

Pierfrancesco Morganti

Abstract: The lack of industrial planning and management of the purchase, storage, preparation, and reuse of food and food packaging, as well as consumers' food wastage habits, are at the heart of the environmental waste problem. Thus, there is a need to increase public and private investments in R&D for the bionanotechnological valorization of marketable ingredients that are available in food waste, in addition to a need for the introduction of better environmental regulations, scholarly teaching, and mass media communications. In conclusion, an innovative policy to reduce, reuse, and recycle waste will ameliorate the environment and better our quality of life.

1. Introduction: The Environment and the Waste Problem

According to the United Nations Environment Programme (UNEP) [1], human health as well as economic and social progress are closely linked to the quality of the environment. Thus, "investing in improved natural resources", such as plant biomass and fishery waste, in addition to reducing carbon emissions and better planning of urban areas, could "generate high social rates of return" [2].

In fact, better environmental regulations and innovative policies can generate new forms of capital, pushing industry to make green technological progress. This is why UNEP research suggests that "an investment scenario of allocating 2% of global GDP to greening economic sectors will produce a higher global Gross Domestic Product (GDP), compared to the business-as-usual scenario, within 10 years." Moreover, reduced production with organized waste recycling, as well as the production and use of goods made of local, renewable materials, will have increased strategic importance to assure the availability of raw materials and drastically reduce the environmental impact of production.

Thus, it is necessary to reduce the production of waste by the imporoved design of goods and packaging, which also favours global recyclability. In so doing, it will be possible to increase products' shelf-life and gear people's mentality towards reuse, through the mass media and R&D involvement.

During the period 1964–2014, global plastic production increased from 15 to 311 million tons, the major portion of which was used for disposable packaging, with a lost material value of US$80–120 billion per year [3]. Thus, it has been estimated that only 5% of plastic packaging is retained for subsequent use, being used one time only [3,4]. For these reasons, plastics and their by-products are littering cities, oceans, and waterways, also contributing to health problems because of the durability of the involved polymers that accumulate as debris (Figure 1). Additionally, it is important to underline that only 14% of plastics are actually recycled by melting and the adopted methodologies are not only harmful to the environment, but also present health threats to people who come in contact with this material [3–5].

Figure 1. Plastic debris accumulated on land and in the ocean.

Plastic melting, in fact, generates volatile organic compounds (VOC) and fumes which are harmful to plant and animal life, near industrial sites. Moreover, the relative carbon emissions contribute to global warming. Finally, since plastic is not classified as a hazardous material, its recycling does not come under any international regulations, further complicating the efforts to solve the waste problem [3–5]. Thus, there is a necessity to produce and use more bioplastics (Figure 2), which unfortunately represent only 1% of the global plastic production (Figure 3) [6].

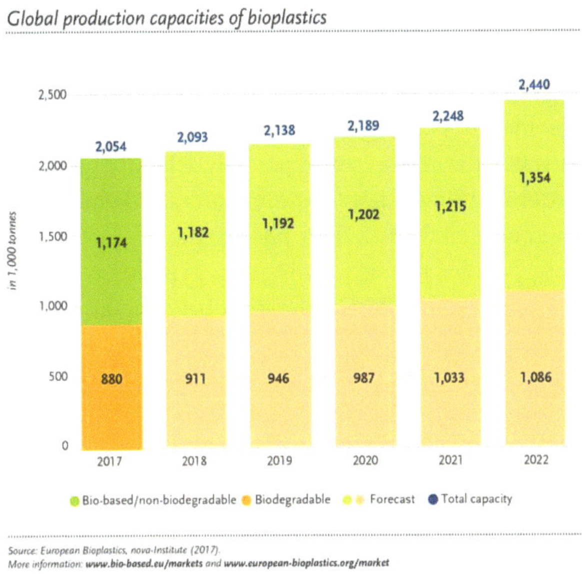

Figure 2. Bioplastics production.

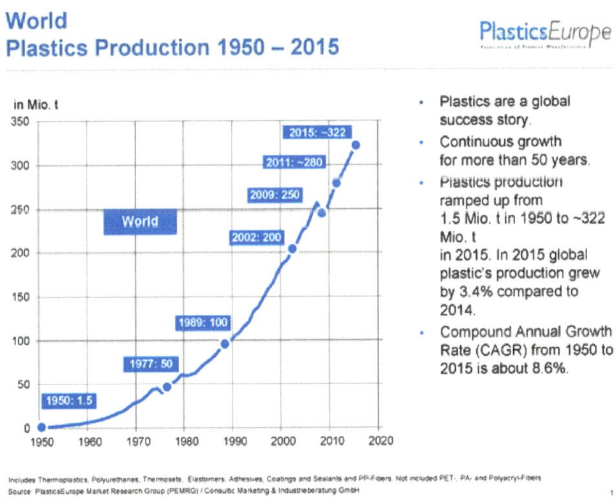

Figure 3. The global production of plastics.

R&D is therefore a fundamental pillar for the creation of novel products that, based on the use of new bionanotechnological science, are capable of profoundly changing today's economy and improving our standards of living. This is why the development of a bionanotechnological product normally starts with an innovative way of thinking that can offer better goods for human welfare.

On the one hand, it will be necessary to avoid food waste by improving consumers' household food management behavior, handling its know-how together with the knowledge of all the stakeholders and the supply chains involved [7]. On the other hand, the industry has to optimize the distribution chains and extraction conditions of food waste through environmentally-friendly biochemical and physical technologies, in order to utilize all of the potentially marketable components present in these discarded goods, as well as their packaging.

Continuous support for public funding and private investments are needed to further develop and increase the R&D means necessary to ameliorate the production, consumption, processing, storage, recycling, and disposal of food and food waste [8]. Towards this purpose, we must remember that 20–25% of food wasted is due to packaging matters. As an example, waste material from coffee-pods is considered to be represented by more than 20 billion pieces/year worldwide, with an annual growth of more than 18% per year over 2011–2016 [9]. Moreover, it must be underlined that in order to produce 1 kg of these capsules, 4 kg of water and 2 kg of petrol must be used together with 22 kW of energy!

In conclusion, along with North America, Western Europe is the chief region for where coffee drinking is a main driver. Thus, to address the problem of sustainability, it is necessary to produce innovative biodegradable pods by increasing the R&D studies on this topic [9].

2. Culture of Knowledge, R&D and Nanobiotechnology

It is fundamental to diffuse the culture and knowledge of these innovative, environmentally friendly nanomaterials and nanotechnologies by scholarly education, and by the use of newspapers, television and any other means of mass media communication.

It is necessary to remember that nanotechnology and bionanotechnology, based upon molecular self-assembly with nanoscale dimensions (a nanometer is a billionth of a meter), deal with the study and application of biological and biochemical activities from the elements of nature to make new and advanced products.

Nature, which self-assembles molecules by biological processes to create complex structures with nanoscale precision, is in fact the best nanotechnology platform, having developed a large array of different materials [10]. They range from feathers to shells, wood, bone and many other macro-, micro- and nanostructures that confer specific properties to materials, such as colour, permeability, elasticity, strength, etc. Inside the cell, for example, there are numerous self-assembled structures that encapsulate specific biomolecules to be released as a consequence of molecular signaling. The human body is composed of soft and hard structures such as skin or bone, made of different kinds of polymers and block copolymers, hydrophilic or hydrophobic, assembled into ordered architectures in the nanoscale regime. Thus, depending on the concentration and the volume ratio between insoluble and soluble blocks, they can form nanoparticles and thin films, used as signalling molecules or cell membranes. Other natural structures, made of chitin crystals, are capable of mimicking different colours found in bird plumage, butterfly wings, or rose flowers (Figure 4) [10,11].

Figure 4. Chitin crystals in birds, butterflies and flowers are capable of mimicking different kinds of colours.

The chitin polymers, organized in a nanostructure, give rise to photonic crystals that reflect sunlight in the presence of water so that the colours that appear depend on the angle from which we observe the animal. This phenomenon inspired the production of biomimetic nanoparticles in cosmetics and drug delivery, as well as the use of crystal chitin to produce innovative smart textiles [10,11].

3. Nanotechnology and Nanoscale

Chitin and chitosan have recently [12–14] found applications in cellular technology for one-, two- and three-dimensional matrices, which have shown good adhesion to stem cells, favouring cell proliferation and differentiation. Additionally, chitosan is a very promising material for the production of absorbable surgical sutures or threads because of its strength and flexibility. However, these properties have to be well preserved upon contact with aqueous media. Thus, a better kinetic investigation of the chitosan fibre resorption in vivo is required to best predict the properties of this type of product.

For all of these reasons, nanomaterials, as cornerstones of nanoscience and nanotechnology, have the potential to revolutionize the ways in which goods and products are created [14,15]. Today they can be found in cosmetics, food, sporting goods, stain-resistant clothing, tires and electronics, and they are also used in medicine for the purposes of diagnosis, imaging, and drug delivery.

Industrially, nanomaterial synthesis and processing may be by two different approaches: a bottom-up method based on the assembling of atoms or a top-down one that breaks or dissociates bulk solids into finer pieces, constituted of a few atoms [15,16]. Milling is a typical top-down method of making nanoparticles, whereas colloidal dispersion is a good example of the bottom-up approach in the synthesis of nanoparticles. These innovative materials, with structural features between those of atoms and bulk materials, possess special characteristics due to their nanosize, higher surface area to volume ratio, high surface energy, spatial confinement and reduced imperfections.

Other fascinating and useful aspects of nanomaterials are their optical and electrical properties, which depend on parameters such as feature size, shape, surface characteristics and other variables necessary to interact with the surrounding environment or other nanostructures.

In conclusion, nanobiotechnology is to be considered a marriage between biological science and engineering methodologies. It seeks to study and apply the biological and biochemical activities of elements in nature to fabricate new goods. The use of waste materials is an important way to protect the environment [17]. Environmental biotechnology, a system based on developing our scientific and engineering knowledge, is related to the use of microorganisms in the prevention of environmental pollution [18].

However, the fabrication and assembly of nanoscale materials, goods and devices capable of conferring health and safety benefits while maintaining the global ecosystem and biodiversity, still represent a major challenge [19]. With this book we have tried to propose some solutions to this challenge.

Acknowledgments: We thank the EU for the economic support given to the European Research Projects: n-Chitopack grant no. 315233, and Biomimetic grant no. 282945.

Conflicts of Interest: I work as Head of the R&D Centre of Nanoscience, Mavi Sud, s. r. l, Italy.

References

1. UNEP (United Nations Environment Programme). *Building Inclusive Green Economy for All*; Joint Agency Paper; UNEP: Geneva, Switzerland, 2012.
2. World Bank. *Inclusive Green Growth: The Pathway to Sustainable Development*; World Bank: Washington, DC, USA, 2012; Available online: www.worldbank.org/inclusivegreengrowth (accessed on 2 May 2018).
3. Hopewell, J.; Dvorak, R.; Kosior, E. Plastics recycling: Challenges and opportunity. *Philos. Trans. R. Soc. B* **2009**, *364*, 2115–2126. [CrossRef] [PubMed]
4. Mc Arthur, D.E.; Waughray, D.; Stuchtey, M.R. *The New Plastics Economy: Rethinking the Future of Plastics*; Ellen McArthur Foundation: Chicago, IL, USA, 2016.
5. Hartman, D. *The Disadvantages of Recycling Plastics*; Sciencing: Chicago, IL, USA, 2017; Available online: https://sciencing.com/disadvantages-recycled-plastics-7254476.html (accessed on 2 May 2018).
6. European Bioplastics. Bioplastics Market Data 2017. Available online: www.european-bioplastics.org/market (accessed on 2 May 2018).

7. Aschemann-Witzel, J.; De Hooge, I.; Aman, P.; Bech-Larsen, T.; Oostindjer, M. Consumer-Related Food Waste: Causes and Potential for Action. *Sustainability* **2015**, *7*, 6457–6477. [CrossRef]
8. Barra, M. *The Global Coffee Pods Market in 2017: The End of an Era?* Euromonitor Report. 2017; Euromonitor International: London, UK, 2017.
9. Baiano, A. Recovery of Biomolecules from food Waste—A Review. *Molecules* **2014**, *19*, 14821–14842. [CrossRef] [PubMed]
10. Morganti, P. Nanoparticles and Nanostructures Man-Made or Naturally recovered: The Biomimetic Activity of Chitin Nanofibrils. *J. Nanomater. Mol. Nanotechnol.* **2012**, *1*, 1–4. [CrossRef]
11. Park, B.K.; Kim, M.-M. Application of Chitin and its Derivatives in Biological Medicine. *Int. J. Mol. Sci.* **2010**, *11*, 5152–5164. [CrossRef] [PubMed]
12. Elieh-Ali-Komi, D.; Hamblin, R. Chitin and Chitosan: Production and Application of Versatile Biomedical Nanomaterials. *Int. J. Adv. Res.* **2016**, *4*, 411–427.
13. Tuyshime, P.; Lee, B.H.; Nsanzabera, F. Current Status and New Perspectives on Chitin and Chitosan as Functional Biopolymers. *Appl. Biochem. Biotechnol.* **2016**, *181*, 1314–1337. [CrossRef]
14. Kinoshita, S. *Structural Colors in Realm of Nature*; World Scientific Press: Singapore, 2008.
15. Ozin, G.A.; Arsenault, A.C. *Nanochemistry: A Chemical Approach to Nanomaterials*; 20 Royal Society of Chemistry Pub: Cambridge, UK, 2005.
16. Cao, G.; Wang, Y. *Nanostructures and Nanomaterials*, 2nd ed.; World Scientific: London, UK, 2014.
17. Alamkhan, F. *Biotechnology Fundamentals*; CRC Press: New York, NY, USA, 2012; ISBN 978-1-4398-2009-4.
18. Jeyabalan, S.; Devarajan, T.; Muniswamy, D.; Mohd, A.A. *Environmental Biotechnology*; AAP Academic Press: New York, NY, USA; CRC Press: New York, NY, USA, 2016; ISBN 9781771883627.
19. Ramalingam, M.; Ramakrishna, S.; Best, S. *Biomaterials and Stem Cells in A Regenerative Medicine*; CRC Press: Boca Raton, FL, USA, 2012.

PART II

Biomass Renewable Raw Materials for a Green Economy

Sustainable Products: The Innovation Strategy for the European Chemical Industry

Amilcare Collina

Abstract: Sustainability will be the key factor which will dictate the progress of the European Chemical Industry in the next decade. Looking at process sustainability—i.e., sustainable manufacturing technologies—we can observe a high level of consciousness as well as behavior aimed at environmental sustainability. As far as product sustainability is concerned, the pace of change is still too slow, and the understanding of what is required for companies to become genuinely sustainable enterprises is often lacking. The European chemical industry has the challenge to become the most sustainable industry in the world by choosing the road which leads to product innovation considering sustainability as the guiding star for this path. Sustainability may represent a factor for competitiveness, as long as the characteristics of sustainability are explicitly illustrated in the performances of the products, and are acknowledged by the market as a value.

1. Introduction

The environment in which today's chemistry-using industries operate has changed enormously in recent years and will continue to change in the future [1]. The challenges of sustainability take many forms: the globalization of markets and of manufacturing infrastructures; the growth in demand for chemical products in developing countries; climate change and the energy debate; a shift in society's concerns, and those of individual consumers, from the production of chemicals to chemical product concerns relating to human health and environmental impact; a growing awareness throughout society of the resource limitations of our planet; the increasing shift in developed countries from bulk chemicals to specialty chemicals to custom-formulated products.

2. Sustainability: The Contribution of the Chemical Industry

Chemical industry stakeholders recognize and demand all the benefits that this industry provides—in health, nutrition, clothing, housing, transportation, communication—but they want all these benefits with none of the negatives.

Major retailers all report heightened consumer interest in the potential impacts on their health and on the environment of chemical residues in food, packaging materials, hygiene and personal care products, and so on.

In this scenario, the chemical industry claims to be science and technology, enabling a number of innovations to provide solutions for the overall objective of conserving natural resources and the environment and protecting people's health: to sum up, aiming to improve the quality of life.

In other words, the chemical industry claims to be part of the solution, which is true, and does not fully recognize that it is part of the problem, which is also true. As a consequence, the chemical industry is not perceived by the general public and by policymakers, as "sustainable" and people are wondering why the claimed capabilities of this industry are not focused first on improving its own sustainability.

Sustainability must be the key factor which will dictate the progress of the European chemical industry in the next decade. Sustainability depends upon maintaining and where possible, increasing stocks of certain capital assets, so that we can learn to live off the income from that capital asset without depleting the capital asset itself [2]. There are five types of capital needed by our economy (and each and every organization operating within that economy) in order to function properly [3].

Natural capital is any stock or flow of energy and matter that yields valuable goods and services. It includes resources, some of which are renewable (timber, grain, fish and water), whilst others are not (fossil fuels); sinks that absorb, neutralize or recycle wastes (for example, forests sequestrating carbon dioxide; and services, such as fertility building or climate regulation. Natural capital is the basis not only of all production in the human economy, but of life itself.

Human capital consists of people's health, knowledge, skills and motivation, all of which are required for productive work. Enhancing human capital (for instance, through investment in education and training) is central to a flourishing economy.

Social capital comprises all the different cooperative systems and organizational frameworks people use to live and work together, such as families, communities, governments, businesses, schools, trade unions, voluntary groups, and so on. Although they involve very different types of relationships and organizations, they are all structures or institutions that add value to human capital, and tend to be successful in doing so if based on mutual trust and shared purpose.

Manufactured capital comprises material goods or fixed assets—tools, machines, buildings and other forms of infrastructure—which contribute to the production process, but do not become embodied in its output.

Financial capital plays an important role in our economy, by reflecting the productive power of the other types of capital, and enabling them to be owned and traded. However, unlike the other types, it has no intrinsic value; whether it is in the form of shares, bonds or banknotes, its value is purely representative of natural, human, social or manufactured capital.

These five forms of capital, judiciously combined by entrepreneurs, are the essential ingredients of all economic activity. Natural capital, despite the sophistication of our modern economy, is still absolutely necessary to maintain a functioning biosphere, to supply resources to the economy, and to help dispose of its wastes.

Human capital provides the knowledge and skills that create manufactured capital, and operate it effectively. Social capital creates the institutions that provide the stable context and conditions within and through which economic activity can take place, and which enables individuals to be vastly more productive. Financial capital provides the lubricant to keep the whole system operating.

The chemicals industry is central to the pursuit of a sustainable society; without it, the prospects of sustainably meeting the needs of nine billion people by the second half of this century are zero.

In operational terms, this combination of responsible care, eco-efficiency and corporate social responsibility has provided many of the world's leading chemical companies with a solid starting point for addressing the challenge of sustainable development.

Looking at process sustainability—i.e., sustainable manufacturing technologies—we can observe a high level of consciousness as well as a behavior aimed at environmental sustainability. On the basis of the significant results obtained, we can conclude that the European chemical industry is on the right track.

As far as product sustainability is concerned, the pace of change is still too slow, and the understanding of what is required for companies to become genuinely sustainable enterprises is often lacking.

This is the heart of the matter.

3. Product Sustainability: The Heart of the Matter

Europe can gain a leadership position for the following reasons:

The market is ready to accept new sustainable products; moreover it is asking for such a product.

A huge market is there.

For a few years, REACh—Regulation on Registration, Evaluation, Authorization and Restriction of Chemicals—has been in force.

The main aim of REACh is to ensure a high level of protection of human health and the environment from the risks that can be posed by chemicals. REACh makes the industry responsible for assessing and managing the risks posed by chemicals and providing appropriate safety information to their users [4].

Europe has a lead over other countries. Without a doubt, the new regulations will force the chemical industry to carry out a severe, costly assessment on how to implement them.

It is an opportunity to take advantage of, not a threat.

The European chemical industry is at a crossroads. It may choose the road which leads to resisting the changes, by trying to postpone the obligations required by the new standards and to ignore the demand from the market for as long as possible, but this is only a short-sighted, losing approach.

Short-sighted in that, even if it managed to postpone the changes to a date yet to be defined, it would never be able to elude them; they are required by the general public and they are clear market needs.

A losing choice because, by focusing its efforts in this direction, the industry would lose sight of the crux of the problem: namely, reduced competitiveness and lower prospects compared with newly-industrialized countries, where the production costs are considerably lower, and will remain so for a number of years.

What, then, is the alternative?

To bravely choose the road which leads to product innovation, considering sustainability as the guiding star for this path. The current state of awareness of the environment and safeguarding our health imposes behavior which is aimed at sustainability, and encourages a wide-reaching innovation program that regards the sustainability of products as its main priority.

There are four key actions that the chemical industry has to take in order to implement this strategy:

Define with the stake-holders (including NGO) a charter of sustainability for a chemical product. Commit to avoiding the introduction into the market of new products unless they comply with the defined charter. Commit to substituting, in a reasonable timeframe which has to be clearly defined, all the existing products that do not comply with the defined charter (starting with Persistent, Bioaccumulative and Toxic substances) with new sustainable products. Commit to manufacturing its products using sustainable manufacturing processes and technologies.

If the chemical industry shows concrete, credible signs that it wishes to follow this road, support from the stakeholders will not be lacking, to guarantee a transparent, scientifically-based risk assessment program, along with support from the policymakers for the high investments which the road to innovation brings.

Sustainability may represent a factor for competitiveness, as long as the characteristics of sustainability are explicitly illustrated in the performance of products, and are acknowledged by the market as a value, which means information and education strategies aimed at consumers and industrial users, the definition of a coherent standards program and a legislative program which provides incentives for sustainable solutions.

In so doing, the European chemical industry may construct legitimate barriers against the entry of products from other markets with sustainability characteristics that do not meet the European standards.

The European chemical industry has the challenge to become the most sustainable industry in the world.

Europe has a solid base of tradition, culture, knowledge, business capacity and resources to take up this road successfully. One should not forget that the chemical industry was born in Europe, and that Europe still claims a high position for the value of the production of chemicals in the world.

Conflicts of Interest: The author declares no conflict of interest.

References

1. Forum for the Future. Available online: https://www.forumforthefuture.org (accessed on 6 November 2017).
2. Chemistry Leadership Council. *A Vision for the Sustainable Production & Use of Chemicals*; Forum for the Future: London, UK, 2008. Available online: https://www.forumforthefuture.org//chemistry-leadership-council (accessed on 6 November 2017).
3. Five Capitals Model. Available online: https://www.forumforthefuture.org//five-capitals-model.pdf (accessed on 6 November 2017).
4. REACh. Available online: https://ec.europa.eu/environment/chemicals/reach/reach_intro.htm (accessed on 6 November 2017).

Biomass Waste—A Source of Raw Materials

Matjaž Kunaver

Abstract: Biomass represents an immense and renewable source for the production of bio-fuels and valuable chemicals. A little amount of this is used in industry and the remaining is leftover in huge quantities. Agricultural crop residues, such as straw, corn stover, and wood and wood wastes such as leftovers from timber cutting, broken furniture, sawdust, residues from paper mills etc. contain appreciable quantities of cellulose, hemicelluloses, and lignin. Much effort has been devoted to converting these types of biomass into useful industrial and commercially viable products. In recent years, some effective processes have been found, such as thermochemical conversion, producing several platform chemicals from these renewable resources. Lignocellulosic biomass contains cellulose, hemicellulose, and lignin in various proportions. The first step towards the conversion of these basic ingredients into different basic chemicals is their separation. Each of these three biomass components can then be converted into different platform chemicals. Different reaction pathways and final products with high added value will be presented. One of possibilities of converting biomass is the liquefaction in glycols using mild acid catalysts. During the liquefaction reaction, lignocellulosic components are depolymerised to low molecular mass compounds with high reactivity, high hydroxyl group content, and can be used in many useful applications. The liquefied biomass can be used as a feedstock in the synthesis of polyesters, polyurethane foams, and adhesives with properties similar to those produced from crude oil feedstock. Nanocellulose, a carbon neutral and sustainable material with outstanding mechanical properties, is one of the most promising materials and has been a subject of recent interest. An overview of some applications and methods are presented in this contribution. The utilization of liquefied lignocellulosic materials can at least partially reduce the crude oil consumption, thus increasing the use of the renewable resources in large extent.

---

1. Introduction

Biomass-based materials and wood in particular are among the more abundant renewable resources. Much effort has been devoted to converting these types of biomass into useful industrial and commercially viable products. Recently, considerable attention has been given to the preparation of environmentally friendly products from liquefied biomass materials. Some typical conversion processes are depicted in Figure 1. However, much remains to be done in synthesis, analysis, and characterisation if these new materials are to be exploited to their full advantage. Cai et al. [1] published a review of physicochemical properties and characterisation methods of lignocelleulosic biomass. The knowledge of biomass properties and main components is essential for effective process development.

With respect to the liquefaction of wood, wood wastes, and other lignocellulosic biomass, earlier applications were mainly based on their use as fuels. A direct liquefaction of wood at high temperatures and pressures in order to convert it into a more usable fuel was achieved using different technologies, such as pyrolysis, plasma techniques, supercritical water gasification, and enzymatic conversion, thus providing a longer-term solution to the need to decrease the crude oil dependency [2–5]. With respect to the liquefaction of wood and wood wastes and their use as a liquid fuel, Rezzoug and Capart [6] used solvolysis in ethylene-glycol and catalytic hydrogenation as the second stage for the production of oils that possessed a high heating value. Seljak et al. [7] reported the results of a successful combustion of liquefied wood prepared in an acid catalysed reaction with polyhydroxy alcohols.

Since petrochemicals may soon be limited, considerable attention has been given to the selective conversion of biomass into chemicals such as 5-hydroxymethylfurfural and dimethylfuran by selective chemical catalysts [8–10]. The conversion of biomass by depolymerisation and fermentation is presently the approach with the most potential. Recent studies of acid catalysed depolymerisation with organic solvents has allowed to increase yields of basic chemicals such as furfural, hydroxymethylfurfural, and levulinic acid [11–13].

Although biomass-based lignocellulosic materials are complex and the liquefaction process of their components is not fully understood, three major transformation pathways are more commonly used.

- Fast pyrolysis at high temperatures and subsequent conversion into the liquid products and gaseous products that are used for energy production;
- Thermochemical conversion, at elevated temperatures and with the use of different liquefying agents and catalysts. Here, solvolysis and depolymerisation take place simultaneously with the liquid final product being rich in hydroxyl groups;
- Biochemical modifications (enzymatic conversion into ethanol and other depolymerisation products).

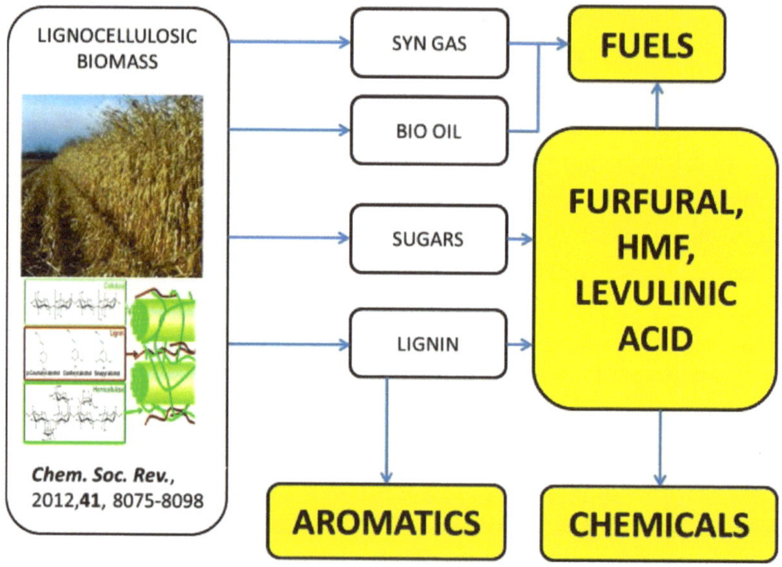

Figure 1. Biomass conversion pathways.

This review provides some references on conversion methods of biomass into some important chemicals and polymers. Because of numerous chemical products that can be manufactured from biomass, the interest in biomass conversion to chemicals has been a subject to many studies during the last few years. Researchers worldwide have developed methods to provide bio-based products from a wide selection of biomass resources and some processes are now suitable for scale-up production in industry. The industrial production of levulinic acid from tobacco wastes is one of the first applications and the isolation of 5-hydroxymethylfurfural, isosorbide, γ-valerolactone, nanocellulose, and other specialty chemicals will follow. Interests in the conversion of biomass to chemicals are strongly connected to the use of renewable resources instead of crude oil and the need to decrease greenhouse gas emissions. Therefore, due to the abundance and availability of biomass with no impact on food production, its utilization in the production of fuels and chemicals is highly promising.

This article presents some pathways for the production of chemicals from biomass. Special attention was given to thermoconversion methods, such as the liquefaction of biomass due to its simple chemical conversion process and equipment needed.

1.1. The Liquefaction Process

The liquefaction of wood and other lignocellulosic materials in the presence of phenol or polyhydroxy alcohols has been intensively studied by several authors [14] and has been described in detail [15]. Such a liquefaction process is usually carried out at elevated temperatures and in the presence of an acid catalyst [16–18]. The liquefied wood contains depolymerized products from the β 1–4 glucosidic bond cleavage of the cellulose and hemicelluloses molecules and from low molar mass oligomers. The complex structure of the lignin molecule is broken into smaller fragments. Several attempts to elucidate the mechanisms associated with the liquefaction of lignin, of cellobiose, and of cellulose have been published. The liquefaction of lignin with phenol has been studied on the basis of the behaviour of model substances, such as guaiacylglycerol-β-guiacyl ether. According to Lin et al., the dominant products were guaiacylglycerol-α-phenyl-β-guaiacyl ethers, followed by guaiacol, triphenylethanes, diphenylmethanes, benzocyclobutanes, and phenylcoumaranes [19,20].

The behaviour of cellulose in acidic ethylene glycol and ethylene carbonate was studied by Yamada and Ono [21]. These authors proved the formation of ethylene glycol—glucosides in the early stage of liquefaction. These later decomposed into the 2-hydroxyethyl levulinate. Jasukaityte et al. [22] studied the liquefaction of different types of cellulose by monitoring the molar mass decrease and changes in degree of crystallinity during the reaction. They found that the less disordered regions degraded in the initial minute of liquefaction. The highly ordered cellulose regions remained relatively stable for a longer time. Kobayashi et al. [23] have undertaken research into the liquefaction of cellulose powder, of steamed lignin, and of mixtures of these two components in order to characterise the reaction process on the basis of the polyhydric alcohol nature of the materials. They found that the condensation reaction occurred only in mixtures of cellulose and lignin. Under these conditions, dimethyl formamide insoluble residue was formed, being similar in nature to wood. They suggested that a condensation reaction occurred between the depolymerised cellulose residues and the aromatic derivatives of lignin. Yan Shi et al. [24] studied kinetics of biomass solvolysis and found that glycerol was very efficient solvent due to higher polarity.

While a large number of papers have been published describing the liquefaction process, the characterisation of the product, and the reaction pathways, little information is available on the application of the more specialised techniques in biomass liquefaction. One of these is microwave-driven wood liquefaction, where microwave heating has been used in a very efficient and fast liquefaction of wood. The efficiency of liquefaction and the time needed for the complete liquefaction was reduced significantly [25,26]. Complete liquefaction was achieved in 20 min under the prevailing processing conditions. The use of ionic liquids for wood liquefaction is also considered to be specific due to the application of reagents and their recycling [27]. Zavrel et al. very efficiently dissolved wood chips in ionic liquids, an environment-friendly alternative to conventional methods [28].

A novel approach to very efficient energy input during the thermochemical conversion of lignocellulosic biomass into liquefied depolymerised products is the use of ultrasound technologies [29–31]. High-energy ultrasound has been used in liquefaction reactions with different wood waste materials. The most common wood waste materials were chosen in order to prove the efficiency of the ultrasound process and to establish a different way for recycling wood wastes found in large quantities in municipal waste deposits. The reaction times were shortened up to nine times when using the ultrasound process, with smaller residual particles and with no influence on the hydroxyl number of the final products.

Yamazaki et al. have studied the liquefaction of beech wood using supercritical alcohols with 100% yields [32]. Liquefied wood has a high reactivity due to a large amount of phenolic and alcoholic hydroxyl groups that are present. The hydroxyl value of the liquefied wood has been determined by several authors and is generally determined to be higher than 200 mg KOH/g. The value depends on the liquefaction time and on the ratio between the wood content and the liquefying reagents.

Although the liquefaction of biomass was studied in detail and its further use in polymer synthesis, its direct utilization in film formation was published by Budja et al. [33] and Briones et al. [34]. Authors produced polymer films from liquefied biomass with promising mechanical properties and with good future perspective in film production for agricultural applications.

1.2. Chemicals from Biomass

Biomass is primarily composed of cellulose (40–50%), hemicellulose (25–35%), and lignin (15–20%). The composition varies considerably depending on types of biomass, part of the plant etc. Cellulose is the major structural component of plant cell walls and is organized into fibrils, which are aligned parallel to each other, surrounded by a matrix of hemicellulose and lignin.

1.2.1. Cellulose

The use of biomass as a natural source for industrial applications and for the production of chemical are today regarded as an alternative to petrochemical processes. In particular, cellulose as the most abundant renewable polymer has become an object of studies on how to obtain many valuable chemicals from biomass [11].

A multitude of different types of biomass has been tested, including algae, cultivated crops, agricultural waste, forest resources, etc. The term that is usually associated with such conversion is biorefinery. A variety of technologies have been used including fermentation, pyrolysis, thermochemical conversion, hydrothermal liquefaction, catalytic conversion, etc.

Some final and intermediate products, derived from cellulose are shown in Figure 2. Chemicals, such as 2,5-Dimethylfuran, 2,5-Dimethyltetrahydrofuran, are considered as the target chemicals for fuel. Isosorbide (1,4:3,6-Dianhydrohexitol) is a desired chemical that allows further chemical modifications through the conversion of the hydroxyl groups into other functional groups, which leads to monomers suitable for the production of polymers [35,36].

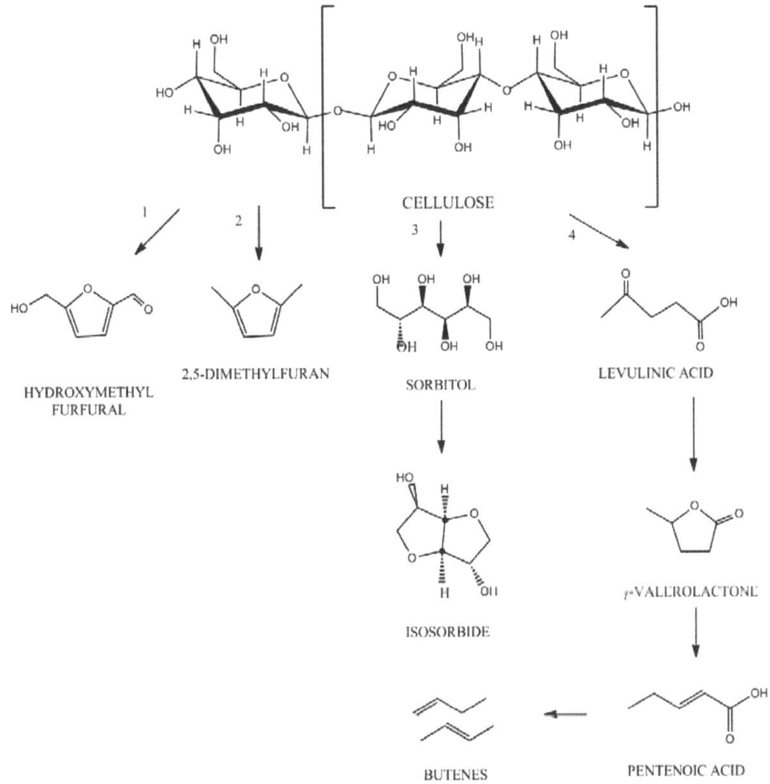

Figure 2. Cellulose conversion into different chemicals by biorefinery.

1.2.2. Hemicelluloses

Hemicelluloses, hetero polymers that are present along with cellulose in plant cell walls, contain xylose, mannose, galactose rhamnose, and arabinose. They can easily be isolated from plants by different purification methods using acids, organic solvents, and alkaline reagents. Other methods include steam explosion, ultrasonication, extrusion, and microwave irradiation. Further conversion into hydroxymethyl furfural, furfural, and even propionic acid can be achieved by reaction in a catalysed ionic liquid, fermentation processes, or hydrothermal processes [37–39].

1.2.3. Lignin

Lignocellulosic biomass is mostly cell wall material, and lignin is one of three principal components beside cellulose and hemicelluloses. Lignin serves as a binding agent, adding strength to the cell walls. It is a cross-linked amorphous polymer synthesized from aromatic alcohol precursors, p-coumaryl, coniferyl, and sinapyl alcohols and has a very complex structure. These precursors are linked together via radical coupling reactions to form a complex three-dimensional molecular architecture that contains a great variety of bonds with typically around 50% β-*O*-4 ether linkage [40]. The characterization and structure determination of lignin has been a challenge to scientists for many years.

Since it represents 30% of all non-fossil organic carbon on Earth, it is logical to explore the potential use of this natural polymer as a source of valuable chemicals, as shown in Figure 3. Lignin is extracted from biomass by physical, chemical, and biochemical treatments. Pulping processes that utilize extraction methods are based on ester and ether bonds cleavage and thus change the structure of lignin macromolecule into more simple and smaller ones. The product is different than the original lignin present in the plant cell. In general, we can distinguish two different extraction processes: sulphur and sulphur-free process. The product of the first one is the lignosulfonate and kraft lignin, while the second one gives organosolve and soda lignin.

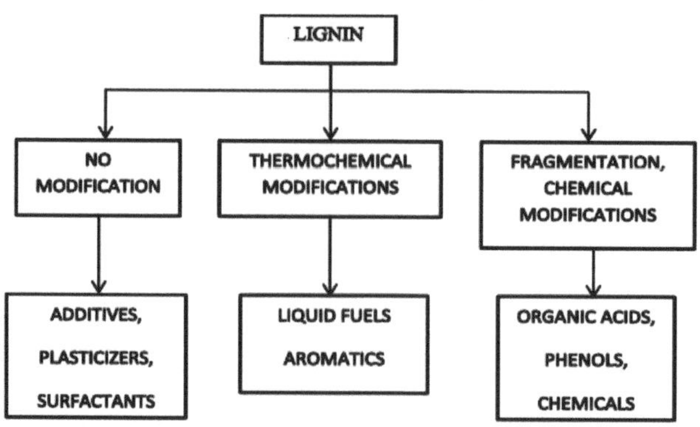

Figure 3. Summary of the uses of lignin.

Lignin without chemical modification can be used as a filler, anti-oxidant, UV-stabilizer, surfactant etc. The chemical modification of lignin [41] has much better potential for multiple applications and can be classified into three main categories:

1. Fragmentation;
2. Modification by creating new chemical active sites;
3. Chemical modification of hydroxyl groups.

Fragmentation of Lignin

The fragmentation methods can be summarized into pyrolysis, hydrogentation, hydrolysis, oxidation, gasification, enzymatic oxidation, and microbial conversion processes. The products are very diverse, from syngas to phenols, cresols, phenol, vanillin, dimethylfuran, different organic acids etc. Lignin fragmentation has had two objectives: the study of the structure and the production of useful materials from waste lignin.

Lignin pyrolysis has been studied for many years with the abovementioned goals and is based on heating biomass at around 500 °C without oxygen. The degradation products are solid (tar), liquid (methanol, acetic acid, acetone, and water), and gaseous fraction (CO, CO_2, CH_4, etc.). Pyrolysis has been used predominantly for bio-oil production which can be used as a fuel.

Lignin oxidation with one of most common oxidants such as nitrobenzene, some metallic oxides, air, and oxygen is another way to obtain phenolic derivatives. Of course, the proper reaction conditions must be met to obtain products with high added value, such as vanillin, which is mainly used in the food processing industry. The lignin oxidation process can also provide some very useful basic chemicals for the chemical industry, such as vanillic acid, hydroxybenzaldehyde, syringic aldehyde, syringic acid etc.

One of the most promising methods is the thermochemical conversion—liquefaction of lignin. This method utilizes simultaneous solvolysis and depolymerisation. The resulting products are in hydroxyl groups and can be used as a feedstock in the preparation of polyurethanes, polyesters, and epoxy resins [42,43].

Modification of Lignin

In order to increase the reactivity of these groups, several modifications have been studied, such as nitration, amination, alkylation, dealkylation, carboxylation, and halogenation. The common goal was to achieve better solubility in organic solvents, to increase chemical reactivity, and to improve mechanical properties and processing of lignin.

Chemical Modification of Hydroxyl Groups

Special attention was given to the functionalization of phenolic and hydroxyl groups in lignin since they can be converted through esterification, etherification, oxidation/reduction, phenolation, reactions with isocyanates, or sylilation. Such modifications improve the solubility of lignin. Demethylation is well-known method for dimethyl sulfoxide (DMSO) production. Here, two methyl groups are transferred from lignin to sulfur, which is further oxidized to DMSO. Esterified lignins can be used as a precursor of polyesters, epoxy resins, and elastomeric materials. Etherification leads to polyols that can be used for the production of polyurethane foams. Hydroxymethylated lignin can substitute a certain part of phenol in phenol-formaldehyde resins. Variations with the use of glyoxal, furfural, and glutaraldehyde have also been published [44].

1.3. Adhesives

Hydroxyl groups in liquefied lignocellulosics can also react with different reactive sites in thermosetting systems as well as in two component systems. Lee and Liu prepared a resol type of resin from liquefied bark that was used in particle board preparation [45]. Similarly, a phenol formaldehyde adhesive was prepared by Shenyuan et al. from liquefied bamboo [46]. Zhang et al. used Chinese fir and poplar in this context [47]. They tested the bonding strength of such adhesive in plywood, showing that the product successfully met the requirements. Gagnon et al. have undertaken research into the softwood bark pyrolysis oils and their use as adhesives for particle boards in combination with isocyanates [48]. Dos Santos et al. have used liquefied cork and wood to produce polyurethane adhesive [49]. Juhaida et al. liquefied kenaf core and prepared polyurethane adhesive for wood laminates [50]. Kunaver et al. applied a mixture of liquefied wood and melamine—urea—formaldehyde resin as an adhesive in wood particle board production [51].

The hydroxyl groups of the liquefied wood can react with epichlorohydrin, thus introducing the epoxy functionality. Kobayashi et al. prepared such epoxy compounds from the heartwood meal of Japanese cedar [52,53]. Kishi et al. prepared a wood-based epoxy resin from wood meal of German spruce [54]. The authors cured their epoxy resin precursors with suitable amino precursors. Such two-component systems were used as adhesives in plywood preparation, and good mechanical and physical properties were obtained.

1.4. Polyurethanes

Liquefied lignocellulosic materials are considered to be an alternative feedstock for polymer synthesis. Recently, considerable attention has been given to the preparation of environmentally friendly polymeric products from liquefied biomass materials and their derivatives.

In the liquefaction process, the hydroxyl group–containing species in the wood components can be used as polyols for several different purposes. The polyhydric units that are present in the resulting liquefied/derivatised wood can be used in the creation of polyurethane foams, polyurethane resin precursors [55,56], and in the recently developed wood polyalcohol–based urethane adhesives [57]. Several authors have found a combination of PEG-400 and glycerine to be the most effective liquefying reagent [58,59]. Claims are made that greater than

95% liquefaction has been achieved in most experiments. Liquid wood can be used directly without any additional treatment. It is claimed that mixtures of such liquefied/derivatised mixtures can be directly used for the preparation of polyurethane foams [55]. Chen and Lu prepared rigid polyurethane foams from liquefied wheat straw [60]. The products are rich in ethylene oxide units and, due to the presence of extensive hydrogen bonding, have a highly hydrophilic character. Hu and Li produced rigid and semi-rigid polyurethane foams by using crude glycerol with high organic impurities [61]. The foams had suitable properties comparable to those derived from petrochemical-based chemicals. Lee et al. [62] have studied the thermal stability, biodegradability, and genotoxicity of a range of polyurethane foams that were produced from polyols made from waste paper. They found that the foams possessed the same thermal stability as those made from liquefied starch or wood. The foams were biodegradable to some extent. No mutagens or carcinogenic substances were found in water extracts from the foams. These materials are technologically simple and inexpensive to produce. It is stated that they can be used in agricultural and horticultural applications. D'Souza et al. [63] have liquefied bark at different reaction conditions and have found that the foams exhibited a high amount of open cell content when utilizing the polyol synthesized at lower temperature.

1.5. Polyesters

The hydroxyl value of the liquefied wood has been determined by several authors and is generally determined to be between the equivalent of 200 and 500 mg KOH/g. The actual value depends on the liquefaction time. This large number of hydroxyl group sites means that the products have applications in those areas in which the hydroxyl component can be used in complex polyester synthesis. Yu et al. [64] liquefied corn stover in a mixture of ethylene glycol and ethylene carbonate and used the liquefied product as a polyol in a condensation reaction with different organic dibasic acids. The product was a completely crosslinked solid polyester, insoluble in most solvents.

Kunaver et al. [65] used the liquefied wood as a polyol in the polyester synthesis. Saturated polyesters are defined as polyesters whose components contain no polymerizable double bonds, the opposite of what occurs with alkyd resins and unsaturated polyesters. Saturated polyesters are synthesized by the condensation of bifunctional or polyfunctional monomers containing hydroxyl or carboxylic groups. The hydroxyl groups in the current investigation originate from liquefied wood. The polycondensation process takes place at temperatures of 150 °C to 260 °C. Water, as the main reaction byproduct, is removed from the system by the creation of a slight vacuum, gas stream condensation, or an azeotropic process. Čuk et al. [66] combined recycled PET polyester with liquefied wood polyester for the production of polyurethane/polyisocyanurate foams. Glass transition temperature, density, and water absorption of the foam increased with increasing amount of glycerol in liquefied wood, and compressive stress also increased, while thermal conductivity was not affected.

Some benefits could be expected from such systems. These include the incorporation of the biomass components into the polymeric compositions and, consequently, the provision of a certain degree of biodegradability. Esterification of a proportion of the hydroxyl groups reduces the reactivity of the liquid wood, a feature that is sometimes desired in polyurethane synthesis.

1.6. Nanocellulose

1.6.1. What Is Nanocellulose?

In recent years, there has been increasing interest in the development of biodegradable composite materials, so-called green composites. Natural fibers lack the high performance characteristics of many synthetic fibers. Cellulose acts as a framework in lignocellulosic materials and is based on highly oriented cellulose fibrils. Within cellulose fibrils, there are highly oriented regions, namely nanocrystalline cellulose, as presented in Figure 4.

Highly crystalline cellulose nanofibers (CNCs), present in natural plant bodies, have unique properties and have been identified as a new reinforcing agent in nanocomposites due to their low cost, availability, renewability, light weight, nanoscale dimension, and high aspect ratio [67,68]. The CNC has high strength due to its dense and ordered crystalline structure and therefore its axial Young's modulus is extraordinary, potentially stronger than Kevlar.

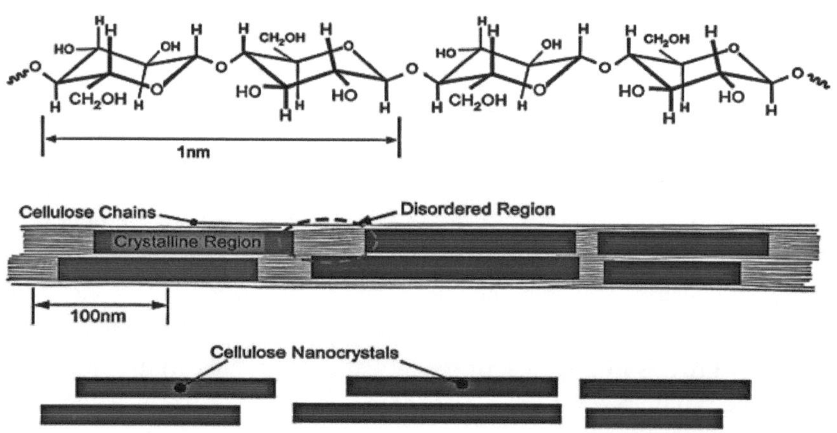

Figure 4. Cellulose nanocrystall regions in plant structure [69].

1.6.2. Nanocellulose Isolation

CNC is usually produced from native cellulose by isolation of its crystalline regions [70]. Different processes have been developed for the isolation on nanocrystalline cellulose from biomass. These include

- Chemical pulping where lignin is removed by a Kraft process;
- Steam explosion followed by hydrolysis;
- Acid hydrolysis;
- Microwave assisted hydrolysis;
- Mechanical structure breakdown.

The typical procedure currently employed for the production of CNC consists of subjecting the cellulosic material to strong acid hydrolysis under strictly controlled conditions of temperature, agitation, and time. Since it is quite difficult to achieve CNC with stable and controlled size (length) in a short duration of time. Kos et al. [71] developed an optimized process for rapid preparation of CNC using microwave heating, while Kunaver et al. [72] modified the process of biomass liquefaction in polyhydroxy alcohols and acid catalyst for the rapid and non-expensive isolation of CNC from different lignocellulosic sources. Ultrasound can be used as an additional source of energy.

1.6.3. Nanocrystalline Cellulose (NCC) Applications

Many new nanocomposite materials with attractive properties have been prepared by physical incorporation of CNC into a natural or synthetic polymeric matrix and also holds promise in many different applications such as in nanopaper, coatings, adhesives, optical sensors, biomedical scaffolds, filtration membranes, electronic devices, foams, aero gels etc., which make NCC of significant interest from a scientific and industrial perspective [73–76]. Major studies over the last decades have been dedicated to use nanocellulose as a filler in nanocomposites to improve mechanical and barrier properties. It can be easily chemically modified due to the abundance of OH groups on the surface of the nanocellulose through the acetylation reaction, TEMPO oxidation, silylation, polymer grafting, all with the purpose of modifying the polarity of the nanocellulose surface and improving the compatibility with different polymer matrices. Due to nanocellulose's high functionality and so many possible chemical modifications, the material applications for nanocellulose are almost limitless.

2. Characterisation Techniques

A number of methods have been used for the process control and for the characterisation of products. Here, only the most commonly used are described.

2.1. *Scanning Electron Microscopy (SEM) and Transmission Electron Microscopy (TEM) Observations*

Wood meal and wood residue, after the liquefaction reaction, can be micrographed on a scanning electron microscope. Samples must be sputter coated with a thin layer of gold to avoid electrostatic charging during scanning. Samples of nanocellulose must first be diluted and sonicated to prevent agglomeration. Both SEM and TEM micrographs give the information about the particle dimensions and the degree of agglomeration. An example of micrograph of wood residue after the liquefaction and nanocrystalline particles, isolated from wood meal, are shown in Figure 5.

Figure 5. Scanning electron microscopy (SEM) micrograph of wood residue after the liquefaction process (**left**). SEM micrograph of nanocrystalline cellulose isolated form wood meal (**right**).

2.2. *Viscosity Measurements*

The viscosity of polyesters and of the liquefied wood is measured using a rotational viscometer. Standard ISO 3219-1993 describes the measurement method.

2.3. *Measurement of Hydroxyl Number and Acid Number of Liquid Biomas and of Liquefied Biomass Polyester Derivatives*

The hydroxyl numbers of liqefied biomass and its derivatives can be determined using standard ASTM Standard D4274-05 [77]. The acid number is determined using standard ASTM D974-08 [78]. Both measurements are needed to control the process of polyester formation and for polyurethane foam preparation.

2.4. *Fourier Transform Infrared (FTIR) Spectroscopic Analysis*

Spectra are usually obtained using an FTIR spectrophotometer. The liquid/ derivatised wood samples and the polyesters are each supplied on a NaCl crystal as a thin film. Transmittance spectra are measured in the wavelength range from 4000 cm^{-1} to 500 cm^{-1}. The method helps to identify the characteristic functional groups in liquefaction products.

2.5. GPC Measurements

Gel permeation chromatographic (GPC) representations are obtained on a chromatographic system, consisting of isocratic pump, a refractive index detector and a diode array detector. Polystyrene standards are used for the calibration curve. The mobile phase is sometimes prepared from dimethyl acetamide containing 0.05 M LiBr. The latter is added in order to prevent interaction between the sample molecules and the stationary phase. A typical GPC curve of liquefied wood and of liquefied wood polyester is presented in Figure 6.

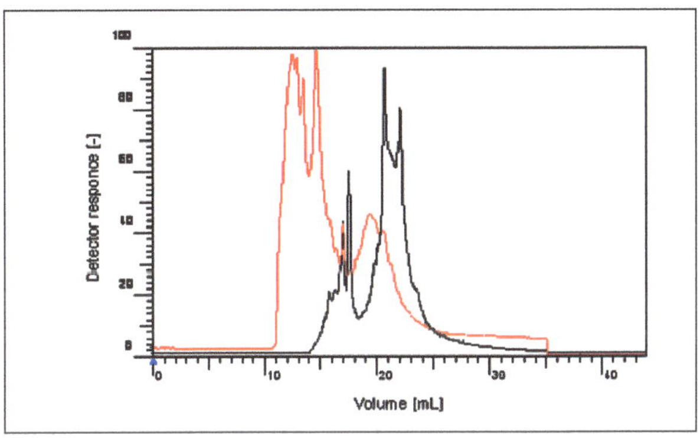

Figure 6. Gel permeation chromatographic (GPC) chromatogram of liquefied wood (black) and polyester synthesized from liquefied wood (red).

2.6. Particle Size Distribution from DLS Studies

The dynamic light scattering (DLS) technique is usually employed to find the statistical distribution of the particles present in nanocellulose or other biorefinery products.

2.7. X-Ray Diffraction

The crystallinity index of nanocellulose is usually calculated from X-ray diffractograms (Figure 7). The diffraction data are collected on a wide-angle X-ray diffraction instrument, equipped with Cu anode at wavelength Cu Kα. Diffractograms are measured in the 2-theta range from 5 °C to 40 °C. The most used method for the crystallinity index calculation is the Segal method [79] although this method suffers some inaccuracy due to the inaccurate amorphous peak position. The crystallite size can be estimated by using a well-known Scherrer equation [80].

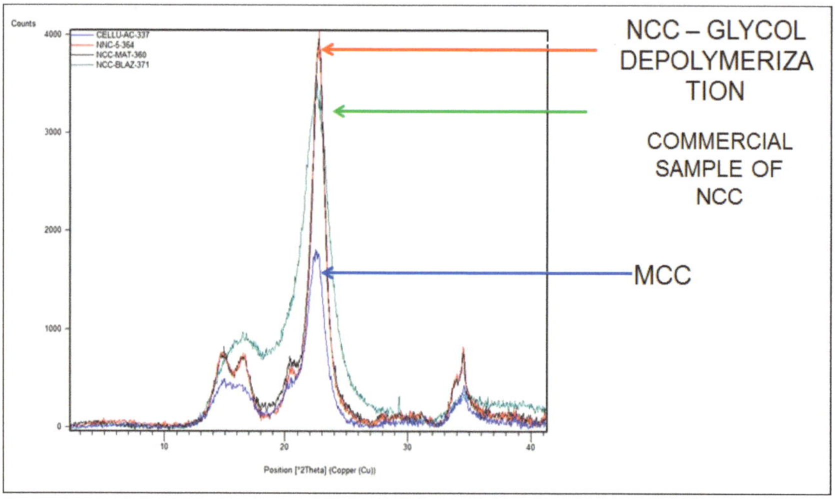

Figure 7. X-ray diffraction (XRD) diffractograms of nanocrystalline cellulose and microcrystalline cellulose.

2.8. *Differential Scanning Calorimetry (DSC)*

The thermal behaviour, the melting temperature, and the crystallization temperature is measured using differential scanning calorimeter. Samples are heated from 30 °C to 350 °C under a nitrogen atmosphere.

2.9. *Thermogravimetric Analysis (TGA)*

Thermogravimetric analysis (TGA) analyses are used for the determination of an initial weight loss at low temperatures, the thermal degradation temperature, and the maximum decomposition temperature. Samples are heated at constant heating rate under nitrogen purging.

3. Conclusions

The possibilities of converting lignocellulosic biomass and biomass waste in particular into valuable chemicals and raw materials for further use in polymer chemistry are numerous. Intensive studies of this challenge have been governed for the last two decades. The biorefinery concept has been firmly established, and the only danger that has to be guarded carefully is that the production does not compete with food production. It provides new prospects of utilizing renewable resources for the production of energy, organic chemicals, and polymers through complex processing technologies. Lignocellulosic biomass, the most abundant sustainable feedstock on earth, contains cellulose, hemicellulose, and lignin in various proportions. The first step toward the conversion of these basic ingredients into different basic chemicals is their separation. Each of these three biomass components can then be converted into different platform chemicals. One of the newest challenges is the isolation and application of nanocellulose, an outstanding material that is carbon neutral, sustainably recyclable, and non-toxic. It has the potential to become a true green nanomaterial that will be used in high-performance applications. The final cost of bioproducts depends upon the processing costs. Therefore, high tonnage, high yields, and the reduction of the synthesis steps are the most important development trends today and in the future.

All the products can be regarded as higher added-value materials, decreasing the dependency to the crude oil production, and at least not increasing the greenhouse gas emissions.

Acknowledgments: The author wishes to gratefully acknowledge the support for the presented work received from the Ministry of Higher Education, Science and Technology of the Republic of Slovenia within the Program P2-0145-0104.

Conflicts of Interest: The author declares no conflict of interest.

References

1. Cai, J.; He, Y.; Yu, X.; Banks, S.C.; Yang, Y.; Zhang, X.; Yu, Y.; Liu, R.; Bridgwater, A.V. Revew of physicochemical properties and analytical characterization of lignocellulosic biomass. *Renew. Sust. Energy Rev.* **2017**, *76*, 309–322. [CrossRef]
2. Zhang, L.; Xu, C.; Champagne, P. Overview of recent advances in thermos-chemical conversion of biomass. Energy Conversion and Management. *Energy Convers. Manag.* **2010**, *51*, 969–982. [CrossRef]
3. Anwar, Z.; Gulfraz, M.; Irshad, M. Agro-industrial lignocellulosic biomass a key to unlock the future bio-energy: A brief review. *J. Radiat. Res. Appl. Sci.* **2014**, *7*, 163–173. [CrossRef]
4. Molino, A.; Chianese, S.; Musmarra, D. Biomass gasification technology: The state of the art overview. *J. Energy Chem.* **2016**, *25*, 10–25. [CrossRef]
5. Demirbas, M.F.; Balat, M.; Balat, H. Biowastes-to-biofuels. *Energy Convers. Manag.* **2011**, *52*, 1815–1828. [CrossRef]
6. Rezzoug, S.A.; Capart, R. Liquefaction of wood in two succesive steps: Solvolysis in ethylene-glycol and catalytic hydrotreatment. *Appl. Energy* **2002**, *72*, 631–644. [CrossRef]
7. Seljak, T.; Oprešnik, S.R.; Kunaver, M.; Katrašnik, T. Wood, liquefied in polyhydroxy alcohols as a fuel for gas turbines. *Appl. Energy* **2012**, *99*, 40–49. [CrossRef]
8. Maity, S.K. Opportunities, recent trends and challenges in integrated biorefinery: Part 1. *Renew. Sustain. Energy Rev.* **2014**, *43*, 1427–1445. [CrossRef]
9. Wettstein, S.G.; Alonso, D.M.; Gürbüz, E.I.; Dumesic, J.A. A roadmap for conversion of lignocellulosic biomass to chemicals and fuels. *Curr. Opin. Chem. Eng.* **2012**, *1*, 218–224. [CrossRef]

10. Gallezot, P. Conversion of biomass to selected chemical products. *Chem. Soc. Rev.* **2012**, *41*, 1538–1558. [CrossRef] [PubMed]
11. Dutta, S.; Pal, S. Promises in direct conversion of cellulose and lignocellulosic biomass to chemicals and fuels: Combined solvent—Nanocatalysis approach for biorefinery. *Biomass Bioenergy* **2014**, *62*, 182–197. [CrossRef]
12. Gandini, A.; Silvestre, A.J.D.; Neto, C.P.; Sousa, A.F.; Gomes, M. The Furan counterpart of Poly(ethyleneterephthalate): An alternative material based on renewable resources. *Polym. Sci. A Polym. Chem.* **2009**, *47*, 295–298. [CrossRef]
13. Binder, J.B.; Raines, R.T. Simple chemical transformation of lignocellulosic biomass into furans for fuels and chemicals. *J. Am. Chem. Soc.* **2009**, *131*, 1979–1985. [CrossRef] [PubMed]
14. Qiuhui, Z.; Guangjie, Z.; Jinpeng, C. Effect of inorganic acid catalysts on liquefaction of wood in phenol. *Front. For. China* **2006**, *2*, 214–218.
15. Zhang, Y.; Ikeda, A.; Hori, N.; Takemura, A.; Ono, H.; Yamada, T. Characterization of liquefied product from cellulose with phenol in the presence of sulfuric acid. *Bioresour. Technol.* **2006**, *97*, 313–321. [CrossRef] [PubMed]
16. Jiulong Xie, J.; Chung-Yun Hse, C.Y.; Shupe, T.F.; Qi, J.; Pan, H. Liquefaction Behaviors of Bamboo Residues in a Glycerol-Based Solvent Using Microwave Energy. *J. Appl. Polym. Sci.* **2014**, *131*, 40207. [CrossRef]
17. Alma, M.H.; Maldas, D.; Shiraishi, N.J. Liquefaction of several biomass wastes into phenol in the presence of various alkalis and metallic salts as catalysts. *J. Polym. Eng.* **1998**, *18*, 162–177. [CrossRef]
18. Lin, L.; Yao, Y.; Yoshioka, M.; Shiraishi, N.J. Liquefaction mechanism of cellulose in the presence of phenol under acid catalysis. *Carbohydr. Polym.* **2004**, *57*, 123–129. [CrossRef]
19. Lin, L.; Yao, Y.; Shiraishi, N.J. Liquefaction Mechanism of β-O-4 Lignin Model Compound in the presence of Phenol under Acid Catalysis, Part 1. Identification of the Reaction Products. *Holzforschung* **2001**, *55*, 617–624. [CrossRef]
20. Lin, L.; Nakagame, S.; Yao, Y.; Yoshioka, M.; Shiraishi, N.J. Liquefaction Mechanism of β-O-4 Lignin Model Compound in the presence of Phenol under Acid Catalysis, Part 2. Reaction Behavior and Pathways. *Holzforschung* **2001**, *55*, 625–630. [CrossRef]
21. Yamada, T.; Ono, H. Characterization of the products resulting from ethylene glycol liquefaction of cellulose. *J. Wood Sci.* **2001**, *47*, 458–464. [CrossRef]
22. Jasiukaitytė, E.; Kunaver, M.; Strlič, M. Cellulose liquefaction in acidified ethylene glycol. *Cellulose* **2009**, *16*, 393–405. [CrossRef]
23. Kobayashi, M.; Asano, T.; Kajiyama, M.; Tomito, B. Analysis on residue formation during wood liquefaction with polyhydric alcohol. *J. Wood Sci.* **2004**, *50*, 407–414. [CrossRef]
24. Shi, Y.; Xia, X.; Li, J.; Wang, J.; Zhao, T.; Yang, H.; Jiang, J.; Jiang, X. Solvolysis kinetics of three components of biomass using polyhydric alcohols as solvents. *Bioresour. Technol.* **2016**, *221*, 102–110. [CrossRef] [PubMed]
25. Kržan, A.; Kunaver, M. Microwave heating in wood liquefaction. *J. Appl. Polym. Sci.* **2006**, *101*, 1051–1056. [CrossRef]

26. Xie, J.; Hse, C.-Y.; Shupe, T.F.; Qi, J.; Pan, H. Liquefaction Behaviors of Bamboo Residues in a Glycerol-Based Solvent Using Microwave Energy. *J. Appl. Polym. Sci.* **2014**, *131*, 40207:1–40207:8. [CrossRef]
27. Honglu, X.; Tienju, S. Liquefaction of wood (Metasequoia glyptostroboides) in allyl alkyl imidazolium ionic liquids. *Wood Sci. Technol.* **2010**, *44*, 119–128.
28. Zavrel, M.; Bross, D.; Funke, M.; Büchs, J.; Spiess, A.C. High-throughput screening for ionic liquids dissolving (ligno-)cellulose. *Bioresour. Technol.* **2009**, *100*, 2580–2587. [CrossRef] [PubMed]
29. Kunaver, M.; Jasiukaitytė, E.; Čuk, N. Ultrasonically assisted liquefaction of lignocellulosic materials. *Bioresour. Technol.* **2012**, *103*, 360–366. [CrossRef] [PubMed]
30. Mateus, M.M.; Acero, N.F.; Bordado, J.C.; Dos Santos, R.G. Sonication as a foremost tool to improve cork liquefaction. *Ind. Crops Prod.* **2015**, *74*, 9–13. [CrossRef]
31. Karimi, M.; Jenkins, B.; Stroeve, P. Ultrasound irradiation in the production of ethanol from biomass. *Renew. Sustain. Energy Rev.* **2014**, *40*, 400–421. [CrossRef]
32. Yamazaki, J.; Minami, E.; Saka, S. Liquefaction of beech wood in various supercritical alcohols. *J. Wood Sci.* **2006**, *52*, 527–532. [CrossRef]
33. Budja, F.; Tavzes, Č.; Zupančič-Kralj, L.; Petrič, M. Self-crosslinking and film formation ability of liquefied black poplar. *Bioresour Technol.* **2009**, *100*, 3316–3323. [CrossRef] [PubMed]
34. Briones, R.; Torres, L.; Saravia, Y.; Serrano, L.; Labidi, J. Liquefied agricultural residues for film formation. *Ind. Crops Prod.* **2015**, *78*, 19–28. [CrossRef]
35. Kurian, J.K.; Nair, G.R.; Hussain, A.; Raghavan, G.S.V. Feedstocks, logistics and pre-treatment processes for sustainable lignocellulosic biorefineries: A comprehensive review. *Renew. Sustain. Energy Rev.* **2013**, *25*, 205–219. [CrossRef]
36. Kajaste, R. Chemicals from biomass e managing greenhouse gas emissions in biorefinery production chains e a review. *J. Clean. Prod.* **2014**, *75*, 1–10. [CrossRef]
37. Peleteiro, S.; Garrote, G.; Santos, V.; Parajo, J.C. Furan manufacture from softwood hemicelluloses by aqueous fractionation and further reaction in a catalyzed ionic liquid: A biorefinery approach. *J. Clean. Prod.* **2014**, *76*, 200–203. [CrossRef]
38. Peng, F.; Peng, P.; Xu, F.; Sun, R.C. Fractional purification and bioconversion of hemicelluloses. *Biotechnol. Adv.* **2012**, *30*, 879–903. [CrossRef] [PubMed]
39. Wei, L.; Shrestha, A.; Tu, M.; Adhikari, S. Effects of surfactant on biochemical and hydrothermal conversion of softwood hemicellulose to ethanol and furan derivatives. *Process Biochem.* **2011**, *46*, 1785–1792. [CrossRef]
40. Laurichesse, S.; Avérous, L. Chemical modifications of lignins: Towards biobased polymers. *Prog. Polym. Sci.* **2014**, *39*, 1266–1290. [CrossRef]
41. Azadi, P.; Inderwildi, O.R.; Farnood, R.; King, D.A. Liquid fuels, hydrogen and chemicals from lignin: A critical review. *Renew. Sustain. Energy Rev.* **2013**, *21*, 506–523. [CrossRef]
42. Jasiukaitytė-Grojzdek, E.; Kunaver, M.; Crestini, C. Lignin structural changes during liquefaction in acidified ethylene glycol. *J. Wood Chem. Technol.* **2012**, *32*, 342–360. [CrossRef]

43. Anguillesi, I.; Lazzeri, A.; Cinelli, P. Green synthesis of flexible polyurethane foams from liquefied lignin. *Eur. Polym. J.* **2013**, *49*, 1174–1184.
44. Lee, W.J.; Chang, K.C.; Tseng, I.M. Properties of phenol-formaldehyde resins prepared from phenol-liquefied lignin. *J. Appl. Polym. Sci.* **2012**, *124*, 4782–4788. [CrossRef]
45. Lee, W.; Liu, C. Preparation of liquefied bark based resol resin and its application to particle board. *J. Appl. Polym. Sci.* **2002**, *87*, 1837–1841. [CrossRef]
46. Shenyuan, F.; Lingfei, M.; Wengzhou, L.; Shuna, C. Liquefaction of bamboo, preparation of liquefied bamboo adhesives, and properties of the adhesives. *Front. For. China* **2006**, *2*, 219–224.
47. Zhang, Q.; Zhao, G.; Yu, L.; Jie, S. Preparation of liquefied wood-based resins and their application in molding material. *For. Stud. China* **2009**, *9*, 51–56. [CrossRef]
48. Gagnon, M.; Roy, C.; Riedl, B. Adhesives made from isocyanates and pyrolysis oils for wood composites. *Holzforschung* **2004**, *58*, 400–407. [CrossRef]
49. Dos Santos, R.G.; Carvalho, R.; Silva, E.R.; Bordado, J.C.; Cardoso, A.C.; Costa, M.R.; Mateus, M.M. Natural polymeric water-based adhesive from cork liquefaction. *Ind. Crops Prod.* **2016**, *84*, 314–319. [CrossRef]
50. Juhaida, M.F.; Paridah, M.T.; Hilmi, M.M.; Sarani, Z.; Jalaluddin, H.; Mohamadzaki, A.R. Liquefaction of kenaf (*Hibiscuscannabinus* L.) core of wood laminating adhesive. *Bioresour. Technol.* **2010**, *101*, 1355–1360. [CrossRef] [PubMed]
51. Kunaver, M.; Medved, S.; Čuk, N.; Jasiukaitytė, E.; Poljanšek, I.; Strnad, T. Application of liquefied wood as a new particle board adhesive system. *Bioresour. Technol.* **2010**, *101*, 1361–1368. [CrossRef] [PubMed]
52. Kobayashi, M.; Hatano, Y.; Tomita, B. Viscoelastic properties of liquefied wood/epoxy resin and its bond strength. *Holzforschung* **2001**, *55*, 667–671. [CrossRef]
53. Kobayashi, M.; Tukamoto, K.; Tomita, B. Application of liquefied wood to a new resin system- synthesis and properties of liquefied wood/epoxy resins. *Holzforschung* **2000**, *54*, 93–97. [CrossRef]
54. Kishi, H.; Fujita, A.; Miyazaki, H.; Matsuda, S.; Murakami, A. Synthesis of wood based epoxy resins and their mechanical and adhesive properties. *J. Appl. Polym. Sci.* **2006**, *202*, 2285–2292. [CrossRef]
55. Wei, Y.; Cheng, F.; Li, H.; Yu, J. Synhesis and properties of polyurethane resins based on liquefied wood. *J. Appl. Polym. Sci.* **2004**, *92*, 351–356. [CrossRef]
56. Hakim, A.A.A.; Nassar, M.; Emmam, A.; Sultan, M. Preparation and characterization of rigid polyurethane foam prepared from sugar-cane bagasse polyol. *Mater. Chem. Phys.* **2011**, *129*, 301–307. [CrossRef]
57. Tohmura, S.; Li, G.; Qin, E. Preparation and characterization of wood polyalcohol-based isocyanate adhesives. *J. Appl. Polym. Sci.* **2005**, *98*, 791–795. [CrossRef]
58. Hassan, E.M.; Shukry, N. Polyhidric alcohol liquefaction of some lignocellulosic agricultural residues. *Ind. Crops Prod.* **2008**, *27*, 33–38. [CrossRef]
59. Kurimoto, Y.; Takeda, M.; Doi, S.; Tamura, Y.; Ono, H. Network structures and thermal properties of polyurethane films prepared from liquefied wood. *Bioresour. Technol.* **2001**, *77*, 33–40. [CrossRef]

60. Chen, F.; Lu, Z. Liquefaction of wheat straw and preparation of rigid polyurethane foam from liquefaction product. *J. Appl. Polym. Sci.* **2009**, *111*, 508–516. [CrossRef]
61. Hu, S.; Li, Y. Production and characterization of biopolyols and polyurethane foams from crude glycerol based liquefaction of soy bean straw. *Bioresour. Technol.* **2012**, *103*, 227–233. [CrossRef] [PubMed]
62. Lee, S.H.; Teramoto, Y.; Shiraishi, N.J. Biodegradable polyurethane foam from liquefied waste paper and its thermal stability, biodegradability, and genotoxicity. *J. Appl. Polym. Sci.* **2002**, *83*, 1482–1489. [CrossRef]
63. D'Souza, J.; Camargo, R.; Yan, N. Polyurethane foams made from liquefied bark-based polyols. *J. Appl. Polym. Sci.* **2014**, *131*, 40599–40609. [CrossRef]
64. Yu, F.; Liu, Y.; Pan, X.; Lin, X.; Liu, C.; Chen, P.; Ruan, R. Liquefaction of corn stover and preparation of polyester from liquefied polyol. Appl. *Biochem. Biotechnol.* **2006**, *130*, 574–585. [CrossRef]
65. Kunaver, M.; Jasiukaitytė, E.; Čuk, N.; Guthrie, J.T. Liquefaction of wood; synthesis and characterization of liquefied wood polyester derivatives. *J. Appl. Polym. Sci.* **2010**, *115*, 1265–1271. [CrossRef]
66. Čuk, N.; Fabjan, E.; Grželj, P.; Kunaver, M. Water-Blown Polyurethane/Polyisocyanurate Foams Made from Recycled Polyethylene Terephthalate and Liquefied Wood-Based Polyester Polyol. *J. Appl. Polym. Sci.* **2015**, *132*, 41522. [CrossRef]
67. Habibi, Y.; Lucia, L.A.; Rojas, O.J. Cellulose Nanocrystals: Chemistry, Self-Assembly, and Applications. *Chem Rev* **2010**, *110*, 3479–3500. [CrossRef] [PubMed]
68. Lima, M.M.D.; Borsali, R. Rodlike Cellulose Microcrystals: Structure, Properties, and Applications. *Macromol. Rapid Commun.* **2004**, *25*, 771–787. [CrossRef]
69. Dodson, B. Wood Pulp Extract Stronger than Carbon Fiber or Kevlar. Available online: http://www.gizmag.com/cellulose-nanocrystals-stronger-carbon-fiber-kevlar/23959 (accessed on 20 December 2017).
70. Brinchi, L.; Cotana, F.; Fortunati, E.; Kenny, J.M. Production of nanocrystalline cellulose from lignocellulosic biomass: Technology and applications. *Carbohydr. Polym.* **2013**, *94*, 154–169. [CrossRef] [PubMed]
71. Kos, T.; Anžlovar, A.; Kunaver, M.; Huskić, M.; Žagar, E. Fast preparation of nanocrystalline cellulose by microwave-assisted hydrolysis. *Cellulose* **2014**, *21*, 2579–2585. [CrossRef]
72. Kunaver, M.; Anžlovar, A.; Žagar, E. The fast and effective isolation of nanocellulose from selected cellulosic feedstocks. *Carbohydr. Polym.* **2016**, *148*, 251–258. [CrossRef] [PubMed]
73. Peng, B.L.; Dhar, N.; Liu, H.L.; Tam, K.C. Chemistry and applications of nanocrystalline cellulose and its derivatives: A nanotechnology perspective. *Can. J. Chem. Eng.* **2011**, *89*, 1191–1206. [CrossRef]
74. Moon, R.J.; Martini, A.; Nairn, J.; Simonsen, J.; Youngblood, J. Cellulose nanomaterials review: Structure, properties and nanocomposites. *Chem. Soc. Rev.* **2011**, *40*, 3941–3994. [CrossRef] [PubMed]

75. Klemm, D.; Kramer, F.; Moritz, S.; Lindström, T.; Ankerfors, M.; Gray, D.; Dorris, A. Nanocelluloses: A New Family of Nature-Based Materials. *Angew. Chem. Int. Ed.* **2011**, *50*, 5438–5466. [CrossRef] [PubMed]
76. Lam, E.; Male, K.B.; Chong, J.H.; Leung, A.C.W.; Luong, J.H.T. Applications of functionalized and nanoparticle-modified nanocrystalline cellulose. *Trends Biotechnol.* **2012**, *30*, 283–290. [CrossRef] [PubMed]
77. American Society for Testing and Materials (ASTM) International. *Standard Test Method for Testing Polyurethane Raw Materials: Determination of Hydroxyl Numbers of Polyols*; ASTM D4274-05; ASTM International: West Conshohocken, PA, USA, 2005.
78. American Society for Testing and Materials (ASTM) International. *Standard Test Method for Acid and Base Number by Color-Indicator Titration*; ASTM D974-08; ASTM International: West Conshohocken, PA, USA, 2008.
79. Segal, L.; Creely, J.J.; Martin, A.E.; Conrad, C.M. An empirical method for estimating the degree of crystallinity of native cellulose using the X-ray diffractometer. *Text. Res. J.* **1959**, *29*, 786–794. [CrossRef]
80. Klug, H.P.; Alexander, L.E. (Eds.) *X-ray Diffraction Procedures: For Polycrystalline and Amorphous Materials*; John Wiley & Sons: New York, NY, USA, 1974; ISBN 978-0-471-49369-3.

Green-Bio-Economy and Bio-Nanotechnology for a More Sustainable Environment

Pierfrancesco Morganti, Hong-Duo Chen and Yuan-Hong Li

Abstract: There is growing recognition that the transition to a green bioeconomy can generate more sustainable growth. This is the reason why the EOCD and EU economic platform until 2030 is based on the use of natural raw materials obtained from plant biomass and fishery waste, in substitution to fossil-derived ones. From this new economic vision, fundamentally focused on innovative agriculture and bionanotechnological systems, great potential for delivering economic growth with environmental protection and social inclusion is expected. Thus the green bioeconomy will have a positive impact on our quality of life, maintaining in equilibrium the planet's ecosystem and biodiversity. Since the realization of a global sustainability depends on renewable sources of materials and energy, bio-based polymers and products to replace petroleum-based ones must become a mainstay of our society. The use of chitin nanofibrils (obtained from crustacean waste) to produce goods and innovative nanocomposites is a step in this direction, as reported in this chapter.

1. Introduction

The bio-economy, synonym for a green economy or ecological economics [1–3], can be described as an interdisciplinary field of academic research that, strictly connected with the bio-nanotechnologies, is based on building blocks of materials, chemicals, and energy derived from renewable industrial byproducts such as plant biomass and fishery waste [4,5].

This new branch of economy, based on the use of natural bio-energy instead of fossil fuels and focused on socioeconomic, agricultural, and technical systems, will represent a fundamental platform of the 2030-EU and OECD economy [6,7]. This new economic vision will meet the requirements for industrial sustainability and environmentally friendly social and economic impacts. At the same time, it is essential not only to increase human well-being and decrease environmental pollution and climate change, but also to stop the depletion of natural resources [1–7]. To this purpose, both biotechnologies and nanotechnologies will be indispensable to stimulate the increase of a sustainable economy, fundamental for developing jobs and the industrial production of tools and energy at low cost. This will make it possible to maintain the actual standard of living, ameliorating its quality and considering human health and wellness as the real goal for an advanced economy and a safe industrial development [6–8].

From this knowledge-based bio-economy (KBBE) potential benefits are expected, such as a reduction of greenhouse gas emissions, a decrease in dependence on fossil "resources, wiser management of natural resources, and improved food security" [6–9].

2. Maintaining Biodiversity through an Ecosystem in Equilibrium

For all these reasons, a must of our future society should be the amelioration of the Earth's habitat, developing a sustainable economy based on the use of materials and processes of biological origin [10].

This new way to produce goods and services has to be based on the same methodologies adopted from nature, such as the use of enzymatic reactions instead of the classical chemical reactions. Thus the bio-economy, driving the growth of smart agriculture and forestry sustainability by innovative processes of bio-engineered technologies, must produce enzymes, amino acids, and active ingredients for pharmaceutical, food, and cosmetic products, as well as make biopolymers and bio-fibers from renewable industrial resources and biofuel produced by bio-refineries, necessary for transporting raw materials and goods [4,10].

It is interesting to underline that, in this different economic vision, the use of industrial byproducts derived from plant biomass and fishery waste (Figure 1) [11,12] should be considered an alternative way to produce goods without impoverishing the earth of its natural resources, which is indispensable to maintain the ecosystem' equilibrium and species biodiversity [13].

Figure 1. Plant and fishery waste.

Thus, the green bio-economy could solve the majority of the aspects required for industrial sustainability, which, by respecting the biodiversity of our planet, will safeguard the health of humans, animals, and plants together (Figure 2).

Figure 2. Biodiversity and socio-assets of the planet.

3. The Bio-Economy and Industrial Sustainable Development

The green bio-economy is considered a way to ensure all the social, economic, and environmental prerequisites to meet a sustainable development. However, it is also increasingly recognized as a greater economic opportunity for removing barriers and enabling poor and disadvantaged groups to participate in, contribute to, and benefit from the transition [14]. To obtain these results it will be necessary to develop a new economy by meeting different economic, social, environmental, and political parameters.

To this purpose, the following will be necessary to realize: (a) an economic growth capable to support the efficient use of resources with low carbon emissions, and favoring the adoption of green and high-tech nano-biotechnologies with the use of raw materials, obtained from byproducts and poor countries; (b) a social development that can improve health and well-being, especially among the poor, thus promoting social equity; (c) an environmental sustainability based on the increase of productivity and efficiency of the natural resources used, with a simultaneous reduction of pollution obtained by the right investments necessary to sustain ecosystem health, and resilience in management of environmental risks; (d) a governance that can empower citizens through access to information, justice, and participation in decision-making, particularly among marginalized groups, contemporary improving transparency and accountability in the public and private sectors, including a better regulation of the market [1].

According to EU, OECD, and USA reports [6,7,15], it seems that the economic, environmental, and social benefits ,obtained from the use of nanotechnologies

(Figure 3), could represent the optimum route towards sustainable development at a regional, national, and international level.

Shifting to the European view, the bio-economy is defined as "the production of renewable biological resources and the conversion of these resources and waste into value-added products, such as food, feed, bio-based products and bio energy" [9,16,17]. Environmental assets such as fertile soil, clean water, and biodiversity help drive economic growth, contributing to public health and providing safety nets for the poor [18]. Thus, investing in improved natural resources through proper environmental management in rural and urban areas, such as sustainable forestry and fisheries, can not only reduce carbon emissions but generate also high social rates of return.

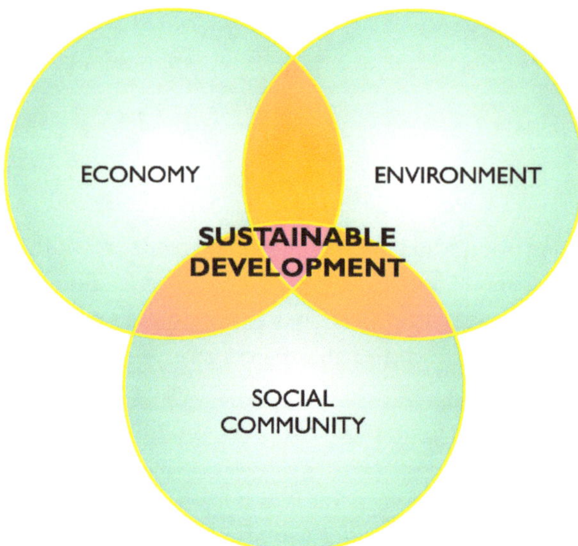

Figure 3. Sustainable development.

However the success of using nano-biotechnologies and byproducts as principal raw materials to make goods and produce clean energy is subject to institutional capacity within implementing agencies and technological organizations from the municipal level up, to meet the requirements with minimal costs. The process of transforming a fossil-based economy into a novel green economy takes time and requires continuous government intervention. The necessary technologies require a strong emphasis to realize a sustainable growth, together with up-front investments in research, development, and innovation.

Thus, the EU has emphasized the research projects by the Seventh Program Framework and are going in this direction with the Horizon 2020 program also [18,19].

To this purpose, our group has realized three interesting projects: Bio-Mimetic (www.biomimetic.eu), coordinated by P&G UK, n-Chitopack (www.n-Chitopack.eu), and Chitofarma (www.chitopharmam.it), both coordinated by MAVI Sud, Italy. By these projects innovative cosmetic products, biodegradable food packaging, and advanced medications have been developed, using chitin nanofibrils and other natural polymers, obtained from fishery waste and plant biomass, respectively, by the use of green industrial processes with low consumption of energy and water [20–22].

In this way new green technologies and new safe products have been developed, with added value for consumers in the fundamental fields of cosmetic, food, wellness, and biomedicine. The food industry is, in fact, one of the largest sectors of the European economy, with a yearly turnover of €800 billion and roughly 4 million employees; the EU cosmetics industry had a turnover of €69 billion in 2012, representing one-third of the global cosmetic market, with direct and indirect employment in EU of 1.7 million people and more than 4000 industry companies, while the global sheet face mask market is expected to reach US$336 million by 2024 (source:Trasparency Market Research) with a year growth of 8.7% from 2016 to 2024 [23].For all these reasons, it is expected that food ,tissue masks and advanced medications with an expected grow of 11%–12% during the forecast period 2014–2019 will play a more prominent role in the future through an increase in innovative diet supplements, therapeutics, and diagnostics.

Nutrition and biomedicine are, in fact, two economic pillars of progress and the key to health prevention, which requires both high-tech technologies and innovative products. Moreover, technological development is hampered by the aging population of the industrialized countries and the necessity of maintaining a high standard of health care, reinforced by the limited availability of raw materials and energy resources, accompanied by global warming also. Thus, the growing demand for a sustainable supply of food, raw materials, and fuel as the major driving force behind the KBBE will probably be based on a combination between plant breeding and industrial nano-biotechnology [7–9,24].

4. The Bioeconomy and R&D Worldwide

The bio-economy, as a key component of the green growth, has a current market in the EU of over 2 trillion euros and provides 22 billion jobs across diverse sectors, representing around 9% of the total labor force of different sectors, including agriculture, forestry, food, chemicals, and bio-energy [7,16,17]. Thus, Europe is the global leader and pioneer in bioscience and related technologies [9,24–26], with many investments dedicated to R&D in nano-biotechnologies, considered the most critical component for the industrial innovation. However, since 2006, the world's four largest spenders on R&D were the United States with US$343 billion, the EU with US$231 billion, China with US$136 billion, and Japan with US$130 billion, respectively, while in terms of percentage of GDP, the order was China, Japan, USA, and EU, with approximate percentages of 4.3, 3.2, 2.6, and 1.8, respectively [27].

It is interesting to underline the changing of the top 10 spender countries in terms of percentage of GDP, in 2011 respect 2006 were: Israel 4.3%, South Korea, 4.03%, Finland 3.78%, Japan 3.39%, Sweden 3.37%, Denmark 3.09%, Taiwan 3.02%, Germany 2.88%, USA 2.77%, and Austria 2.75%, with an average spent of US$104,000 per employee on R&D. However, in 2017 the top 10 innovative economies has been: Switzerland that continues to occupy the first position for the 7th consecutive year, followed from Sweden, Netherlands, USA, UK, Denmark, Singapore, Finland, Germany and Ireland [27]. In the top 25 some economies–such as the Netherlands, Denmark, Germany, Japan, France, Israel and China-move, up with the middle-income countries growing more distant to them. The exception is still China that became the 22th most innovative economy in the world.

To this purpose funding mechanisms have been boosted in the EU by the Horizon 2020 program, which has defined the framework for research and innovation for 2014–2020, continuing the Past Seventh Framework Program for Research and Technological Development (FP7) [18,19].

In these programs the chief purpose of funding has been to increase and innovate the fields of Food, Agriculture, Fisheries, and Biotechnology as part of the Bio-economy (FABS) [18,19,24–28].

5. Technological Bio-Revolution and the Bio-Economy

According to the OECD [6,28] the bio-economy, considered a new branch of the economy based on innovative technologies (biotechnology and nanotechnology), could substantially contribute to actual economic processes. The bio-economy, in fact, involves three fundamental pillars of biotechnology: (a) a deep knowledge of the biological systems or living organisms necessary to manufacture products or develop processes that benefit humans, such as molecules, enzymes, chemicals, and bio-materials.

As a result, for example, (a) the use of living cells from yeasts, molds, bacteria, plants, and enzymes to synthesize products would require less energy and create less waste; (b) the use of renewable biomasses and efficient industrial bioprocesses with more sustainable technologies for safeguarding humans and the environment; in this case, the main goal of the so-called bio-refinery has to be the organization of sustainable processing of biomass into a spectrum of marketable products, such as food, feed, and chemicals, producing both high-value, low-volume and low-value high-volume products, minimizing and recycling waste streams also [17,29,30]; (c) the integration of the new biotechnological knowledge with all the classic industrial sectors to obtain a real bio-revolution [31,32].

Thus the bio-economy, if designed and implemented intelligently, will have a positive impact on our quality of life, if we maximize the utilization of the animal and plant biomass, simultaneously meeting key sustainable goals and minimizing cumulative pollutants to maintain in equilibrium the Earth's ecosystem [33,34]. It is, in fact, important to remember that pollutants affect both human and climate health, so that the World Health Organization states that 2.4 million people die each year from causes directly attributable to air pollution, with 1.5 million of deaths attributable to indoor air pollution.

In fact, air and water pollution released from industrial, agricultural, and household waste have contaminated the environment to such an extent that, to save our planet, a rescue plan is urgently required. It seems, therefore, fundamental to recognize the economic value of natural capital and ecosystem services, treating them as goods and services to be valued in monetary terms. As a consequence, the bio-economy has to be considered a promising way to reconcile financial and ecological value, estimating how to maintain a stable environment and respect biodiversity before assessing the cost in dollar terms [33,35].

This is the reason why the bio-based economy was outlined, by the EU's public-oriented definition, as a "production paradigm that, relied on biological processes and with natural ecosystems, uses natural inputs, expends minimum amounts of energy and does not produce waste". So "all materials discarded by one process, will be input for another process to be reused in the ecosystem" [9,36].

By the same approach, the European Association for Bio-industries (EuroPaBio), considering industrial nano-biotechnology a key component of the bio-economy, underlines that "the application of biotechnology for sustainable production of chemicals, materials and fuels from biomass, creates an opportunity to reduce significantly our dependence on coal, oil and gas" [9,37].

Consumer attitude towards nanotechnologies is 64% positive with 71% willingness to purchase nano-products, according to a research study by the Nanotechnology Industries Association (NIA) reported by Denis Koltsov in 2011 (Figure 4), so the global manufacturing output of nanotechnologies is estimated to reach US$30 billion in 2015 (Figure 5) [38].

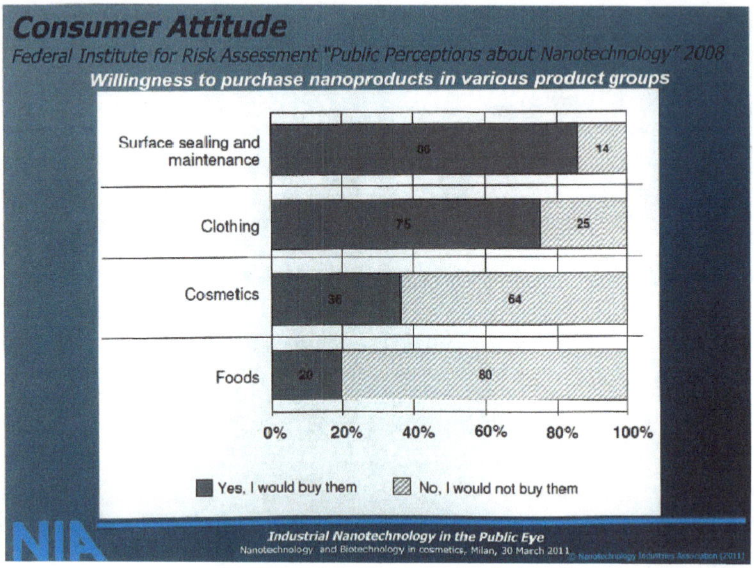

Figure 4. The willingness of consumers to buy nanotechnological products (courtesy of NIA).

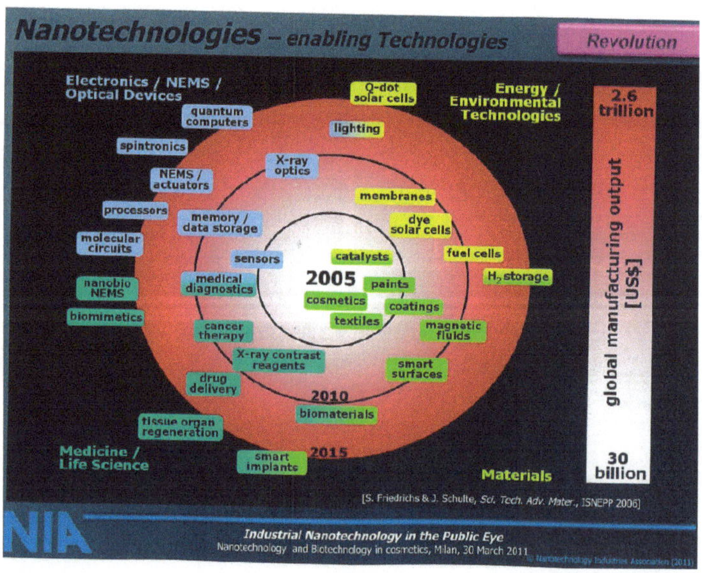

Figure 5. Global manufacturing output of nanotechnologies (courtesy of NIA).

6. Bionanotechnology in Health Care

Bio-nanotechnology has found a wide range of applications in medicine, health care, and cosmetic dermatology, including prevention, diagnosis, and cure of diseases. It is a multidisciplinary field that covers a vast array of devices and machines derived from engineering, physics, material science, chemistry, and biology and deals with the study and application of biological and biochemical activities from elements of nature to target drug delivery, bio-imaging, sensing, and diagnosis of pathologies at early stages [27]. On the other hand, naturally occurring nano-structures in biology have been a source of inspiration for new nanotechnological hybrid nano-structures made of biological and non-biological, organic and inorganic building blocks.

These bio-technological systems involve the fabrication of nano-devices or nano-particles on the order of 100 nm or smaller that, when made of organic polymers, colloids or molecules, including DNA, proteins, lipids, and polysaccharides, are classified as soft nanotechnology [39,40]. The physicochemical properties of nano-particles, in fact, can be engineered at the molecular level; their shape, size, and charge can be controlled, and the surface density of the eventual targeting ligand can be optimized for specific applications [40–43], so that charged particles may create electrostatic interactions with charged skin elements in the interstitial matrix, such as positively charged collagen or negatively charged glycosaminoglycans [44]. This is the reason why "many efforts to reduce material dimensions are motivated by the attractive properties and functions unique to their nanometer regime".

"At nanoscale, certain properties of matter become scale-dependent, including capillary forces, optical effects/color, conductivity, electron affinity", surface reactivity, and so on [45]. In fact, as material size decreases, the surface properties of the atoms increasingly dominate, producing significant changes in material reactivity [45,46]. Hence, surface effects are a unique and very significant functional nano-property that requires both control and careful characterization for exploitation in specific nanotechnologies.

According to the different and controlled technologies adopted by our research group, it is possible to produce block co-polymeric micro/nano-lamellae or nano-particles (NPs) (Figure 6) made of chitin nanfibrils (electropositive polymers) and hyaluronan (an electronegative polymer), which, by entrapping different active ingredients, may have their surface covered by positive or negative charges [29,47–51].

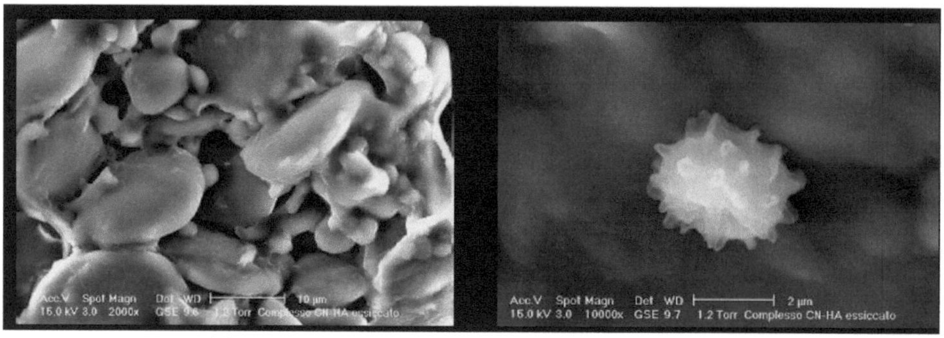

(a) (b)

Figure 6. Micro/nano-lamellae (**a**) and nanoparticle at SEM (**b**).

When the surface is covered with positive charges these NPs show an ability to disturb lipid lamellae of the stratum corneum, enabling better diffusion of the entrapped active ingredients through the skin layers; when their surface is covered with negative charges the active ingreduents seem to remain at the level of the more external corneocytes (Figure 7).

In the first case, by entrapping the positive nano-particles with antioxidant ingredients it is possible to design formulations with anti-aging activity, while in the second case the negative nano-particles, entrapped with sunscreen ingredients, have been shown to be effective for sunscreen emulsions to protect the skin against the sun [50–52].

In the same way, it was possible to produce interesting composites by the use of natural fibers for making non-woven tissues made by electrospinning (Figure 8a) or casting technology (Figure 8b), for cosmetic or medical utilization.

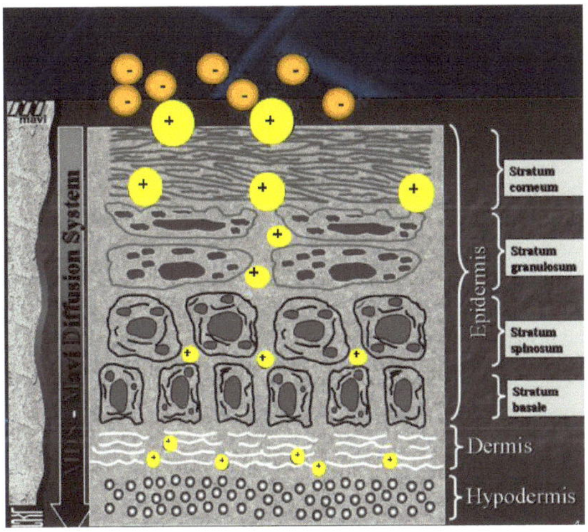

Figure 7. Negatively charged nano-particles remain on the skin surface; meanwhile, positively charged ones penetrate through the different skin layers depending on their dimension.

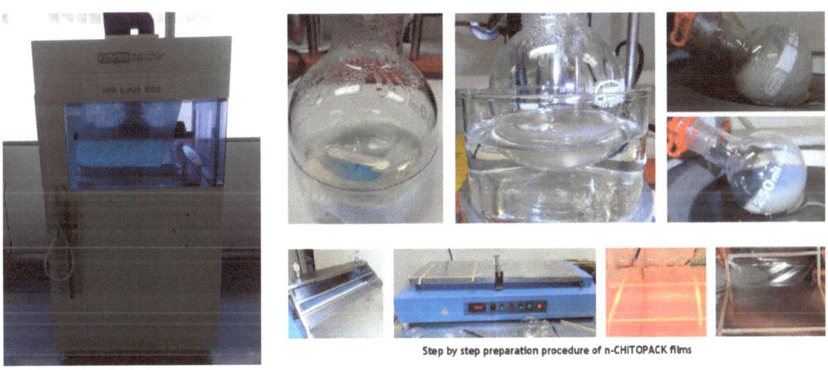

(a) (b)

Figure 8. (**a**) The electrospinning technology at lab level; (**b**) the casting technology at lab level.

It is important to remember that a composite may be defined as a physical mixture of two or more different materials, having properties that are generally better than those of any one of the materials used.

Different from synthetic fibers, natural ones possess more desirable properties such as biodegradability, renewability, and a lower price, although they may have poorer mechanical properties. Composites are made of a strong load-carrying material (known as the reinforcement filler) embedded in a weaker material (known as the matrix). Reinforcement provides strength and rigidity, helping to support the structural load. The matrix or binder (organic or inorganic) maintains the position and orientation of the reinforcement. Other advantages of natural fibers lie in their low density, high toughness, comparable specific strength, reaction in tool wear, ease of separation, and low energy of fabrication.

Naturally, the type of the final natural composite is dependent upon the variations in the characteristics and amount of their components, such as cellulose, lignin, pectin, or chitin nano-fibrils, as well as their chemical structures and the production processes. While natural fibers are an interesting option for wider application in composite technology, natural ingredients are of increased use in the cosmetic and food fields [53].

Chitin nano-fibrils (CN), obtained from fishery and crustacean waste, and ligno-cellulosic polymers from plant biomass provide powerful toolboxes for innovative nano-biotechnological processes. These natural polymers offer, in fact, characteristics with interesting properties for various purposes (energy, textiles, cosmetics, biomedicine, etc.) [54].

In cosmetic dermatology CN, complexed or non-complexed with different active ingredients, have been embedded into numerous emulsions characterized by anti-aging (Figure 9) [55], anti-inflammatory (Figure 10) [56,57], anti-acne (Figure 11) [58], or photo-protective activity (Figure 12) [59].

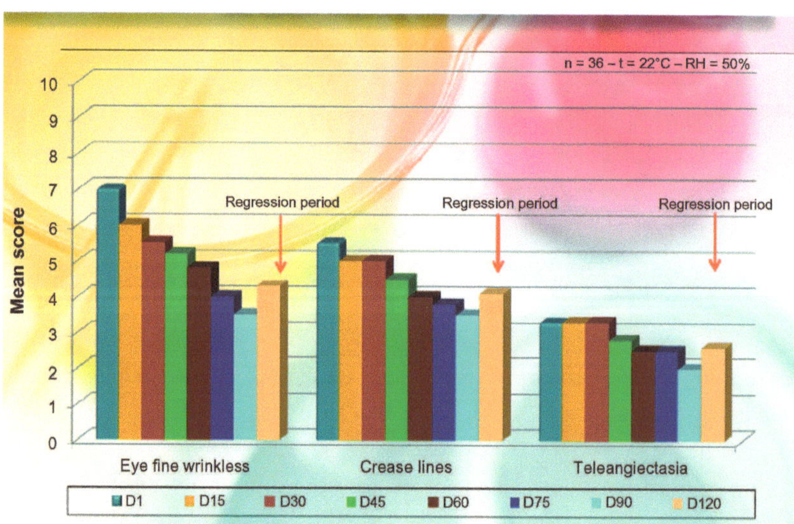

Figure 9. Dermatological mean evaluation on sign of photo-aging after injective treatment with phosphatidylcholine-hyaluronic acid-chitin nano-fibrils encapsulating active compounds (BPN). Note: all p values are highly significant in comparison to the baseline ($p < 0.005$). Abbreviations: BPN block-polymer nano-particles; RH: Relative Humidity; D: day.

Moreover, CN have been bound with other natural polymers to produce non-woven tissues for developing wound dressings by electrospinning (Figure 13), or to make films for food packaging by the casting technology (Figure 14) (unpublished data).

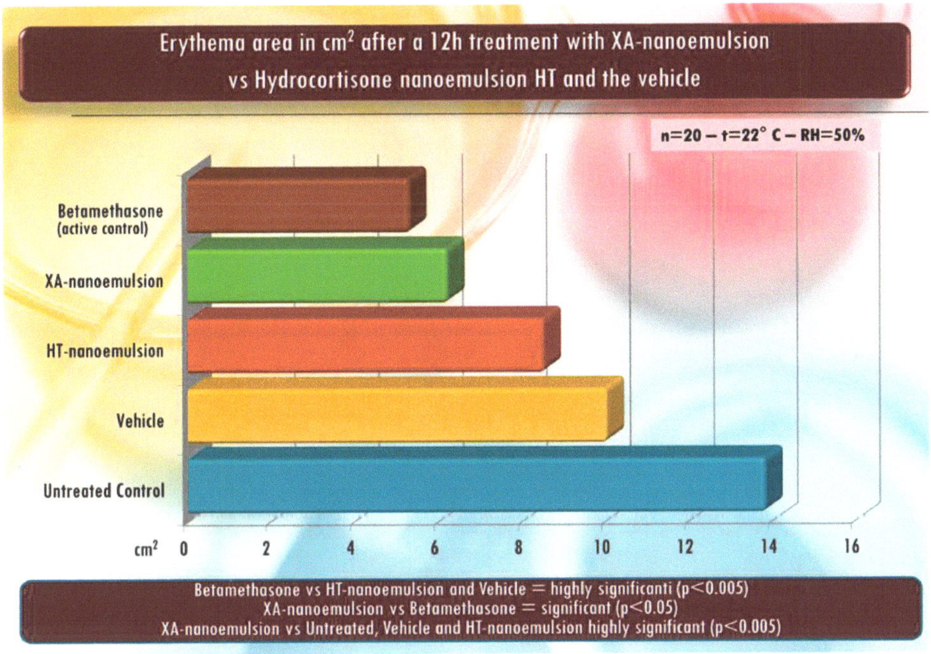

Figure 10. Anti-inflammatory activity of a CN-HA emulsion entrapping active ingredients (XA-nano-emulsion) compared to betamethasone and hydrocortisone (HI) emulsion.

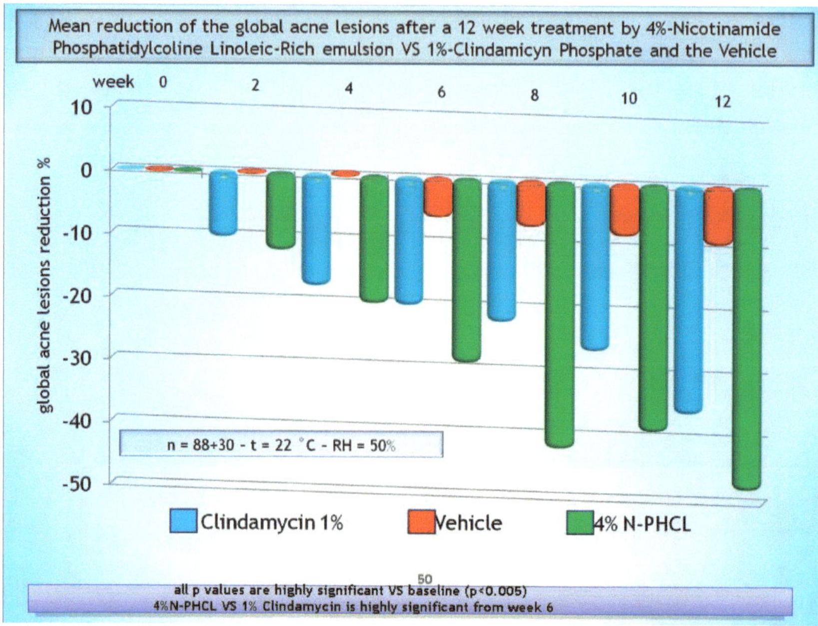

Figure 11. Reduction of acne lesions by a CN-phosphatidylcholine emulsion compared to the clindamycin phosphate emulsion activity.

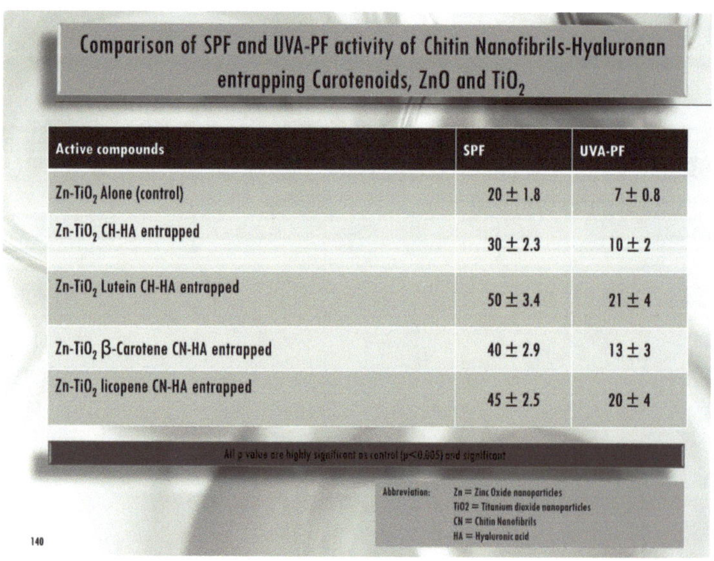

Active compounds	SPF	UVA-PF
Zn-TiO_2 Alone (control)	20 ± 1.8	7 ± 0.8
Zn-TiO_2 CH-HA entrapped	30 ± 2.3	10 ± 2
Zn-TiO_2 Lutein CH-HA entrapped	50 ± 3.4	21 ± 4
Zn-TiO_2 β-Carotene CN-HA entrapped	40 ± 2.9	13 ± 3
Zn-TiO_2 licopene CN-HA entrapped	45 ± 2.5	20 ± 4

Figure 12. Sun protective activity of CN-HA block co-polymers entrapping different sunscreen agents.

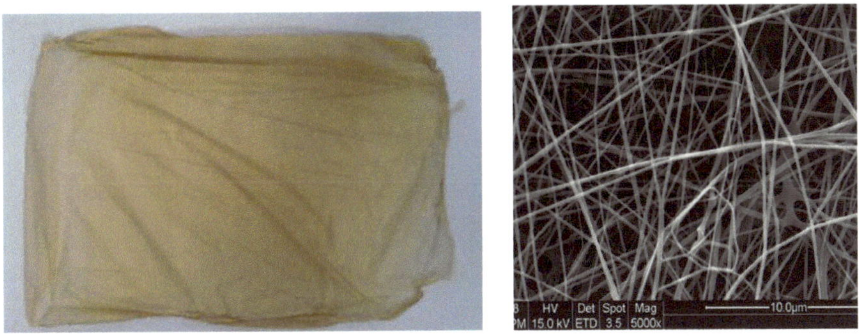

(a) (b)

Figure 13. (**a**) is the wound dressing made by electrospinning. (**b**) is the same tissue at SEM.

The most important features of wound dressing are bio-adhesion to the wounded site and appropriate humidity, which are effective against the so called burden-microorganisms. Due to the difficulty of obtaining the ideal non-woven tissue, wound therapy represents a challenging area in drug product development. In the USA alone, more than 6.5 million patients are affected annually by burns and wounds, while the costs of treatment are estimated to be US$25 billion per year [59,60]. Moreover, wound infections are the most serious complications related to burn injuries and up to 1% of the world population requires medical treatment each year for burn injuries [60–62].

Figure 14. CS/CN film produced by the casting technique.

For all these reasons, modern wound dressings should preserve a humid environment, creating a protective barrier against both mechanical stress and secondary infections. Therefore, the dressing has to enable absorption of wound exudate and elimination of pathogen microorganisms, be safe, non-irritant, and acceptable to the patient with a low cost per unit, which would reduce the need for dressing changes [63,64]. These are the challenges solved by the Chitofarma research project.

7. Concluding Remarks

Global sustainability depends on finding renewable sources of materials and energy, so there is an ever-increasing need to develop bio-based polymers and products able to replace petroleum-based ones. Research in this field has shown the strong potential of generating high-performance functionalized polymers and nano-particles from plant and animal biomass. With the anticipated large-scale production of lignin, cellulose, and hemicellulosic polysaccharides from plant biomass as well as of chitin, chitosan, and oligosaccharides from fishery waste, renewable feedstock for nano-particles, biopolymers, and bio-composites will be available, having physicochemical properties that match or exceed those of petroleum-based compounds [65].

Thanks to our growing knowledge, bioinformatics, more secure energy supply, and the interaction of engineering and life sciences, it will be possible to find new foodstuffs, innovative cosmetics, and drugs opening avenues for developing novel crops and plants, as well as new diagnostic and therapeutic options. Thus, improved management of ecosystem goods and services, carried out by socially sustainable national and international institutions in both the public and private sectors, will increase the health and productivity of all the environmental assets, expanding and securing the green bio-economy and harnessing innovation to ensure well-being for all (Figure 15).

Figure 15. Green economy to protect the environment and human well-being.

In conclusion, the transition from the normal economy to the innovative, green bio-economy is a real future challenge for politicians and all scientists operating in the fields of cosmetology and dermatology who wish to promote and support equitable and pedagogic skills at all levels, respecting different cultural values, supporting equity between and within countries and between generations, and ensuring the conservation of natural resources with optimal water and energy use, thus restoring lost biodiversity, striving for zero emissions and zero waste, and promoting the quality of life over the long term.

Acknowledgments: We thank the EU for the economic support given to the European Research Projects: n-Chitopack grant no. 315233, and Biomimetic grant no. 282945.

Author Contributions: This chapter has been written. with the same contribution by all the authors

Conflicts of Interest: We declare that Hong-Duo Chen and Yuan-Hong Li have no conflict of interests. Pierfrancesco Morganti works as Head of R&D Centre of Nanoscience, Mavi Sud, s.r.l, Italy.

References

1. Building an Inclusive Green Economy for All. Available online: http://www.wbcsd.org/Clusters/Social-Impact/Resources/Building-an-Inclusive-Green-Economy-for-All (accessed on 27 October 2017).
2. Shmelev, S.E. *Ecological Economics: Sustainability in Practice*; Springer: New York, NY, USA, 2012.
3. Spash, C.L. Social ecological economics: Understanding the past to see the future. *Am. J. Econ. Soc.* **2012**, *70*, 340–375.
4. Converting Waste Agricultural Biomass into a Resource (2009). Available online: www.unep.org/publications/ (accessed on 27 Spetember 2017).
5. Fishery and Aquaculture Statistics. Available online: http://www.fao.org/docrep/015/ba0058t/ba0058t.pdf (accessed on 27 October 2017).
6. The Bioeconomy to 2030: Designing a Policy Agenda-OECD. Available online: http://www.oecd.org/futures/long-termtechnologicalsocietalchallenges/thebioeconomyto2030designingapolicyagenda.htm (accessed on 27 October 2017).
7. Innovating for Sustainable Growth: A Bioeconomy for Europe. Available online: https://wbc-rti.info/object/event/11271/attach/05_Bio_economy_for_Europe.pdf (accessed on 27 October 2017).
8. En Route to the Knowledge-Based Bio-Economy. Available online: http://www.bs-witzenhausen.de/cms/Download/2007-cologne-paper_property_pdf.pdf (accessed on 27 October 2017).
9. McCormick, K.; Kautto, N. The Bioeconomy in Europe: An overview. *Sustainability* **2013**, *5*, 2589–2608. [CrossRef]
10. Langeveld, J.W.; Sanders, J.P.M.; Meesen, M. *The Biobased Economy: Biofuels, Materials and Chemicals in the Post-Oil Era*; Earthscan: London, UK, 2010; pp. 3–17.

11. Morganti, P.; Del Ciotto, P.; Carezzi, F.; Morganti, G.; Chen, H.-D. From Waste Material a New and Safe Anti- aging Compound: A Chitin Nanofiber Complex. *SOEW J.* **2012**, *138*, 128–138.
12. Morganti, P.; Li, Y.H. Healthy products from waste materials. *Eur. Cosmet. China Spec. Issue* **2012**, *20*, 60–63.
13. UNEP. *Biodiversity Barometer 2013, United Nations Environment Programme 19 April*; UNEP: Montréal, QC, Canada, 2013.
14. The Economist. *Biotechnology: Chemistry Goes Green*; The Economist Newspaper Ltd.: London, UK, 2010.
15. White House. *National Bioeconomy Blueprint*; White House: Washington, DC, USA, 2012.
16. European Commission Research DG. *New Perspective on the Knowledge-Based Bio-Economy*; European Commission: Brussels, Belgium, 2005.
17. Bio-Economy Technology Platforms (BECOTEPS). *The European Bioeconomy in 2030: Delivering Sustainable Growth by Addressing the Grand Societal Challenges*; Bio-Economy Technology Platforms: Brussels, Belgium, 2011.
18. Europe 2020: Commission Proposes New Economic Strategy in Europe. Available online: http://csdle.lex.unict.it/docs/labourweb/Europe-2020-Commission-proposes-new-economic-strategy-in-Europe-/369.aspx (accessed on 27 October 2017).
19. Horizon 2020 Italia. Available online: https://www.researchitaly.it/en/horizon-2020 (accessed on 27 October 2017).
20. Morganti, P. Saving the Environment by Nanotechnology and Waste Material: Use of Chitin Nanofibrille by EU Research Projects. *J. Appl. Cosmetol.* **2013**, *31*, 89–96.
21. Morganti, P. Innovation, Nanotechnology, and Industrial Sustainability by the Use of Underutilized By-Products: The EU Support to SMEs. *J. Mol. Biochem.* **2013**, *2*, 137–141.
22. Morganti, P. To Improve Quality of Life Minimizing the Environmental Impact. *SOFW J.* **2013**, *139*, 66–72.
23. IB 2025: Maximizing UK Opportunities from Industrial Biotechnology in a Low Carbon Economy. Available online: http://webarchive.nationalarchives.gov.uk/20090609032547/http://www.berr.gov.uk/files/file51144.pdf (accessed on 27 October 2017).
24. Sheet Face Masks Market (Fabric-Non-Woven, Cotton, Hydrogel, and Bio-Cellulose; Category-Premium Sheet Face Masks and Mass Sheet Face Masks)-Global Industry Analysis, Size, Share, Growth, Trends and Forecast 2016–2024. Available online: https://www.transparencymarketresearch.com/sheet-face-masks-market.html (accessed on 27 October 2017).
25. Clever Consultant. *The Knowledge Based Bio-Economy in Europe: Achievements and Challenges*; Clever Consultant: Brussels, Belgium, 2010.
26. ENDS Europe. Eu Biotech Research Plan. Available online: www.endseurope.com/ (accessed on 22 March 2014).
27. Firdos, A.K. *Biotechnology Fundamentals*; CRC Press: Boca Raton, FL, USA, 2012.
28. The Global Innovation Index 2017. Available online: http://www.wipo.int/edocs/pubdocs/en/wipo_pub_gii_2017.pdf (accessed on 27 October 2017).

29. Fernando, S.; Adhikari, S.; Chandrapal, C.; Amorality, N. Biorefineries: Current status, challenges, and future directions. *Energy Fuel* **2006**, *20*, 1727–1737. [CrossRef]
30. Morganti, P.; Del Ciotto, P.; Fabrizi, G.; Guarneri, F.; Cardillo, A.; Palombo, M.; Morganti, G. Safety and Tolerability of a Chitin Nanofibrils-Hyaluronic acid Nanoparticles Entrapping Lutein. Note I: Nanoparticles Characterization and Bioavailability. *SOFW J.* **2013**, *139*, 12–23.
31. Rothaermal, F.; Thursby, M. The nanotech versus the biotech revolution: Sources of productivity in incumbent firm research. *Regul. Policy* **2007**, *36*, 832–849. [CrossRef]
32. The Third Industrial Revolution. Available online: http://www.economist.com/node/21553017 (accessed on 27 October 2017).
33. Harris, J.; Roach, B. *Environmental and Natural Resource Economics: A Contemporary Approach*, 3rd ed.; ME Sharpe: Armonk, NY, USA, 2013; ISBN 0765637928.
34. Hatta, G.M.; Lovejoy, T.E.; Bennett, C.; Rockstrom, J.; Watson, B.; Steiner, A.; McAlpine, J.; Barbut, M. *Biodiversity and Ecosystem Insecurity: A Planet in Peril*; Djoghlaf, A., Dodds, F., Eds.; Stakeholder Forum: London, UK, 2017.
35. Union for Ethical Biotrade Releases Results of Biodiverdity Barometer. Available online: https://www.iucn.org/fr/node/14132 (accessed on 27 October 2017).
36. Bio-Based Economy for Europe: State of Play and Future Potential—Part 2. Available online: https://ec.europa.eu/research/consultations/bioeconomy/bio-based-economy-for-europe-part2.pdf (accessed on 27 October 2017).
37. Where Next for the European Bioeconomy? Available online: http://ec.europa.eu/research/bioeconomy/pdf/where-next-for-european-bioeconomy-report-0809102014_en.pdf (accessed on 27 October 2017).
38. Koltsov, D. NIA Industrial Nanotechnology in the Public Eye. In Proceedings of the Nanotechnology Conference, Milan, Italy, 30 March 2011.
39. Hanley, I.W. Nanotechnology with soft materials. *Angew Chem. Int. Ed.* **2003**, *42*, 1692–1712. [CrossRef] [PubMed]
40. Whitesides, G.M.; Lipomi, D.J. Soft nanotechnology: Structure vs function. *Faraday Discuss.* **2009**, *143*, 373–381. [CrossRef] [PubMed]
41. Euliss, L.E.; DuPont, J.A.; Gratton, S.; DeSimone, J. Imparting size, shape, and composition control of materials for nanomedicine. *Chem. Soc. Rev.* **2007**, *38*, 1095–1104. [CrossRef]
42. Gratton, S.E.A.; Ropp, P.A.; Pohlhaus, P.D.; Luft, J.C.; Madden, V.J.; Napier, M.E.; DeSimone, J.M. The effect of particle design on cellular internalization pathways. *Proc. Natl. Acad. Sci. USA* **2008**, *105*, 11613–11618. [CrossRef] [PubMed]
43. Gu, F.; Zhang, L.; Teply, B.A.; Mann, N.; Wang, A.; Radovic-Moreno, A.F.; Langer, R.; Farokhzad, O.C. Precise engineering of targeted Nanoparticles by using self-assembled bio integrated block copolymers. *Proc. Natl. Acad. Sci. USA* **2008**, *105*, 2586–2591. [CrossRef] [PubMed]
44. Jain, R.K.; Stylianopoulos, T. Delivering nanomedicine to solid tumors. *Nat. Rev. Clin. Oncol.* **2010**, *7*, 653–664. [CrossRef] [PubMed]

45. Jones, C.F.; Castner, D.G.; Grainger, D.W. Surface Adsorbates on Nanomaterials and Their Possible Roles in Host Inflammatory and Toxicological Processing. In *Handbook of Immunological Properties of Engineered Nanomaterials*; Dobrovolskaia, M.A., McNeil, S.E., Eds.; World Scientific Publishing: Singapore, 2013.
46. Augean, M.; Rose, J.; Botteri, J.K.; Lowry, G.V.; Jolivet, J.P.; Wiesner, M.D. Towards a definition of inorganic Nanoparticles from an environment, health and safety perspective. *Nat. Nanotechnol.* **2009**, *4*, 634–641.
47. Morganti, P.; Fabrizi, G.; Palombo, P.; Palombo, M.; Ruocco, E.; Cardillo, A.; Morganti, G. Chitin-Nanofibrils: A new cosmetic carrier. *J. Appl. Cosmetol.* **2008**, *26*, 113–128.
48. Morganti, P. Chitin Nanofibrils in skin treatment. *J. Appl. Cosmetol.* **2009**, *27*, 251–270.
49. Morganti, P.; Del Ciotto, P.; Gao, X.H. Skin Delivery and Controlled Release of Active Ingredients Nanoencapsulated. *Cosmet. Sci. Technol.* **2012**, *20*, 37–142.
50. Morganti, P. Use of Chitin Nanofibriks from Biomass for an Innovative Bioeconomy. In *Nanofabrication Using Nanomaterials*; Ebothe, J., Ahmed, W., Eds.; One Central Press: Manchester, UK, 2016; pp. 1–22.
51. Morganti, P.; Palombo, P.; Palombo, M.; Fabrizi, G.; Cardillo, A.; Svolacchia, F.; Guevara, L.; Mezzana, P. A phosphatidylcholine hyaluronic acid chitin–nanofibrils complex for a fast skin remodeling and a rejuvenating look. *Clin. Cosmet. Investig. Dermatol.* **2012**, *5*, 213. [CrossRef] [PubMed]
52. Morganti, P.; Del Ciotto, P.; Morganti, G.; Fabien-Soule, V. Application of Chitin Nanofibril and Collagen of Marin Origin as Bioactive Ingredients. In *Marine Cosmeceuticals: Trends and Prospects*; Kim, S.-K., Ed.; CRC Press: Boca Raton, FL, USA, 2012; pp. 267–289.
53. Morganti, P.; Carezzi, F.; Del Ciotto, P.; Morganti, G. Chitin Nanofibrils as Innovative Delivery System. *Person. Care Eur.* **2012**, *5*, 95–98.
54. Morganti, P.; Tishchenko, G.; Palombo, M.; Kelnar, L.; Brizova, L.; Spirkova, M.; Pavlova, E.; Loners, L.; Carezzi, F. Chitin nanofibrils for biomimetic products: Nanoparticles and nanocomposites chitosan films in health care. In *Marine Biomaterials. Isolation, Characterization and Application*; Kim, S.K., Ed.; CRC Press: Boca Raton, FL, USA, 2013; pp. 681–715.
55. Morganti, P.; Palombo, M.; Fabrizi, G.; Guarneri, F.; Slovacchia, F.; Cardillo, A.; Del Ciotto, P.; Carezzi, F.; Morganti, G. New Insights on Anti-Aging Activity of Chitin Nanofibril-Hyaluronan block-copolymers entrapping active ingredients: in vitro and In vivo study. *J. Appl. Cosmetol.* **2013**, *31*, 1–29.
56. Morganti, P.; Chen, H.D. Una nano emulsione innovative per il trattamento della xerosi cutanea e dell'alterazione della barriera. *Cosmet. Technol.* **2011**, *14*, 29–37.
57. Morganti, P.; Fabrizi, G.; Guarneri, F.; Palombo, M.; Palombo, P.; Cardillo, A.; Ruocco, E.; Del Ciotto, P.; Morganti, G. Repair Activity of Skin Barrier by Chitin-Nanofibrils Complexes. *SOFW J.* **2011**, *137*, 10–26.
58. Morganti, P.; Berardesca, E.; Guarneri, F.; Fabrizi, G.; Palombo, P.; Palombo, M. Topical clindamicyn 1% vs phosphatidylcholine-acid rich and nicotianamide 4% in the treatment of acne: A multigenerational-randomized trial. *Int. J. Cosmet. Sci.* **2011**, *13*, 1–10.

59. Morganti, P.; Chen, H.D.; Gao, X.H.; Del Ciotto, P.; Carezzi, F.; Morganti, G. Nanoparticles of Chitin Nanofibril-Hyaluronan block polymer entrapping Lutein as UVA protective compound. In *Carotenoids. Food, Source, Production and Health Benefits*; Yamaguchi, M., Ed.; Nova Science Publishers, Inc.: New York, NY, USA, 2014; pp. 237–259.
60. Sen, C.K.; Gordillo, G.M.; Roy, S.; Kirsner, R.; Lambert, L.; Hunt, T.K.; Gottrup, E.; Gunter, G.C.; Longaker, M.T. Human Skin Wounds: A Major and Snowballing Threat to Public Health and the Economy. *Wound Repair Regen.* **2009**, *17*, 763–771. [CrossRef] [PubMed]
61. Hurler, J.; Skalko-Basnet, N. Potential of Chitosan-based Delivery Systems in Wound Therapy: Bioadhesion Study. *J. Funct. Biomater.* **2012**, *3*, 37–48. [CrossRef] [PubMed]
62. Murphy, K.D.; Lee, J.O.; Herndon, D.N. Current pharmacotherapy for the treatment of severe burns. *Expert Opin. Pharmacother.* **2003**, *4*, 369–384. [CrossRef] [PubMed]
63. Elsner, J.J.; Egozi, D.; Ullman, Y.; Berdicevsky, I.; Shefy-Peleg, A.; Ziberman, M. Novel biodegradable composite wound dressings with controlled release of antibiotics: Results in a guinea pig burn model. *Burns* **2011**, *37*, 896–904. [CrossRef] [PubMed]
64. Gaossens, A.; Cleenewerck, M.B. New Wound dressings: Classification, and tolerance. *Eur. J. Dermatol.* **2010**, *20*, 24–26.
65. Ten, E.; Vermerris, W. Functionalized Polymers from Lignocellulosic Biomass: State of the Art. *Polymers* **2013**, *5*, 600–642. [CrossRef]

Chitin Nanofibrils, a Natural Polymer from Fishery Waste: Nanoparticle and Nanocomposite Characteristics

Pierfrancesco Morganti, Gianluca Morganti and Maria Luisa Nunziata

Abstract: Chitin nanofibrils (CNs), obtained by a patented industrial process, is a pure linear alpha-crystal polysaccharide of acetyl-D-glucosamine and D-glucosamine with a mean dimension of 5 × 7 × 240 nm. By the process a colloidal aqueous suspension, containing ~300 billion of positively charged pure nanofibrils, is obtained. For their physicochemical character CNs form easily block copolymeric nanoparticles (NPs) with macromolecules or natural polymers, negatively charged. These NPs, capable to entrap active ingredients, can be embedded into micro/nano cosmetic pharmaceutical emulsions or into nanocomposite fibers to make non-woven tissues. NPs not only have the capacity to increase the effectiveness of ingredients protect them from environmental aggressions, but also to release them in different skin/mucous layers at different times, depending on the formulation methodology used. Effectiveness and safeness of chitin nanoparticles are reported and discussed in this chapter.

1. Introduction

Chitin nanofibrils (CNs) [1,2], are pure crystalline structures obtained by an industrial patented process free of any waste material [3]. They are made of linear alpha polysaccharides of *N*-acetyl-D-glucosamine and D-Glucosamine with a mean dimension of 5 × 7 × 240 nanometers (nm) and the same backbone of hyaluronic acid (Figure 1) [4,5].

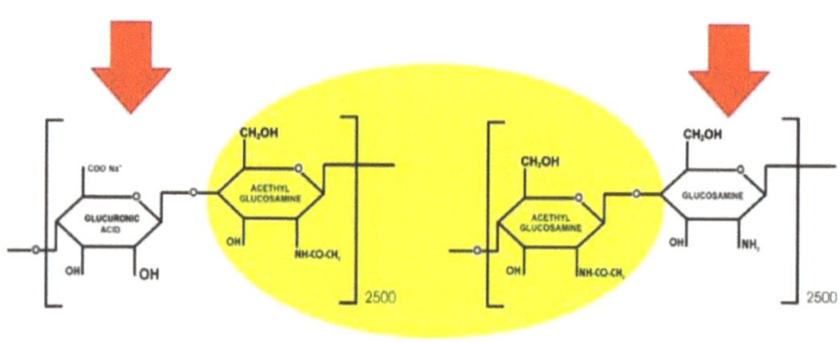

Figure 1. Chitin and hyaluronic acid have the same backbone (courtesy of MAVI SUD, Italy).

By this patented process a colloidal aqueous suspension of CNs are obtained, containing about 300 billion pure nanofibrils per milliliter, as shown elsewhere (Figure 2) [4,5]. A nanometer, with a range equivalent to the billionth of a meter (80,000 times thinner than a human hair), covers sizes smaller than the wavelength range of visible light but bigger than several atoms (Figure 3) [6,7].

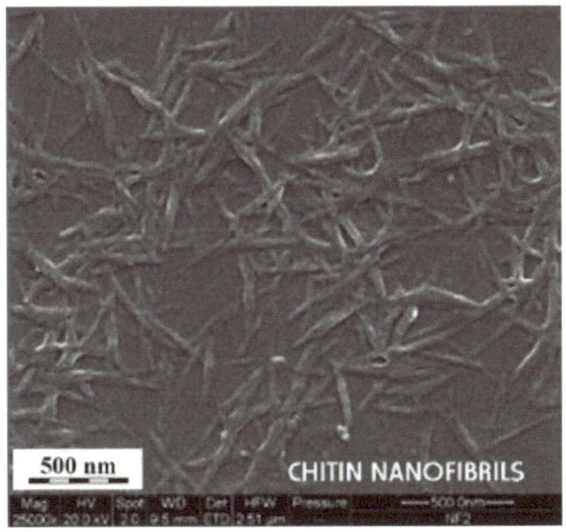

Figure 2. Chitin nanofibrils at SEM.

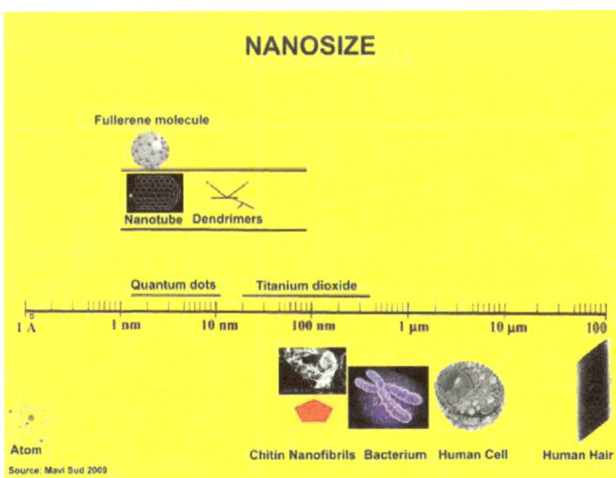

Figure 3. The nanometer dimension [7].

These electropositive nanocrystals with interesting physicochemical characteristics and properties (Figure 4) easily form block copolymeric micro/nanoparticles with electronegative compounds, and give more mechanical strenght to natural or man-made polymers, rendering their activity more effective [8–13].

The higher acetylation degree of chitin can, in fact, contribute to the formation of hydrogen bonds, stabilizing the crystalline structure [8], giving not only a greater resistance to the composite fiber made, for example, with PLA or chitosan [9], but also showing antibacterial activity and lower toxicity [10]. On the one hand, nanocrystals could institute a bridge between single molecules and bulk systems, modifying and changing the polymer characteristics [11]. On the other hand, the nanometer range and high-surface-area-to volume ratio of CNs allow them to interact closely with microbial membranes; they display most significant antibacterial and antifungal effects different from those of the bulk counterparts [11,12]. Moreover, these nanocrystals, covered by positive charges on their surface and capable to entrap active ingredients, have the ability to disturb the tight lamellar layers of the Stratum Corneum (SC), enabling better diffusion of the ingredients through the skin layers [13].

On the contrary, when the nanoparticles are negatively charged, the active ingredients remain at the level of the outermost skin. So, it is possible to modify the activity and effectiveness of the CN block-copolymeric nanoparticles not only by selecting the active ingredients, but also by modulating the electrical charges covering their surface. However, these fibrillar bio-nanoparticles are characterized by remarkable properties such as outstanding stiffness and strength, thermostability, barrier properties, degradability, and sustainability, with a global availability from renewable resources and food waste [5,12,13].

Thus, CNs find many applications in foods, advanced medications, cosmetics, smart textiles, waste water treatments, and other biotechnological products [14–16] (Figure 5).

For this purpose it is interesting to underline that the significant amount of underutilized waste resulting from the industrial processing of seafood and plant biomass became a problem both for the environment and human health.

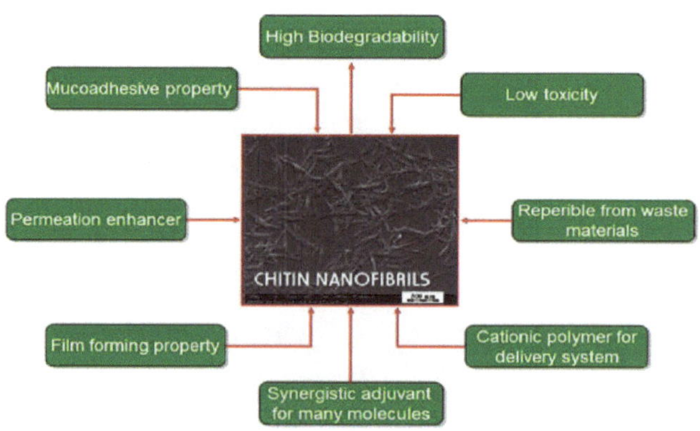

Figure 4. Chitin Nanofibril characteristics. Source: MAVI SUD 2010.

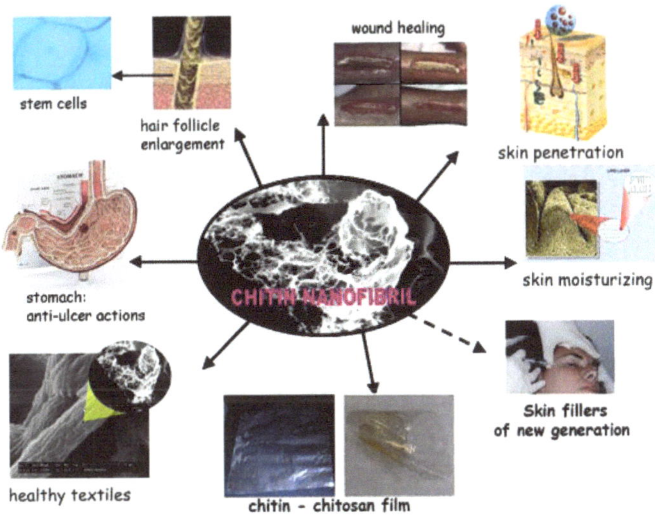

Figure 5. Different uses of Chitin nanofibrils. Source: MAVI SUD 2010.

According to the Food Agricultural Organization (FAO) Yearbook, in fact, the annual total production of crustacean and fishery's, has reached 148.5 million tons in 2010, 45%, producing ~50–70% waste [17]. On the other hand, 140 billion tons of plant biomass are generated every year from agriculture according to the United Nations Environment Program (UNEP) [18]. Therefore fisheries and plant biomass, widely available and renewable materials still largely underutilized byproducts represent an important source of raw material at a low cost (virtually free).

Aside from being carbon-neutral biomass used for energy and innovative goods it could reduce greenhouse gas (GHG) emissions and dependency on fossil fuels, closing the carbon cycle loop and contributing to climate change mitigation. Thus, the main industrial objectives are to select raw materials (possibly obtained from wastes and byproducts), identifying and assessing environmentally sound technologies to convert chitinous and lignocellulosic biomass into energy and innovative goods.

The use of chitin nanofibrils to produce nanoparticles and nanocomposites is the object of the following section.

2. Chitin, Chitosan and Chitoolisaccharides

While chitosan is the low acetyl substituted form of chitin, composed mainly of Beta (1–4)-2-deoxy-2-amino-D-glucopyranose residues [19], chitoolisaccharides are the smallest chitosan oligomers, characterized by different molecular weights with unusual biological activities [20] (Figure 6).

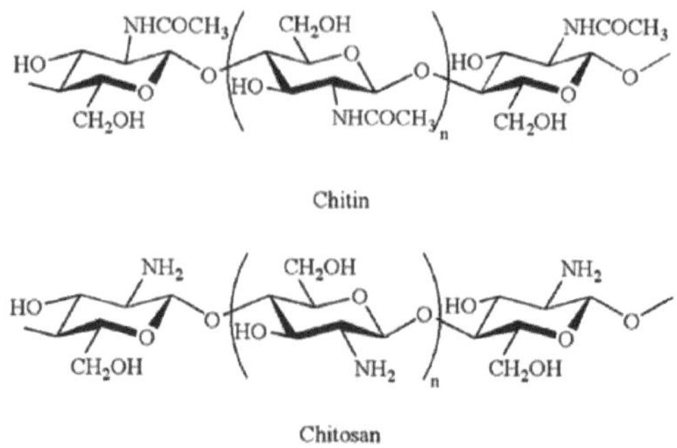

Figure 6. Chemical structures of chitin and chitosan. When the degree of *N*-acetylation (DA) is greater than 50%, the polysaccharide is considered to be chitin. When the DA is less than 50%, the polysaccharide is considered to be chitosan.

These natural polymers, which elicitate defence responses in mammals and plant tissues, because of their mucoadhesive property, low toxicity and antifungal/antibacterial activity, have found a variety of applications in different fields such as the biomedical, pharmaceutical, food and environmental industrial [21], as previously reported. Moreover, the presence of amino groups makes chitin and its derivatives easier to modify by chemical reactions than cellulose [22]. In addition, CN, for its crystalline structure, nanosize dimension and natural origin, has been shown to be not only an interesting active carrier for innovative pharmaceutical and cosmetic products, but also a good candidate for reinforcing polymer nanocomposite fibers to be used for advanced medications and food packaging films [23,24].

Specifically in the case of composites and films, the use of biopolymers has emerged as an interesting alternative to fossil fuel-based products. With the impending fossil fuel crisis, the search for and development of alternative chemical/material substitutes is pivotal to reduce human dependence on fossil resources. The various advantages of natural fibers over man-made glass and caron fibers include their low cost, low density, comparable specific tensile properties, reduced energy consumption, lower health risk, renewability, recyclability, and biodegradability, all necessary to safeguard both the environment and human health [25]. Moreover, owing to the increase in the concept of ecological safety and utilization of renewable materials towards a greener society, the application of natural fibers in industry as bio-filler/reinforcement materials in composite has considerably increased in recent years [26]. These composites are, in fact, eco-friendly to a greater degree and offer a lower density, better matrix–fiber compatibility, and recyclability compared to conventional ones [27,28].

For all these reasons, CN has shown interesting properties for both its crystallinity and nanodimension. In any case, chitin nanofibrils and chitosan possess interesting characteristics qualifying their use in wound healing for their capacity to (a) attract and activate macrophages and neutrophils initiating the healing process; (b) promote granulation and re-epithelialization of the tissue; (c) limit scar formation and retraction; (d) show analgesic and haemostatic activity; (e) activate the immunocompetent cells function; and (f) stimulate the cellular activities by the release of monomers and oligomers of glucosamine and *N*-acetyl glucosamine, used as building blocks in the synthesis of the natural Extra Cellular Matrix (ECM) [29–34]. Moreover, they show interesting skin protective activity and antimicrobial/antifungal effectiveness [10,15].

In addition, CN and chitosan, as polymers characterized by their good adsorption and attraction ability towards transition metal ions and electronegative polymers, easily form stable chelate compounds and block co-polymeric nanoparticles, respectively (Figure 7). This complex-forming ability, useful to clean air and water or produce specialized pharmaceutical and cosmetic carriers, mostly depends on polymer parameters such as de-acetylation degree, polymer chain length and crystallinity [35,36], as well as on its own absorptive properties and physical form of the adsorbent selected, connected also with its composition, pH, and ionic strength.

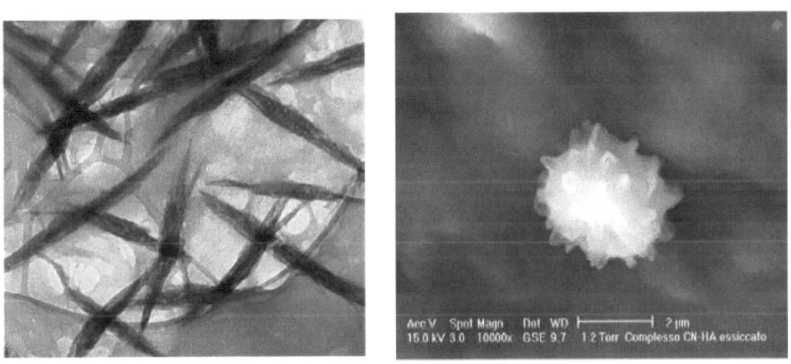

Figure 7. Left: Chitin nanofibrils bonded to metal ions (courtesy of G. Tischenko); Right: single CN–Hyaluronan nanoparticle (courtesy of Morganti et al. [36]).

3. CN Nanoparticles and Nanocomposites

On the one hand nanoparticles, obtained combining the electropositive CN with the electronegative hyaluronic acid [36], have been used recently to design cosmetic products characterized, for example, by their antiaging effectiveness on skin folds has been shown (Figure 8) [37], anti-acne activity treatment with an evident decrease of lesion counts (Figure 9) [38] and melanin synthesis inhibition verified on melanocytes cultures (Figure 10) [39], or for their elastic activity on damaged hair exposed to UV (Figure 11) [40] and wound healing activity (Figure 12) [32]. On the other hand, they are also considered as high-potential filler material for the improvement of the mechanical and physical properties of the nanocomposite polymer matrix (Figure 13) [41,42].

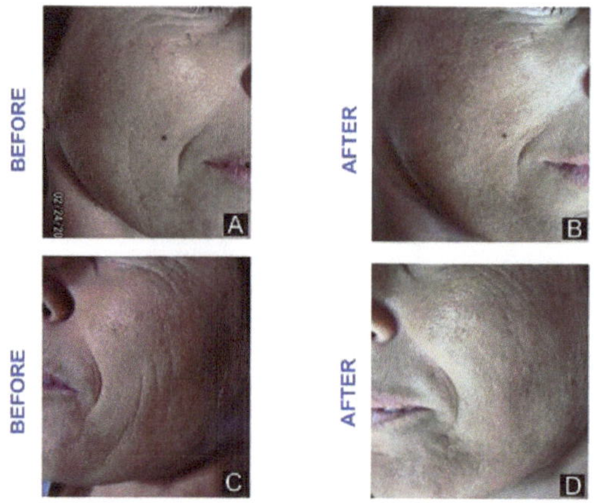

Figure 8. Antiaging effectiveness on nasolabial folds and mid-cheek before (**A-C**) and after (**B-D**) treatment by the active block-polymer-nanoparticles (BPN) at 30 days (courtesy of Morganti et al. [37]).

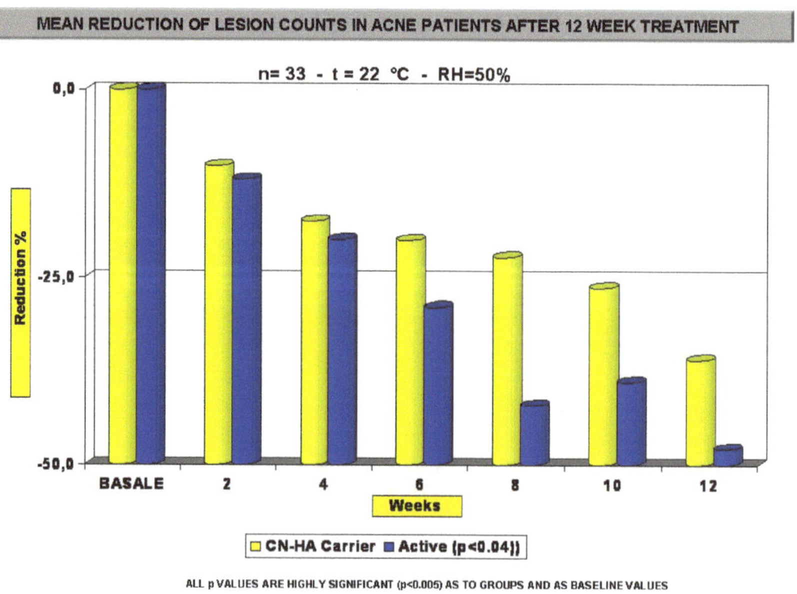

Figure 9. Anti-acne treatment effectiveness by active block-polymer-nanoparticles (courtesy of Morganti et al.).

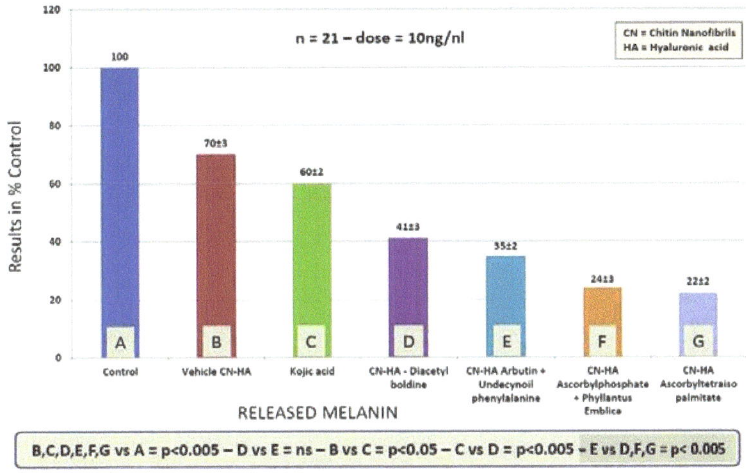

Figure 10. % Melanin synthesis inhibition on melanocyte cultures treated by CN–HA entrapping active ingredients vs. control and vehicle [39] (courtesy of Morganti et al.).

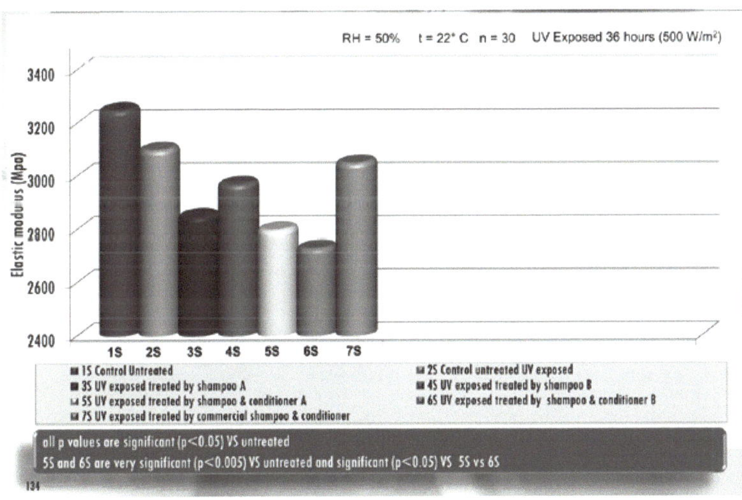

Figure 11. Decrease in elastic modulus of hair exposed to UV and treated with a Zn–CN shampoo and conditioner [40] (courtesy of Morganti et al.).

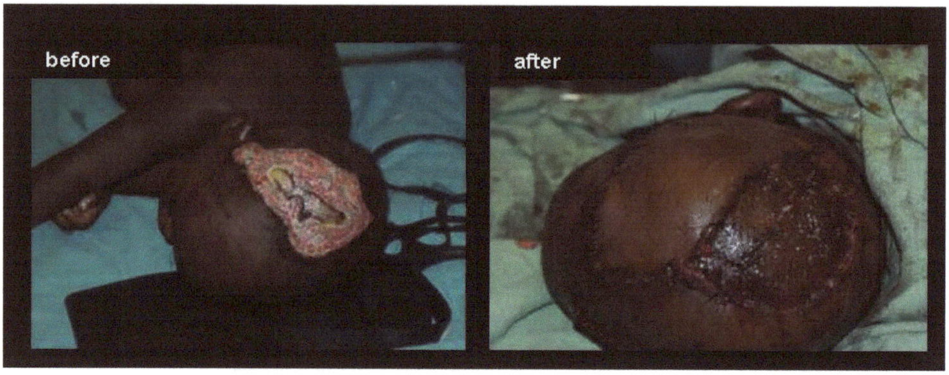

Figure 12. Wound healing activity of CN–HA nanoparticles (courtesy of Dr. P. Mezzana).

Composite Fibers	Tensile Strength MPa	Tensile Modulus MPa	Elongation at break, %
Chitosan 100%	199	7950	7.2
+0.5% Chrysotile	292	13740	7.6
+1.0% Chrysotile	220	8510	8.5
+1.5% Chrysotile	241	11370	8.5
+2.0% Chrysotile	187	7310	4.8
+3.0% Chrysotile	214	9200	4.7
+ 5% CN	**391**	**18240**	**6.2**
+ 20% CN	**411**	**20700**	**3.9**

Figure 13. Mechanical properties of chitosan/CN on composite fibers [40] (courtesy of V.E. Yudin).

4. Nanoparticles and Regenerative Medicine

4.1. Nanoparticles

Nanoparticles are used in nanomedicine as diagnostic imaging agents and therapeutic delivery vehicles to treat different disorders such as cancer, infection diseases, neurological modifications, etc. [42–49].

The primary amine groups of Chitin Nanofibril (CN) [50], assessing the special properties to this crystalline compound as material of choice for developing micro/nanoparticles, make it very useful, especially in biomedical applications [51]. The CN nanoparticles, in fact, have the ability to control the release of active ingredients, avoiding the use of hazardous organic solvents, since they are produced by the gelation method in aqueous solution by the use of electronegative, natural polymers (Figure 14) [36].

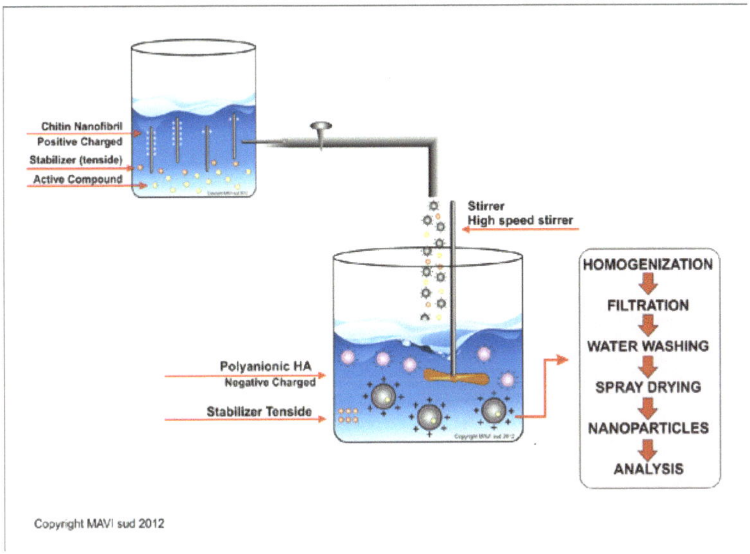

Figure 14. The gelation method. Hyaluronic Acid = HA. Source: MAVI SUD, Italy.

As an electropositive polymer, CN, when in contact with electronegative compounds, easily forms block copolymeric nanoparticles, also having the ability to entrap different active ingredients during the gelation process (Figure 15).

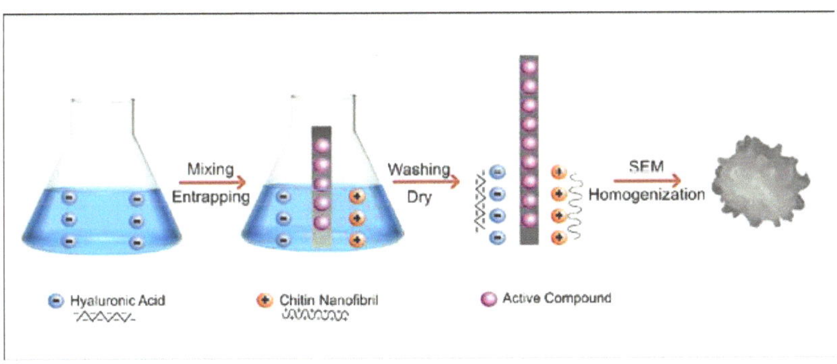

Figure 15. Entrapping active ingredients by CN–HA during the gelation method. Source: MAVI SUD, 2013.

Both CN and CN-derived nanoparticles are biocompatible with living tissues, breaking down to harmless amino sugars by the activity of the chitotriosidase (families of chitinases), secreted by humans [52]. This specific enzyme degrades chitin and chitosan primarily by the endo-processive mechanism, showing an absolute preference for acetylated sites compared with deacetylated ones. Thus CN is more easily degraded than chitosan because of its higher content of acetylated glucosamine, probably acting as a template for both the regular synthesis of hyaluronan and glucosaminoglicans, inducing the normal and regular dispositions of the fibers into the ECM (Figure 16). The facility to modulate the collagen synthesis could explain why its use may reduce the risk of hypertrophic formation of scars and keloids, also slowing down the adhesion of intra-peritoneal and intestinal structures [29,30,53,54].

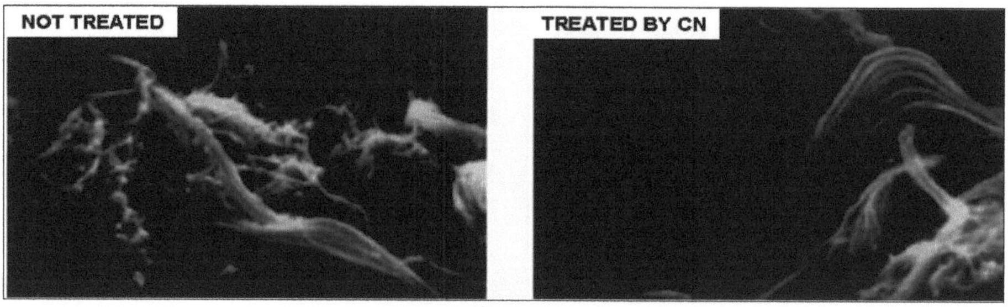

Figure 16. Disposition of collagen fibers on wounded tissue treated (**right**) by a CN-enriched gel emulsion compared to the control (**left**). Courtesy of M. Tucci et al. [29].

Thus, when used as carrier, this natural polymer could modulate the penetration through the skin layers of the entrapped/encapsulated active ingredients and the healing process by its own metabolized components (e.g. glucosamine and acetyl glucosamine) [29,30].

During the wound healing process, in fact, glucosamine and *N*-acetyl glucosamine groups should serve as a reinforcement substratum of the epidermis for keratinocyte activity at level of the wounded tissue. At dermis level, fibroblasts can be activated to produce the right quality and quantity of fine collagen fibers, necessary in the early period of the tissue rebuilding. Thus, CN could constitute a micro-environmental stimulus for the cell that, by influencing its correct trophism, could ameliorate and modulate the skin granulation process, enhancing the activity of defensines. Moreover, they modulate the activity of both metalloproteinases and angiogenesis favouring the regular deposition of the collagen fibers necessary to repair the dermo–epidermal lesions [55,56].

4.2. Regenerative Medicine

The aim of regenerative medicine is to repair and replace damaged tissues or organs by mean of natural regeneration processes. In fact, million people worldwide had an increase in life expectancy, while millions of patients are suffering from the skin and cartilage defects caused by trauma, injury and age-related degeneration. For example, the main cause of hospital admission and death in Brazil is injuries caused by accidents. In 2004 alone, the National Health Care System has spent US$585 million on orthotics and prosthetics [57,58].

According to The WHO Report on Disability [47], there are more than 1 billion people with disability worldwide (i.e., about 15% of the global population) who would benefit from prosthetics and orthotics services. The presence of disability is rising because of the aging population and the global increase in chronic diseases. Thus, 30 million people in Africa, Asia and South America require 180,000 rehabilitation professionals and devices whose cost is unsustainable for poor people. However, the mass production of these devices can lower their cost, using universal and innovative ingredients that are administered widely. Therefore, it becomes imperative to develop biomaterials, such as CN, with the main purpose of regenerating tissues and organs, possibly at low cost and without side effects.

In a regeneration strategy, biomaterials have to promote new tissue formation by providing adequate porosity and an appropriate surface to foster and direct cellular attachment, migration, proliferation, favouring the desired differentiation of specific phenotypes throughout scaffolds where new tissue formation is needed [59]. Thus films, fibers and bulk materials, based on the natural polysaccharide CN, could represent an elective biomaterial combining bioresorption properties, absence of cytotoxicity and low environmental impact during processing and use [60,61].

5. Nanocomposites

Nanocomposites, as the most advanced and adaptable engineering material, are considered to belong to the group called nanomaterials, where a nanoparticle (nanofiller) is distributed into a matrix [62].

Generally, a nanocomposite is a multiphase dense material in which at least one of its phases has either one, two or three measurements lower than 100 nm [63].

The perfect combination of the right polymeric matrix and reinforcing natural fibers produces composites possessing the finest properties of each component. However, while the term natural fiber-reinforced composite usually refers to natural fibers in any sort of polymeric matrix (natural or man-made), the nanofillers (nanoparticles) in a nanocomposite material are the component constituted of inorganic/inorganic, inorganic/organic or organic/organic sources.

Polymer/inorganic nanoparticles find applications in diverse areas, including biomedical applications. While on the one hand the polymer component has structural functions, also tuning the mechanical features and processability of the final material, on the other hand inorganic components such as TiO_2, ZnO, or the metals Ag, Cu and Bi, can not only reinforce the mechanical and thermal properties, but can characterize the final product by their anti-inflammatory, antibacterial and anti-fungal activities (Figure 17), necessary, for example, to accelerate skin regeneration by the use of advanced medications, appositely designed (Figure 18).

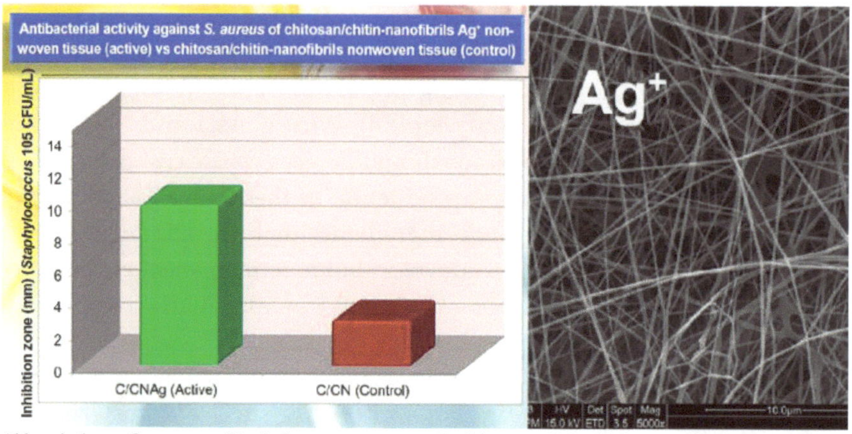

Abbreviations: S. aureus; Staphylococcus aureus; C: chitosan; CN: chitin-nanofibrils; Ag+: silver ions; CFU: colony forming unit.

Figure 17. Antibacterial activity of Ag+ treated chitosan/chitin fibers. Courtesy of Morganti et al. [55].

Thus, before starting the electrospinning and/or casting process to make non-woven tissues, Ag nanoparticles have been bound to CN fiber by our group [16]. This methodology allows the inorganic particle to be either physically trapped within the matrix or ionically bound to the polymer [64], depending on the productive process selected. However, both electrospinning and casting are technologies for fabricating fibers at nanometer dimension, which for their specific high surface areas and their ability to mimic the native ECM are very useful for developing nanofibrous cellular scaffolds for human tissue engineering.

All the tissues of human organs, such as bone, cartilage, tendon, ligaments, skin, nerve, and blood vessels, in fact, are hierarchically organized into fibrous structures with fiber dimensions down to the nanometer scale [65].

Advanced medications made by the use of electrospinning or casting technology can provide an environmental or physical cell scaffold, promoting cell growth and function towards the synthesis of ECM over time. This is the reason why natural, biodegradable, nanofibrous materials that mimic the ECM nanostructure have been investigated as ideal components for many human tissues.

Therefore, nanostructured biomaterials such as nanoparticles, nanofibers and nanocomposites made by the use of CN have gained increasing interest in regenerative medicine, so that recently they have been investigated for the capacity to emulate the nanofibrous features of ECM components and modulate the regeneration of burned skin in a faster way (Figure 18) [53,66].

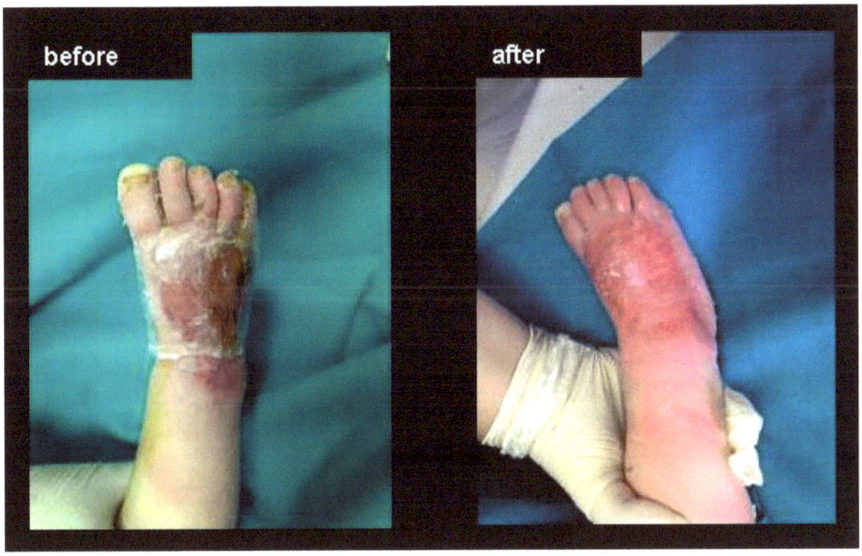

Figure 18. Scar-forming and antibacterial activity of chitosan/chitin fibers treated with Ag+ on burned baby skin treated for six days. Courtesy of M. Palombo et al. [53].

The general idea of nanocomposites is based on the concept of creating a very large interface between nanosized building blocks and the natural/manmade polymer matrix [67]. In any case, the performance of natural composites is influenced by several factors, such as the fiber's microfibrillar angle and its architectural structure, the physicochemical properties and composition, as well as the cell dimensions, and the mechanical properties connected with the interaction between the fiber and the polymer matrix. For these reasons, in recent years reinforced composites containing natural fibers have received considerable attention. Thus, their use increased rapidly because of their high performance in mechanical properties, significant processing advantages, biodegradability, low cost and low density [68].

Natural fibers are, in fact, renewable and cheaper, and pose no health hazards, providing a solution to environmental pollution by finding new uses for waste materials. At this purpose CN from crustaceans' waste, and lignin from plant biomass, seem to have good potential as byproduct resources for industrial uses, probably for their capacity to form non-woven tissues or films with an extremely high surface-to-volume ratio, and tunable porosity underlined for their non-toxic and skin-friendly character, as reported in other chapters of this book.

As natural compounds, diffuse in nature in great quantity, lignin represents a structural material for plant cells, and chitin (CN) is a defense and protective material for crustaceans. However, to understand the properties of naturally reinforced composite materials, it is essential to recognize and control in advance the mechanical, physical and chemical composition/properties of the fibers selected, together with their naturalness, safeness and effectiveness [69]. Natural polymers such as oligosaccharides, in fact, are hydrophilic, enzymatically degradable compounds capable of retaining the stability of incorporated drugs, thereby increasing their therapeutic effects.

This is why the term bio-nanocomposite has been coined, as well as naturapolyceutics, born from the union of polymer and pharmaceutics [70].

In any case, bio-nanocomposites are regarded as an emerging group of nanostructured materials classified as (a) nanocomposite materials developed from renewable and sustainable nanoparticles like cellulose, chitin and lignin, (b) petroleum-derived polymers like polypropylene (PP), polyethylene (PE) and polyethylene oxide (PEO), and nanocomposites derived from bio polymers like polylactic acid (PLA) and polyhidroxyacids (PHA) or inorganic nanofillers like carbon nanotubes and nanoclay.

In conclusion, nanoparticles, nanofibers and nanocomposites, made by the use of CN and other natural polymers or macromolecules, represent new families of polymeric carriers and matrices able to enhance the mechanical and barrier properties of emulsions, films, and non-woven tissues, used to produce innovative cosmetics, food packaging and advanced medications together with many other composite-based industrial applications (Figure 19).

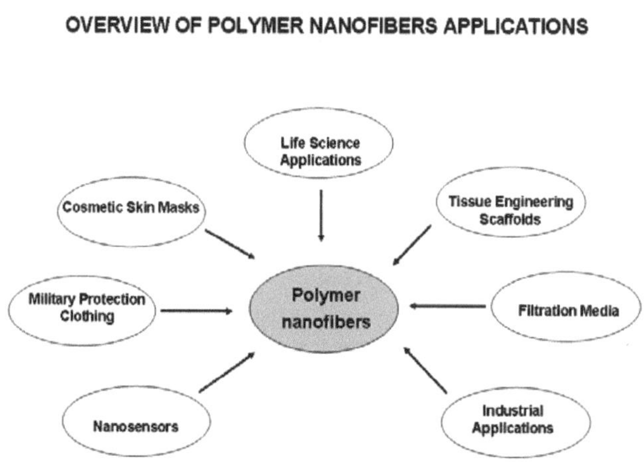

Figure 19. Nanofibers' applications.

Specifically, CN nanoparticles enhance the performance and properties of emulsions and composites, showing great value for both innovative pharmaceutical and cosmetic formulations, and fiber-reinforced composite-based products. This is the reason why bionanotechnology is estimated to contribute at least US$3 trillion to the worldwide economy by 2020 [71].

Furthermore, it has been expected that industries based on nanotechnology might require at least six million workers to sustain them by the end of the decade, hence contributing to a solution to the international economic crisis. Apart from the benefits and safety aspects of bionanotechnology-based products such as nanoparticles and nanocomposites, public opinion and attitudes (Figure 20) towards this new sector are extremely important for the development of this emerging area and business extimated to reach US$172 billion by 2025 [72], according to the research study of Nanotechnology Industries Association (NIA) [73].

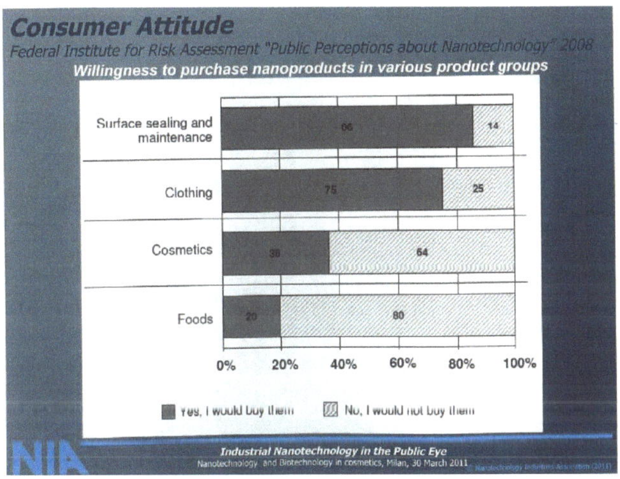

Figure 20. Consumer attitude to purchasing nanoproducts in different industrial areas (courtesy of D. Kolsov) [72].

6. Concluding Remarks

The extraordinary mechanical and biological properties of Chitin nanofibrils seem to be related to their highly layered hierarchical structure, when produced as pure crystals by the right treatment and purification methods [74]. They are natural polymers possessing unique properties like biodegradability and chelating activities, and have been shown to have antibacterial, antifungal, and anti-inflammatory effectiveness also [75]. However, the relationship between its hierarchical structure and the biological properties has to be further explored.

In anyway CN is a biomaterial that seems capable of stimulating the growth of cells in contact with it, interacting with the biological systems.

Finally, its extraction from waste materials of natural origin, the low cost and its wide applicability make it suitable to be used in many industries involved in the production of pharmaceuticals, food, cosmetic products, and packaging materials.

However, more studies must be undertaken in order to recognize in a deeper way all the possibilities these new technologies and chitin nanofibrils have for contributing to effective industrial innovations. It is, therefore, fundamental to underline the necessity of using more waste materials and industrial byproducts to produce goods that will not impoverish our planet and will help maintain its biodiversity [76].

Acknowledgments: The authors thank MAVI SUD for the sample test used for all the reported studies and the EU for the economic support given to finance n-Chitopack research project, grant no.315233, and Biomimetic research project no.282945.

Conflicts of Interest: The authors work at the R&D Centre of Nanoscience, Mavi Sud, s.r.l, Italy.

References

1. Mincea, M.; Negrulescu, A.; Ostafe, V. Preparation, modification, and applications of chitin nanowiskers: A review. *Rev. Adv. Mater. Sci.* **2012**, *30*, 225–242.
2. Neville, A.C.; Parry, D.A.D.; Wodhead-Galloway, J. The chitin crystalline in arthropod cuticle. *J. Cell. Sci.* **1976**, *21*, 531–562.
3. Muzzarelli, C.; Morganti, P. Preparation of Chitin and Derivatives Thereof for Cosmetic and Therapeutic Use. U.S. Patent 8,383,157, 26 February 2013.
4. Muzzarelli, R.A.A.; Morganti, P.; Morganti, G.; Palombo, P.; Palombo, M.; Biagini, G.; Mattioli-Belmonte, M.; Giantomassi, F.; Orlandi, F.; Muzzarelli, C. Chitin Nanofibrils/Chitosan glycolate composites as Wound medicaments. *Carbohydr. Polym.* **2007**, *70*, 274–284. [CrossRef]
5. Morganti, P.; Carezzi, F.; Del Ciotto, P.; Morganti, G.; Nunziata, M.L.; Gao, X.H.; Chen, H.D.; Tischenko, G.; Yudin, V.E. Chitin Nanofibrils: A natural Multifunctional a Polymer: Physicochemical characteristics, effectiveness and safeness. In *Nanobiotechnology*; Phonix, D.A., Ahmed, W., Eds.; One Central Press: Manchester, UK, 2014; pp. 1–31.
6. Kamel, S. Nanotechnology and its applications in lingo-cellulosic composites. A mini review. *Express Polym. Lett.* **2007**, *1*, 546–575. [CrossRef]
7. Morganti, P. Use and Potential of Nanotechnology in Cosmetic Dermatology. *Clin. Cosmet. Investig. Dermatol.* **2010**, *3*, 5–13. [CrossRef]
8. Cui, J.; Yu, Z.; Lau, D. Effect of acetyl group on mechanical properties of chitin/chitosan nanocrystal properties: A molecular dynamics study. *Int. J. Mol. Sci.* **2016**, *17*, 61. [CrossRef] [PubMed]
9. Cinelli, P.; Coltelli, M.B.; Mallegni, N.; Morganti, P.; Lazzeri, A. Degradability and Sustainability of Nanocomposites based on polylactic acid and nanochitin. *Chem. Eng. Trans.* **2017**, *60*, 115–120.

10. Morganti, P.; Del Ciotto, P.; Stoller, M.; Chianese, A. Antibacterial and Anti-inflammatory Green Nanocomposites. *Chem. Eng. Trans.* **2016**, *47*, 61–66.
11. Morones, J.R.; Elechiguerra, J.L.; Camacho, A.; Hollt, K.; Koori, J.B.; Ramirez, J.T.; Yacaman, M.J.C. The bacterial effects of silver Nanoparticles. *Nanotechnology* **2005**, *16*, 2346–2353. [CrossRef] [PubMed]
12. Aghreal, A.; Wejs, T.L.; Schurr, M.J.C.; Faith, N.G.; Czuprynski, C.S.; McAnulty, J.T.; Murphy, C.S.; Abbot, N.L. Surface modified with nanometer -think silver- impregnated polymeric films that kill bacteria but support growth of mammalian cells. *Biomaterials* **2010**, *31*, 680–690.
13. Morganti, P.; Palombi, M.; Tishchenko, G.; Yudin, V.E.; Guarneri, F.; Cardillo, M.; Del Ciotto, P.; Carezzi, F.; Morganti, G.; Fabrizi, G. Chitin-Hyaluronan Nanoparticles: A Multifunctional Carrier to Delivery AntiAging Active Ingredients through the Skin. *Cosmetics* **2014**, *1*, 140–158. [CrossRef]
14. Mantovani, E.; Zappelli, P.; Conde, J.; Sitja, R.; Periales, F. Report on nanotechnology & Textiles-medical textiles, sport/outdoor textiles. *Obs. NANO* **2010**, 44–50.
15. Morganti, P.; Stoller, M. Chitin and Lignin: Natural Ingredients from Waste Materials to make innovative and healthy products for humans and plants. *Chem. Eng. Trans.* **2017**, *60*, 319–324.
16. Morganti, P.; Carezzi, F.; Del Ciotto, P.; Morganti, P.; Nunziata, M.L.; Gao, X.H; Chen, H.-D.; Tishchenko, G.; Yudjn, V.E. Chitin nanofibrils: A Natural Multifunctional Polymer. In *Nanobiotechnology*; Phenix, A.D., Ahmed, W., Eds.; One Central Press: Manchester, UK, 2014; pp. 1–31.
17. Food and Agriculture Organization of the United Nations (FAO). *Yearbook 2012. Fishery and Aquaculture Statistics*; FAO: Rome, Italy, 2012.
18. United Nations Environment Programme (UNEP). *Converting Waste Agricultural Biomass into a Resource*; International Environmental Technology Centre: Osaka/Shiga, Japan, 2012.
19. Rinaudo, M. Chitin and Chitosan: Properties and Applications. *Prog. Polym. Sci.* **2006**, *31*, 603–632. [CrossRef]
20. Mengibar, M.; Ganan, M.; Miralles, B.; Carrascosa, A.V; Martinez-Rodriguez, A.J.; Peter, M.G.; Heras, A. Antibacterial activity of products of depolymerization of chitosans with lysozyme and chitosanase against Campylobacter jejuni. *Carbohydr. Polym.* **2011**, *84*, 844–848. [CrossRef]
21. Park, B.K.; Kim, M.M. Application of Chitin and its Derivatives in Biological Medicine. *Int. J. Mol. Sci.* **2010**, *11*, 5152–5164. [CrossRef] [PubMed]
22. Zheng, J.B.; He, Y.S.; Li, S.L.; Wang, Y.Z. Chitin Whiskers: An Overview. *Biomacromolecules* **2012**, *13*, 1–11. [CrossRef] [PubMed]
23. Morganti, P.; Morganti, G.; Morganti, A. Transforming nanostructured chitin from crustaceans waste into beneficial Health products: A must of our society. *Nanotechnol. Sci. Appl.* **2011**, *4*, 123–129. [CrossRef] [PubMed]
24. Morganti, P.; Li, Y.H. From Waste Materials Skin-Friendly Nanostructured Products to Save Humans and the Environment. *J. Cosmet. Dermatol. Sci. Appl.* **2011**, *1*, 123–129. [CrossRef]

25. Malkapuram, R.; Kumar, W.; Yuvraj, S.N. Recent Development in Natural Fibre Reinforced Polypropylene Composites. *J. Reinf. Plast. Compos.* **2008**, *28*, 1169–1189. [CrossRef]
26. Satyanayama, K.; Arizaga, G.G.C.; Wypich, F. Biodegradable composites based on lignocellulosic fibers -An overview. *Prog. Polym. Sci.* **2009**, *34*, 982–1021. [CrossRef]
27. Matabola, K.P.; Vries, A.R.; Moolman, F.S.; Luyt, A.S. Single polymer composite: A review. *J. Mater. Sci.* **2009**, *44*, 6213–6222. [CrossRef]
28. Azowa, Z.N.; Yousif, B.I.; Manalo, A.C.; Karunasena, W. A review on the degradability of polymeric composites based on natural fibres. *Mater. Des.* **2013**, *47*, 424–442. [CrossRef]
29. Tucci, M.G.; Mattioli-Belmonte, M.; Ricotti, G.; Biagini, G. Polysaccharides: Health-Environment Binomial. *J. Appl. Cosmetol.* **1999**, *17*, 94–101.
30. Morganti, P.; Mattioli-Belmonte, M.; Tucci, M.G.; Ricotti, G.; Biagini, G. New Prospects for Cutaneous Wound Healing and Keloid Treatment. *J. Appl. Cosmetol.* **2000**, *18*, 125–130.
31. Ravi Kumar, M.N.V.; Muzzarelli, R.A.A.; Ssshiwa, C.; Domb, A.J. Chitosan Chemistry and pharmaceutical Perspectives. *Chem. Rev.* **2004**, *104*, 6017–6084. [CrossRef] [PubMed]
32. Mezzana, P. Clinical efficacy of a new nanofibrils-based gel in wound healing. *Acta Chir. Plast.* **2008**, *3*, 5–13.
33. Muzzarelli, R.A.A.; Boudrant, J.; Meyer, D.; Manno, N.; De Marchis, M.; Paoletti, M.G. A tribute to Henri Braconnkt precursor of the carbohydrate polymers Science on the Chitin bicentellian. *Carbohydr. Polym.* **2012**, *87*, 995–1012. [CrossRef]
34. Busilacchi, A.; Gigante, A.; Mattioli-Belmonte, M.; Manzotti, S.; Muzzarelli, R.A.A. Chitosan stabilises platelets growth factors and modulate stem cell differentiate on toward tissue regeneration. *Carbohydr. Polym.* **2013**, *98*, 665–676. [CrossRef] [PubMed]
35. Morganti, P.; Carezzi, F.; Del Ciotto, P.; Morganti, G. Chitin Nanoparticles as Innovative Delivery System. *Personal Care Eur.* **2012**, *5*, 95–98.
36. Morganti, P.; Del Ciotto, P.; Fabrizi, G.; Guarneri, F.; Cardillo, A.; Palombo, M.; Morganti, G. Safety and Tolerability of Chitin Nanofibrils-Hyaluronic acid Nanopartickes Entrapping Lutein. Note I: Nanoparticles Characterization and Bioavailability. *SOFW-J.* **2013**, *139*, 12–23.
37. Morganti, P.; Palombo, M.; Fabrizi, G.; Guarneri, F.; Slovacchia, F.; Cardillo, A.; Del Ciotto, P.; Carezzi, F.; Morganti, G. New Insights on Anti-Aging Activity of Chitin Nanofibril-Hyaluronan Block Copolymers Entrapping Active Ingredients: In Vitro and in Vivo study. *J. Appl. Cosmetol.* **2013**, *32*, 1–29.
38. Morganti, P.; Berardesca, E.; Guarneri, B.; Fabruzi, G.; Palombo, P.; Palombo, M. Topical clindamicyn 1% vs phosphatidylcholine linoleum acid-rich and nicotinamide 4% in the treatment of acne: A multicenter- randomized trial. *Int. J. Cosmet. Sci.* **2011**, *13*, 1–10.
39. Morganti, P.; Del Ciotto, P.; Carezzi, F.; Guarneri, F.; Yip, J.Y. Skin Lightening Efficacy of New Formulations Enhanced by Chitin Nanoparticles Delivery system. Note I. *J. Appl. Cosmetol.* **2014**, *32*, 57–71.

40. Morganti, P.; Palombo, M.; Cardillo, A.; Del Ciotto, P.; Morganti, G.; Gazzaniga, G. Anti-dandruff and anti-oily Efficacy of Hair formulations with a Repairing and Restructuring activity. The positive Influence of the Zn-Chitin Nanofibrils Complexes. *J. Appl. Cosmetol.* **2012**, *30*, 149–159.
41. Saba, N.; Tahir, P.; Jawaid, M. A review on Potentially of Nano Filler/Natural Fiber Filled Polymer Hybrid Composites. *Polymers* **2014**, *6*, 2247–2273. [CrossRef]
42. Yudin, V.E.; Dobrovolskaya, I.P.; Neelov, I.M.; Dresvyanina, E.N.; Popryadukhin, P.V.; Ivan'kova, E.M.; Elokhovskii, V.U.; KAsatkin, I.A.; Okrugin, B.M.; Morganti, P. Wet spinning of fibers made of Chitosan and Chitin Nanofibrils. *Carbohydr. Polym.* **2014**, *108*, 176–182. [CrossRef] [PubMed]
43. Li, K.C.; Pandit, S.D.; Guccione, S.; Bednarski, M.D. Molecular imaging applications in nanomedicine. *Biomed. Microdevices* **2004**, *6*, 113–116. [CrossRef] [PubMed]
44. Ferrari, M. Frontier in cancer nanomedicine: Directing mass transport through biological barriers. *Trends Biotechnol.* **2010**, *28*, 181–188. [CrossRef] [PubMed]
45. Godi, B.; Tasciotti, E.; Liu, X.; Serda, R.E.; Ferrari, M. Multi stage nanovectors: From concept to novel imaging contrast agents and therapeutics. *ACC Chem. Res.* **2011**, *44*, 979–989. [CrossRef] [PubMed]
46. Bae, Y.H.; Lark, K. Targeted drug delivery to tumors: Myths, reality and possibility. *J. Control. Release* **2011**, *153*, 198–205. [CrossRef] [PubMed]
47. Chang, H.I.; Yeh, M.K. Clinical development of liposomes-based drugs: Formulation, Characterization, and therapeutic efficacy. *Int. J. Nanomed.* **2012**, *7*, 49–60.
48. Modulerà, S.; Turanek, J. Liposomal paclitaxel formulations. *J. Control. Release* **2012**, *163*, 322–334.
49. Morganti, P. Chitin Nanofibrils and their derivatives as Cosmeceuticals. In *Chitin Chitosan, Oligosaccharides and Their Derivatives: Biological Activities and Applications*; Kim, S.-K., Ed.; CRC-Press: New York, NY, USA, 2011; pp. 532–541.
50. Morganti, P. The meaning of Nanodimension Involving the Cosmetic Chemist from lab to the Industrial Process. *J. Sci. Res. Rep.* **2014**, *4*, 79–100.
51. Morganti, P.; Febo, P.; Cardillo, M.; Donnarumma, G.; Baroni, A. Chitin Nanofibril and Nanolignin: Natural Polymers of Biomedical Interest. *J. Clin. Cosmet. Dermatol.* **2017**, *1*, 1–7.
52. Eide, K.B.; Norberg, A.L.; Heggset, E.B.; Lindbom, A.R.; Varum, K.M.; Eijsink, V.G.H.; Sørlie, M. Human Chitotriosidase-Catalyzed Hydrolysis of Chitosan. *Biochemistry* **2012**, *51*, 487–495. [CrossRef] [PubMed]
53. Palombo, M. Innovative Solutions for Advanced Medications: Biochemical and Clinical Data. In Proceedings of the 11th International Congress of a Cosmetic Dermatology, Beijing, China, 26–28 September 2014; p. 38.
54. McCarthy, M.F. Glucosamine for wound healing. *Med. Hypotheses* **1966**, *47*, 273–275. [CrossRef]
55. Morganti, P. Use of Chitin Nanofibrils from Biomass for an Innovative Bioeconomy. In *Nanofabrication in Nanomaterials*; Ebothé, J., Ahmed, W., Eds.; One Central Press: Cheshire, UK, 2016; pp. 1–22.

56. Donnarumma, G.; Fusco, A.; Morganti, P.; Palombo, M.; Anniboletti, A.; Del Ciotto, P.; Baroni, A.; Chianese, A. Advanced Medications made by Green Nanocomposites. *Int. J. Res. Pharm. Nano Sci.* **2016**, *5*, 261–270.
57. World Health Organization (WHO). *World Report on Disability*; World Health Organization: Geneva, Switzerland, 2011. Available online: www.who.int/disabilities/world_report/2011/report.pdf (accessed on 17 November 2017).
58. Wei, G. Nanostructured Biomaterials for Regeneration. *Adv. Funct. Mater.* **2008**, *18*, 3566–3582. [CrossRef] [PubMed]
59. World Health Organization (WHO). *Standards for Prosthetics and Orthotics Service Provision. 2015–2017 Work Plan*; International Society of Prosthetics and Orthotics: US ISPO, Dublin, 2015. Available online: http://www.who.int/phi/implementation/assistive_technology/workplan_p-o_standards.pdf (accessed on 17 November 2017).
60. Morganti, P.; Del Ciotto, P.; Morganti, G.; Fabien-Soule', V. Application of Chitin Nanofibrils and Collagen of a Marine Origin as Bioactive Ingredients. In *Cosmeceuticals: Trends and Prospects*; Kim, S.-K., Ed.; CRC-Press: New York, NY, USA, 2012; pp. 267–289.
61. Morganti, P.; Tischenko, G.; Palombo, M.; Kelnar, L.; Brizova, L.; Spirkova, M.; Kkbera, L.; Carezzi, F. Chitin nanofibrils for biomimetic products: Nanoparticles and nanocomposite chitosan films in health care. In *Marine Biomaterials: Isolation, Characterization and Application*; Kim, S.-K., Ed.; CRC-Press: New York, NY, USA, 2013; pp. 681–715.
62. Marquis, D.M.; Guillaume, E.; Chivas-Jaly, C. Properties of nano fillers in polymer. In *Nanocomposites and Polymers with Analytical Methods*; Cuppoletti, J., Ed.; Interch Publishing: Rijcka, Croatia, 2011; pp. 261–284.
63. Ajayan, P.M.; Schadler, L.S; Braun, P.V. *Nan composites Science and Technology*; Wiley-VCH, Verlag GmBH & Co.: Winheim, Germany, 2003.
64. Hood, M.A.; Mari, M.; Munoz-Espi, R. Synthetic Strategies in the Preparation of a polymer/Inorganic Hybrid/Nanoparticles. *Materials* **2014**, *7*, 4057–4087. [CrossRef] [PubMed]
65. Lou, H.; Ding, X.; Zhou, Ag; Li, P.; Wei, X.; Fan, Y. Electrospinning of nanofibers for tissue engineering applications. *J. Nanomater.* **2013**, *2013*. [CrossRef]
66. Morganti, P. A Green Bio-Economy for a Sustainable Cosmetic Dermatology. In Proceedings of the 11th International Congress of Cosmetic Dermatology, Beijing, China, 26–28 September 2014; p. 1.
67. Hari, J.; Pukanzsky, B. NanoComposites: Preparation, structure, properties. In *Applied Plastics Handbook: Processing Materials*; Kutz, M., Ed.; Elsevier Inc.: Waltham, MA, USA, 2011; pp. 109–142.
68. Pathanh, D.; Singh, D. A review on electrical properties of fiber reinforced polymer Composite. *Int. J. Theor. Appl. Sci.* **2009**, *1*, 34–37.
69. Kalia, S.; Kaith, B.; Kaur, I. Pretreatment of natural fibers and their applications as reinforcing material in polymer composites. A review. *Polym. Eng. Sci.* **2009**, *49*, 1253–1272. [CrossRef]
70. Ngwuluka, N.C.; Ochekpe, N.A.; Aruoma, O.I. Naturapolyceutics: The Science of Utilizing Natural Polymers for Drug Delivery. *Polymers* **2014**, *6*, 1323–1332. [CrossRef]

71. Rico, M.C.; Mirkin, C.A.; Hersam, M.C. Nanotechnology research directions for societal Needs in 2020. Summary of International study. *J. Nanopart. Res.* **2011**, *13*, 897–919. [CrossRef]
72. Wood, L. Global Nanocomposites Market Analysis & Trends—Industry Forecast to 2025. *Business Wire*, Dublin, 16 April 2016.
73. Koltsov, D. Industrial Nanotechnology in the Public Eye. In Proceedings of the meeting Nanotechnology and Biotechnology in Cosmetics, Milano, Italy, 30 March 2011; pp. 99–115.
74. Fratzl, P.; Weinkamer, R. Nature's hierarchical materials. *Prog. Mater. Sci.* **2007**, *52*, 1263–1334. [CrossRef]
75. Izumi, R.; Azuma, K.; Izawa, H.; Morimoto, M.; Nagashima, M.; Osaki, T.; Tsuka, T.; Imagawa, T.; Ito, N.; Okamoto, Y.; et al. Chitin Nanofibrils suppress skin inflammation in atopic dermatitis-like skin lesions in NC/Nga mice. *Carbohydr. Polym.* **2016**, *146*, 320–327. [CrossRef] [PubMed]
76. Union for Ethical Biotrade (UEBT). *Biodiversity Barometer*; UEBT: Amsterdam, The Netherlands, 2017. Available online: ethicalbiotrade.org (accessed on 17 November 2017).

Cellulose and Lignin: The Abundant Renewable Polymers from Plant Biomass

Pietro Palmisano and Maria Chiara Piglione

Abstract: The conversion of several biomasses into renewable chemicals and fuels has received great attention in the last decade as a sort of parallel route to the classical petrochemical way of reaching an eco-friendlier and more sustainable economy. Many researchers have further focused their studies towards waste products coming from the agriculture and/or forest industry, and found out the different techniques used for the separation of the main constituents of a classical biomass—cellulose, hemicellulose and lignin—as well as their characterization techniques and analytical studies in order to better identify their physical and chemical characteristics for identifying the best way to transform them into valuable products. An overview of the most important developed processes has been widely coupled with a detailed list of the chemicals and fuels that a potential biorefinery could provide, simply by processing feedstocks that a few years ago were considered just waste.

1. Introduction

Some of the main problems of today's world are climate change due to greenhouse gas emissions (GHG), resource constraints—particularly for fossil fuels—and population fast growth, which causes an increased energy, water and food demand. For these reasons, interest has been focused on studying new technologies and finding sustainable alternatives to the fossil feedstock dependence, developing the biorefinery concept. The idea is very similar to a petroleum refinery, in which petroleum is the starting source for the production of multiple fuels and a wide range of products. In the same way, a biorefinery is a chemical plant fed by biomass and/or renewable feedstocks as inputs, converting them to fuels (e.g., biodiesel and bioethanol), energy and a wide range of chemicals or different kind of materials that can have a multiple range of application fields (e.g., monoethylene glycol, *n*-butanol, xylenes, polyurethane, polyols).

One of the main strengths of a biorefinery is the number of different products and intermediates generated from the whole process that allow the maximization of the value of the biomass itself. Furthermore, big efforts are aimed at the energetic self-sustainability of the biorefinery in order to make their development and diffusion convenient.

It has also been widely demonstrated in the past 30 years of experimental studies that many fossil-derived industrial products could be partially or totally replaced using biomass sources as bio-based final products [1]. In the last decade, many scientific studies have been focused on the conversion of biomasses towards the main block molecules of the chemical and polymer industry, trying to better define the most promising and economically feasible route to get methanol, ethylene, propylene, butadiene, benzene, toluene and xylene from several kinds of biomasses instead of the complementary fossil-derived molecules.

Biorefineries can be fed with different feedstocks, such as perennial grasses, starch crops (wheat, maize, etc.), sugar crops (beet, cane, etc.), lignocellulosic crops as well as oil crops, aquatic biomass and organic residues (industrial, commercial and post-consumer wastes). First-generation biofuels achieved after transformation of food crops, in what we can define as I generation biomass plant, can generate several issues, such as negative energy balance coupled with a negative greenhouse gas emission, and most likely a vigorous increase in food prices. For this reason, the main focus of this section is on II generation lignocellulosic biorefineries, whose industry to date is still under development. Of primary importance in the development of these new technologies is assuring that new bio-based products have lower environmental impacts than their petrochemical equivalent. Another important aspect is the economical one, as in many cases bio-based chemicals and biofuels have higher production costs than their petrochemical equivalents, only partially justifiable with the complexity of their structure and with the heterogeneity and complexity of the starting feedstock (biomass).

If we focus on the historic standpoint, the transformation of the constituent part of woody biomasses is something well known and not new. For instance, cellulose can be used as precursor for textile fibers, paper, and film with different properties deriving from the industrial process. The impact of cellulose crosses different industrial applications as well as domestic life and civil construction [2]. Besides cellulose, lignin has been extensively studied in recent years. Its abundancy is directly connected with the pulp and paper industry (the order of magnitude is roughly millions of tons per year worldwide), where the residual part of the wood is not useful as cellulose fraction (i.e., black liquor coming from the Kraft processes or lignosulfonates coming from the sulfite processes, both of which are the platform for the production of pulp and paper) and it is actually employed as sacrificial fuel or binding agent for concrete. The increasing interest of academia in lignin is ascribable to its aromatic nature; that makes it a possible source of value-added aromatic chemicals, an aspect that will be described more thoroughly in the coming paragraphs.

For both cellulose and lignin, the interest is now focused on new value-added applications that can be combined to form the concept of a II generation platform, in which the valorization of each constituent of the biomass is a condition sine qua non for achieving the self-sustainability of the entire process in terms of revenues (enlarging the range of potential products) and exploiting all the streams and sub-streams generable from the starting woody material trying to make a rather complex technical concept economically sustainable. A possible scheme for a lignocellulosic feedstock biorefinery is shown in Figure 1, in which a thermo-chemical platform is coupled with a biotechnological one. This general scheme highlights the possibility of obtaining fuels, chemicals, and building molecules for the polymer industry, as well as enough energy to achieve energetic self-sustainability. Products can be generated via thermo-chemical conversion or bioprocessing. In both cases, the structure of the natural monomers that constitutes the starting biomass are largely preserved instead of being transformed to obtain the final products.

The aim of this section is to highlight opportunities coming from the use of lignocellulosic feedstocks in order to obtain a number of value-added products through a number of different conversion processes.

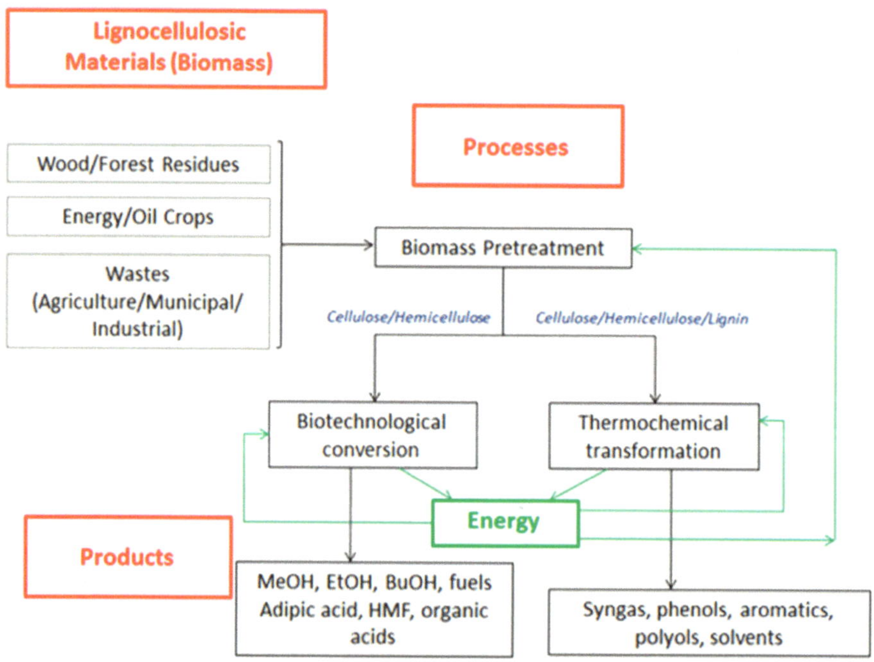

Figure 1. Concept block flow diagram for II generation biorefinery.

2. Lignocellulosic Feedstocks

Lignocellulosic biomass refers to inedible plant material with a complex polymeric generated by photosynthesis, whose main properties are structural strength in combination with flexibility. Lignocellulosic feedstocks are in fact complex natural biopolymers built up by three main constituents—cellulose, hemicellulose and lignin—whose structures and compositions are affected by many factors such as plant species, cultivation regions and subsequently growth conditions [3]. A general composition of lignocellulosic feedstocks is shown in Figure 2.

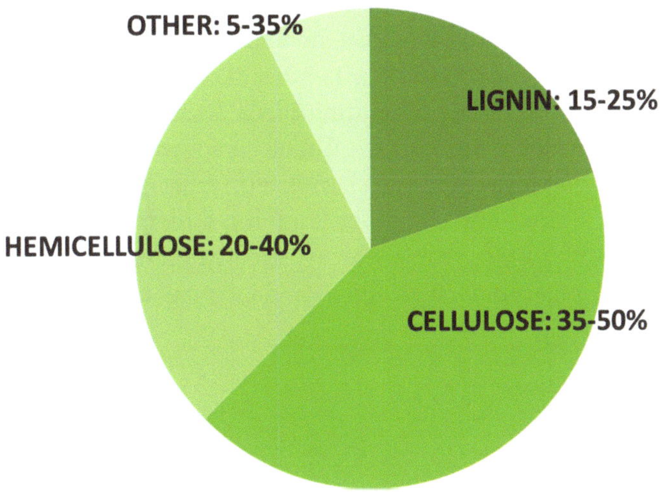

Figure 2. Generic composition for lignocellulosic biomass feedstocks [4].

Cellulose and hemicellulose are carbohydrate polymers, while lignin is a phenolic polymer. Besides these three main constituents, lignocellulosic feedstocks also contain water and minor amounts of extractives and inorganic compounds (ashes). A general overview of lignocellulosic feedstock composition has been collected by Lee DK et al. Table 1 is taken from his monograph [4]. Nevertheless, it is important to remember that the growth in different environments has a significant effect on feedstock composition.

Table 1. Composition of residual lignocellulosic feedstock [4].

	Cellulose % of d.m.	Hemicellulose % of d.m.	Lignin % of d.m	Other % of d.m.
Corn Stover	38	26	19	17
Soybean	33	14	-	53
Wheat Straw	38	29	15	18
Rye Straw	31	25	-	44
Switchgrass	37	29	19	15
Miscanthus	43	24	19	14
Forage Sorghum	34	17	16	33
Sweet Sorghum	23	14	11	52
Bagasse	40	30	20	10

2.1. Cellulose and Hemicellulose

Cellulose is a long-chain linear polymer that contains predominantly crystalline arrangements with smaller amorphous regions. The cellulose polymers are arranged in micro fibrils that are organized in fibrils; these are combined into cellulose fibers which are responsible for the fibrous nature of lignocellulosic biomass cell walls. Hemicelluloses are shorter, or branched amorphous polymers, of five- or six-carbon sugars. Together with lignin, hemicellulose forms the matrix in which the cellulose fibrils are embedded. The hemicellulose acts as a connector between cellulose and lignin [5]. Its basic building block is cellubiose, obtained from the linking of two moleclules of β-D-glucose. Cellulose, shown in Figure 3, presents many polar -OH groups that form hydrogen bonds with adjacent chains and create hard, stable crystalline regions and microfibrils giving strength to the whole biomass structure. The length of the chains can vary greatly, going from a few hundred to over ten thousand sugar units, depending on the type of biomass.

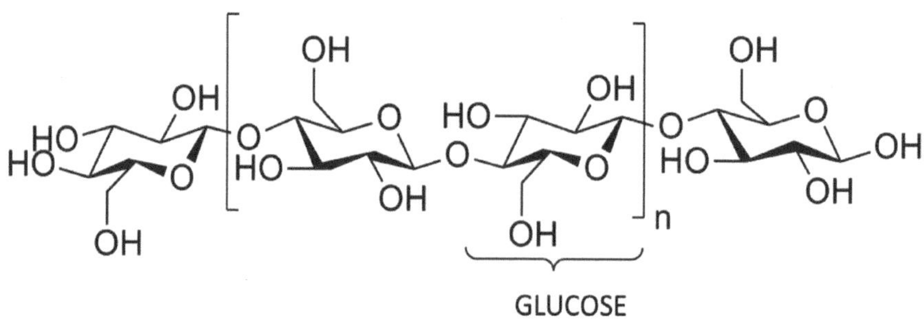

Figure 3. Molecular structure of cellulose.

Referring to Figure 4, it shows that cellulose is arranged into walls that constitute the plant cell.

The hemicelluloses in deciduous woods (hardwoods) and agricultural herbaceous products such as wheat straw, cornstover and switch grass predominantly consist of D-xylose units (xylans), whereas coniferous (softwood) hemicellulose is mainly made up of D-mannose (mannans), L-arabinose (arabinans) and D-galactose (galactans) [5]. Xylans are the main hemicelluloses in hardwood and they also predominate in annual plants and cereals, making up to 30 per cent of the cell wall material. Hardwood xylan (*O*-acetyl-4 methyl-glucuronoxylan) is substituted at irregular intervals with 4-*O*-methyl-α-D-glucuronic acid groups joined to xylose by α-1,2-glycosidic linkages. On average, every tenth xylose unit has a uronic acid group attached at C2 or C3 of xylopyranose [6]. Using hydrolysis, both cellulose and hemicellulose can be disrupted in C5 and C6 sugars. Hydrolysis can be done more readily on hemicellulose than on cellulose, due to its branched, amorphous nature.

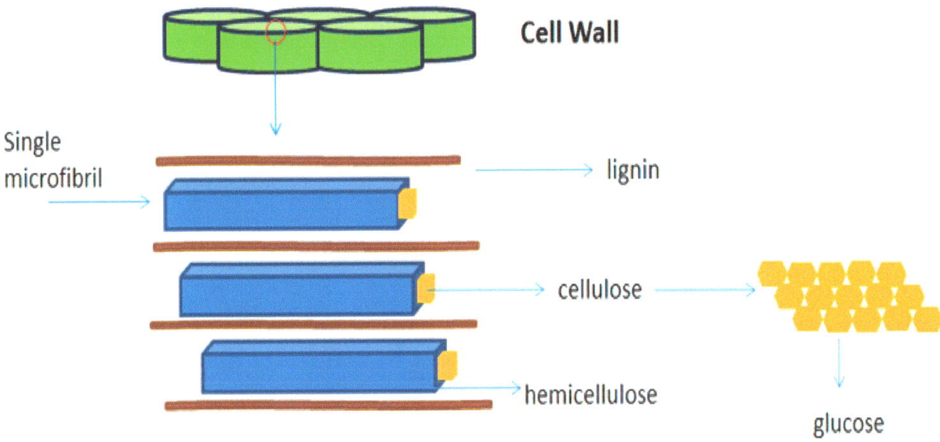

Figure 4. Structure of plant cell walls.

2.2. Lignin

Being one of the most abundant natural polymers, together with cellulose and hemicellulose, lignin forms part of the secondary cell walls of plants and helps maintain the integrity of the cellulose/hemicellulose/pectin matrix. The most complete structural model of lignin was proposed by Adler in 1977 (Figure 5) [7]. The variable molecular mass of this biopolymer is a consequence of the random cross-linked polymerization of phenolic monomeric units, originating from radical-coupling reactions between phenolic radicals [8,9]. It is generally accepted that there are three basic phenol derivatives, the so-called monolignols that make up almost all types of lignin found in nature—*p*-coumaryl alcohol (H-phenolic group)—coniferyl alcohol (G for guayacylic phenolic group), and sinapyl alcohol (S for syringilic phenolic group)] [10]. The scientific literature has defined lignin as a phenolic-propanoid polymer providing structural support for the plant, giving strength, rigidity and resistance to environmental stresses [11]. Of course, as well as cellulose and hemicellulose, the lignin content in biomass varies depending on biomass typology (softwood vs. hardwood vs. herbaceous plants), harvesting time and zone, biomass treatment.

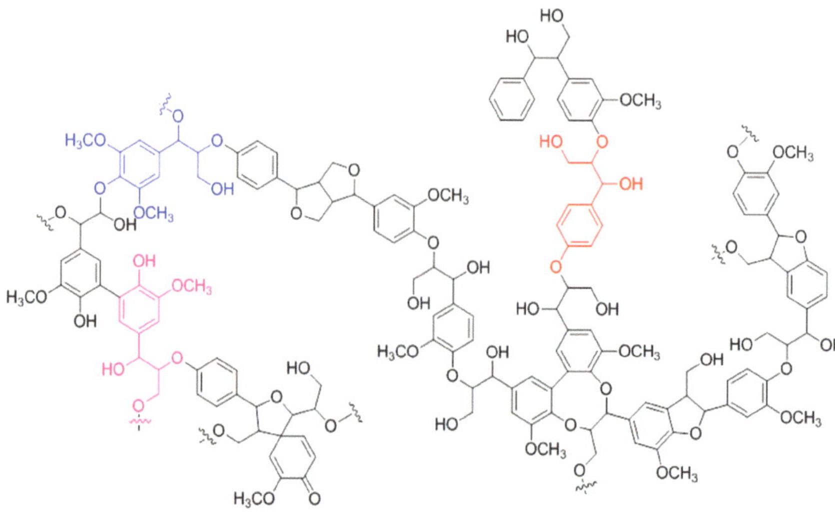

Figure 5. Lignin structure by Adler [7].

If the lignin can retain a generic propanoic polymer, it becomes crucial to identify its monomer distribution and concentration relating them to the three basic precursors mentioned above and shown in Figure 6 [9].

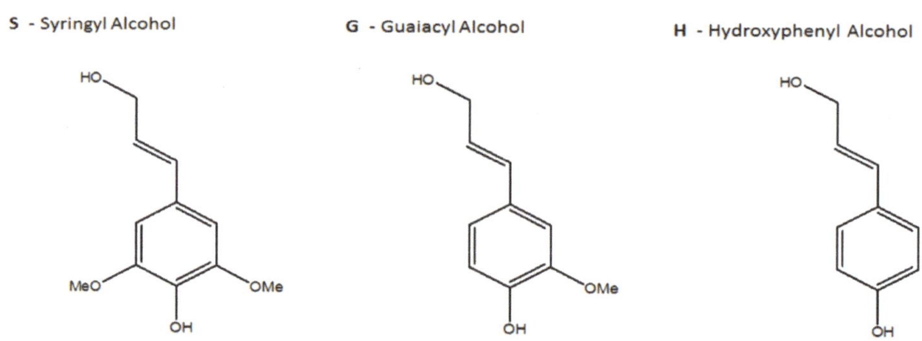

Figure 6. Lignin precursors.

The chemistry of lignin is complex compared to other biopolymers such as proteins or carbohydrates, which are linear chains or, at the most, branched polymers. Lignin contains a range of chemical functional groups, which is partly the result of the extraction method. The main groups in unmodified lignins are hydroxyl (aromatic and aliphatic), methoxyl, carbonyl, and carboxyl [9]. Lignin is composed of a three-dimensional network, lacking the regular and ordered repeating units of other biopolymers such as cellulose. The majority (approximately two thirds) of chemical bonds in the native lignin polymeric network are of the C-O-C ether linkage type between the phenylpropane units, predominantly β-O-4, while about one third consists of C-C bonds between these units. Furthermore, lignin also includes branched and cross-linked structures. Typical bonds and their recurrence in different types of lignin are shown in Table 2 [11].

Table 2. [Typical bonds and their recurrence in different types of lignin each 100 bonds] [11].

Name Bond	Type	Softwood (%)	Hardwood (%)
β-O-4	Etheric	40–50	50–60
β-5	Etheric	10–12	3
4-O-5	Etheric	3	3
5-5	carbon to carbon	13	3
β-β	carbon to carbon	3	3

3. II Generation Biomass Conversion

3.1. Pretreatment Methods

Until recently, the last pilot and demo-scale plant developments have defined pretreatment as the most important step for the conversion of II generation biomass towards chemicals and fuels. The goal of this step is the separation of the cellulose from the other natural polymers as well as the cellulose accessibility to the subsequent process steps such as the enzymatic hydrolysis or any thermic or catalytic chemical transformation. A strong separation from the lignin is needed. At the same time, it is important to allow valorization of the production of a lignin by-product with high purity. For both streams, the purer they are, the better can be retained. From this standpoint, it should also be considered that pretreatment should be tuned compatibility with feedstocks, enzymes and organisms that will be injected later on be in the bioprocessing and/or the needings for thermo/catalytic/chemical operations. Another important aspect that needs to be taken in account is the pretreatment cost, which also includes the costs related to the handling of feedstock and of solid and liquid streams generated, handling of waste and the potential production of co-products. The basic aim of this step of the procedure involves firstly the cleavage of hydrogen typically present in the crystalline cellulose as well as disrupting hemicellulose and secondly lignin matrix disrupting a remarkable increment of porosity as well as surface area of cellulose in order to make the enzymatic/chemical attack more effective [11,12]. The literature presents four pretreatment approaches to reach the goal: physical, physico-chemical, chemical and biological [13].

3.2. *Physical Biomass Pretreatment*

Physical pretreatment consists of mechanical processing for size reduction and specific surface increase. Examples of physical pretreatments are grinding, milling, microwaving and extrusion. Typically, these methods are expensive for their energy requirement [13].

3.3. *Physico-Chemical Biomass Pretreatment*

Physico-chemical pretreatments include steam explosion, catalyzed (SO_2 or CO_2) steam explosion, ammonia fiber explosion (AFEX), liquid hot water and microwave-chemical pretreatment [14,15]. The steam explosion biomass is treated with high-pressure saturated steam and then the pressure is suddenly reduced, which makes the materials undergo an explosive decompression [16]. This kind of explosion allows efficient cellulose separation from the woody matrix as well as making for a good characteristic for the subsequent cellulose hydrolysis, degradation of hemicellulose, an impact modification of native lignin structure. In this last aspect steam explosions usually affect the molecular weight of lignin towards its decrement as well as a reduction of its etheric bonds (C-O-C) despite an increment of biphenilic (C-C types) ones. It requires low capital investment, it has been proven on different feedstocks and it also has a low environmental impact [17]. Ammonia fiber explosion is very similar to steam explosion, but in this case lignocellulosic biomass is exposed to liquid ammonia instead of steam. The AFEX process demonstrates attractive economics compared to several leading pretreatment technologies [18] and the ammonia used during the process can be recovered and reused.

In the case of liquid hot water pretreatment (LHW), biomass undergoes high-temperature cooking in pressurized water [19]. LHW helps enhance cellulose digestibility, sugar extraction, and pentose recovery, and the product contains little or no inhibitor of sugar fermentation [20]. Moreover, this kind of pretreatment does not require biomass size reduction, as particles are broken apart during pretreatment itself. Finally, microwave/chemical pretreatment is very effective as it accelerates reactions during the pretreatment process [16,21].

3.4. *Chemical Biomass Pretreatment*

Today, these types of pretreatments are mainly employed in the pulp and paper industry where the cellulose delignification is a must in order to achieve high quality in terms of pulp pureness (lignin here retains a sort of undesired product or a real poison) and so a high quality of final products (paper). It is important to note that the chemical wood/biomass pretreatment must provide a cellulose having high crystalline degree and molecular weight, and biodegradability of the product.

Among the commercial routes actually used, the Acidic pretreatment foresees the use of concentrated and diluted acids (typically dilute sulfuric acid, H_2SO_4 and Na_2S) with a good hemicellulose removal. Alkaline pretreatment is more focused on lignin separation. The chemicals needed are basically potassium, calcium, and ammonium hydroxide. In any case, a good quality of cellulose is generally achieved. Depending on the starting material origin, type (softwood vs. hardwood) and aging, the tuning of each process parameter (pressure, temperature, residence time, and above all chemical choice and their weight ratio with respect to the biomass) must be defined preliminarily in order to get high cellulose/hemicellulose/lignin separation.

Generally alkaline pretreatment is performed at lower temperature/pressure (almost in environmental conditions) and for a longer residence time than the acid one. In both cases the de-structured biomass needs a washing step where lignin and all the inhibitors for the subsequent hydrolysis must be removed.

Ionic liquids (green solvents) have recently been used for dissolution of several woody biomasses. These pretreatments seem more sustainable in term of energy requirement, less impacting from an environmental standpoint, simple enough in terms of equipment and operating conditions [22], but more study must to be done in order to use them for commercial applications [23].

3.5. Chemical Biomass Pretreatment

Biological pretreatment uses wood-degrading microorganisms to modify the chemical composition and/or structure of the lignocellulosic biomass in order to facilitate enzyme digestion. This kind of pretreatment has multiple limitations but also many advantages, including no chemical requirement, low energy input, mild environmental conditions, and environmentally friendly working parameters [24,25].

4. Cellulose and Hemicellulose Conversion to Sugars

Cellulose and hemicellulose obtained from biomass pretreatment is then hydrolyzed to soluble fermentable sugars using enzyme systems (mainly cellulases and hemicellulases). Several process configurations can be applied: separate hydrolysis and fermentation (SHF), simultaneous saccharification and fermentation (SSF), simultaneous saccharification and co-fermentation (SSCF) and consolidated biomass processing (CBP) [16].

SHF is composed by two steps conducted at different conditions to each other and does not need any washing of pretreated biomass or nutrient supplementation [26]. The first one foresees a biomass hydrolysis to sugars via enzyme injection, followed by a second step where the sugars are fermented to ethanol with the help of various yeasts [27].

During SSF, pretreated biomass hydrolysation to sugars and sugar fermentation to ethanol take place in the same vessel. The difficulty with this solution is the optimizing of all the various conditions governing the two different processes (pH, temperature, substrate concentration, etc.) [28]. On the other hand, its strength is in the removal of end-product inhibition, which causes enhanced hydrolysis rate. The cellulose conversion can be favored only with a washing step where the inhibitor has to be taken out [29].

SSCF is mainly used for xylose-rich lignocellulosic materials and is carried out by genetically engineered microbes.

CBP is characterized by the occurring of the three main bio-transformations involved in lignocellulosic bioprospecting in a single process: production of enzymes, hydrolysis and fermentation. In the first step, cellulase and hemicellulases enzyme production is achieved; simultaneously it is possible to get sugars such as pentose and hexose (from hemicellulose and cellulose) as well as their fermentation. This kind of method allows avoiding the costs associated with cellulose production, but the "engineering" of a culture with many microbial tasks.

5. Lignin Separation Technologies

As already widely discussed in this section, there are distinct lignin separation processes [30] already implemented on the commercial scale for several purposes. Most of them are directly connected to the pulp and paper industry where the lignin is considered a by-product, and in the best case a co-product. Moreover, due to the different properties (molecular weight, dryness, particle size distribution, surface area, presence of sulfonic groups) intensely dependent on the separation processes of origin, we can have different kinds of products and just not a simple lignin with native characteristics. Despite this dilemma, nowadays in the pulp and paper industry 98% of the material coming out from the delignification processes is still burnt in the same factories to generate energy and steam for the energetic sustainability of the entire process. Lignosulfonates are the only products related to the pulp and paper industry that have a market, but they represent no more than 2% of the total amount of lignin product available (in terms of mass, no more than 1 Mton/y). The reduced availability of lignosulfonates on the market is mainly due to the fact that Kraft process has gradually replaced sulfite process in the last 50 years. Today, the amount of lignin extracted and separated from black liquor in the Kraft process is one order of magnitude less than that of lignosulfonates, but a strong effort has been made by some pulp and paper companies in order to obtain high-value products, lowly affected from impurities and sulfur traces [31].

5.1. *Kraft Lignin*

The Kraft process consists of a chemical pretreatment of wood in a range of 155–175 °C, using a solution of $Na_2S/NaOH$ that allows the delignification of the raw material after several hours of decanting time. The solid (cellulose) stream is then separated from the fluid (black liquor) fraction. If isolation of Kraft lignin is desired, the black liquor is precipitated by neutralization with at very low pH by adding some acidic agent. In order to get the solid powder of lignin, a drying process needs to be implemented at a large scale [32]. Kraft lignin could be considered a hydrophobic material, with a molecular mass around 1000 Da, affected from an abundance of biphenilic bonds (C-C′) among propylphenols monomers [33] as reported in Figure 7. This separation process makes lignin water-insoluble, and/or soluble in basic phase.

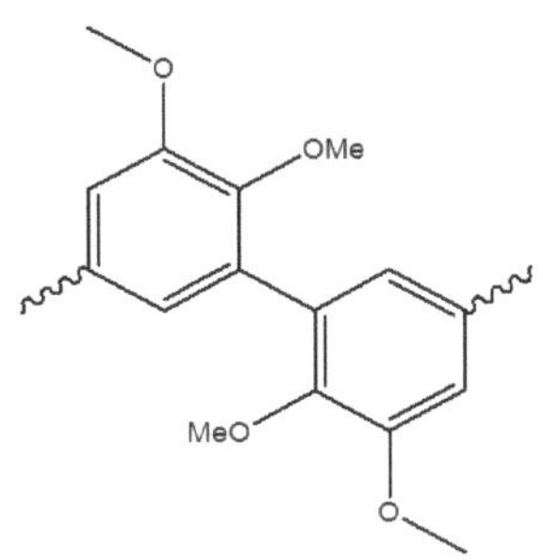

Figure 7. Main structure identified in Kraft extracted lignin [33].

5.2. *Lignosulfonates*

This kind of lignin is generally extracted from liquid residues at the end of the sulfite-pulping process. The raw material generally used for this technology consists of softwoods, and the lignin extraction procedure foresees the use of sulfurous acid (sulfites or bisulfites) followed by the addition of magnesium, ammonium or sodium salts [34] for neutralization and precipitation. The chemical structure of this product presents hydrophobic and hydrophilic properties. In this kind of extraction, the less aggressive and severe conditions ensure a product with a higher molecular weight than Kraft lignin, and a quite important presence of etheric groups among the propylphenolic aromatic monomers. Comparing this polymer matrix with the native lignin, the presence of the sulfonic group attached to the lignin monomers creates a new and completely different physical and chemical behavior with respect to the original one (see Figure 8).

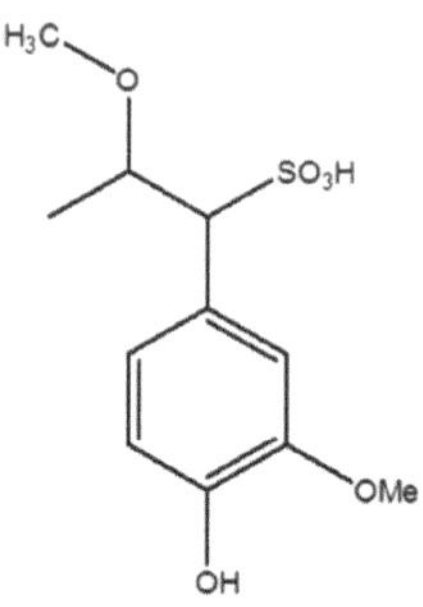

Figure 8. Typical monomer formulation of lignosulfonate.

5.3. *Organosolv Lignins*

Organosolv treatment applied to the raw biomass has been one of the most studied processes in the field of lignin extraction. Many literature papers have highlighted the influence of delignification parameters (temperature, pressure, type of solvent, basically ethanol or acetic acid, solvent concentration in the aqueous phase, addition of a small amount of acid dose) on the fraction of lignin extracted in a batch or in a continuous process. In any case, the aim of the process is solubilizing part of the original lignin. In a subsequent step, lignin separation from the solution is obtained by acidic precipitation (pH ~2), and then solvent recovery is achieved by distillation of the solution. Nowadays there are several processes producing commercial Organosolv Lignin. Most of them are summarized in Table 3, also specifying the solvent adopted in order to achieve them. Usually working temperature is between 180 and 200 °C, working pressure between 2 and 5 bar and solvent concentration in water close to 60% *w*/*w*. Usually lignin yields are quite low and the pureness of the obtained product (sulfur, ash and salt free, abundance of etheric bonds among monomeric units as well as high concentration of guayacilic, siryngilic and phenolic monomers) is combined with low molecular weight (around 1000 g/mol) [35].

Table 3. Organosolv Lignin Commercial/Patented Extraction [10].

Lignin Type	Solvent
Alcell [36]	Ethanol/water
Alcetocell [37]	Acetic acid/water
Acetosolv [38]	Acetic acid/HCl pulping
ASAM [39]	Alkaline sulfite/anthaquinone/methanol
Batelle [40]	Phenol/acid/water
Formacell [41]	Acetic acid/formic acid/water
Milox [42]	Formic acid/hydrogen peroxide
Organocell [43]	Methanol pulping/NaOH/anthaquinone pulping

5.4. Steam Exploded Lignin

If biomasses are treated at high temperature/pressure (i.e., 180–200 °C), followed by a sudden decompression in presence of some chemicals, hydrolysis of lignin is reached. In these conditions a water-insoluble lignin material with a low level of residual carbohydrates and extractives is obtained [10]. Depending on the severity of the explosion and on the acidity level of the hydrolysis, the molecular weight of the product can be affected.

5.5. II Generation Bioethanol Lignin

In recent years the development of 2nd-generation bioethanol plants has made it possible to get a new source of lignin as a by-product of the entire process. In this case, the lignin is a sulfur-free product due to the fact that these processes do not foresee the use of acid or basic attack to the inlet feedstock (mainly agricultures residues like wheat or rice straw or energy crops grown in marginal land). Furthermore, the chemical structure of the obtained lignin is very close to the native one because in this case smart cooking used for biomass pretreatment is less severe in terms of temperature and vapor condition than steam explosion.

6. Lignin Derivatives and Their Applications

As shown in a previous paragraph, lignosulfonate is the only lignoderivative with a wide commercial use due to its capabilities as a binder. Nevertheless, to date, new routes are becoming interesting in terms of application although they remain limited in terms of market dimensions. These routes are the ones of eco friendly renewable materials [44], moisture-retention agent and dust suppressors, gypsum-board manufacture, asphalt stabilizers, dyes, and pigment emulsions [45]. In the last 5 years, many studies and granted projects worldwide [46,47] have been focused on the use of different lignins (Organosolv, Kraft, steam exploded) as co-polymers, additives for making new and functionalized composite polymers, formaldehyde resins, polyester polymers, polyurethane foams, bio-plastics, epoxy resins. The scientific approach to make feasible the realization of these new products with enhanced mechanical and thermal properties (with respect to their petrochemical equivalents) consists in the functionalization of the starting lignins, obtained by chemical modification. This goal can be achieved by the insertion of a specific group into the polymeric matrix by alkylation, dealkylation, amination, sulfonation, silylation, acylation, halogenation, nitration or methylation [48,49].

6.1. Resins

As an alternative to the commercial phenol-formaldehyde resins, lignin-formaldehyde resins seem to be promising materials if used as adhesives for plywood, chipboard, fiberboard, rubber products, and refractory and friction materials. The synthesis of this new type of resin occurs in acidic or basic media [50]. In order to get this new kind of resin, Organosolv or hydrolyzed lignin has been retained as the most performing replacements for phenols coming from fossil sources [51]. The economic and technical evaluation of this new product as adhesive is considered positive for its application and commercialization [52].

Referring to epoxy-resins, a previous modification of the lignin is required before its introduction in the polymer structure as cross-linking agent instead of the bisphenol-A [53]. Generally, the modification foresees a phenolation of the lignin matrix in order to get diphenolic moieties that perform as well as the bisphenols during the epoxidation of the resin precursors [54]. In this type of application, the best performing feedstocks have been identified in lignosulfonates and Organosolv lignin.

6.2. *Polyurethane*

Due to the large use of this material in the construction and building industry, the substitution of these fossil-derivatives with green sources has been widely investigated. In this case, the Organosolv lignin has been demonstrated as the best substitute of the aromatic isociantes (in particular the methyl diphenil isocianate, MDI) for reacting with polyols (PEG or PPG). The products achieved could be considered good in terms of physical properties (both sheets and foams) [55]. Economic evaluations are undergoing in order to ascertain a comparison cost production with the fossil derivatives.

6.3. *Polyolefin Polymers*

The high resistance to biodegradability of polypropylene (PP) and polyethylene (PE) has led to new research towards a blend of both these polyolefin polymers with natural polymer lignin. However, the optimization of good products, especially referring to the HDPE and PP, in terms of mechanical properties, seems to be still far off [56].

6.4. *Bio-Plastics from Lignin*

Remarkable results have been achieved for the production of an alternative thermoplastic material in which sulfur-free lignins can be mixed with natural fiber or additives in order to obtain a plastic granulated called ARBOFORM [57] that can be melted and molded as a petrochemical thermoplastic material that can replace the plastic used in household industries.

6.5. *Carbon Fibres*

The relatively low cost of some lignin sources has oriented many researchers in considering these materials as precursors for preparing carbon fibers by carbonization of dry-spun fibers. A basic aqueous solution needs to be prepared and added during the plasticized treatment with polyvinyl alcohol or glycerol. Due to the high-purity lignin required for melt-spinning (sulfur, sugar and ash free), this application seems to be quite far from an immediate realization [58].

7. Lignin Depolymerization Technologies

In competition with the direct application of lignins in the products/materials listed and described before, there is the transformation of lignin to several families of remarkable chemicals, mainly aromatics, which might substitute products coming from standard petrochemistry. Conversion technologies are based on the depolymerization concept that transforms the starting matrix structure to monomers or dimers having a good relevance in terms of chemical product demand: phenols, cresols, benzene, toluene and xylenes as well as quinones or polyols. From this perspective pyrolysis of lignin can be seen as the oldest and most widely studied depolymerization/cracking technique since the 60s for transforming wood or lignocellulosic feedstocks to chemicals and fuels [59]. Independently from reactor choice (fluidized bed, ablative cone, Auger) [60] the main product achieved by thermocracking (usually 500 °C in inert atmosphere) is an oil with a large amount of substituted phenols (monomeric, dimeric and trimeric), with a huge water, oxygen and acid content, unstable along the time that requires an immediate upgrade process in order to get monomeric phenols/cresols or aromatics like benzene toluene and xylenes. Besides oil, pyrolysis also foresees an important production of char (yields can oscillate between 20 and 40% *w*/*w* of the starting materials) and gases. Both these products need to be burnt in order to make energetically self-sustainable the entire operation. If the thermocracking occurs in an oxidative atmosphere, the oil is richer in vanillin and eugenol than phenols and methoxyphenols (guaiacols and syringols). Both these products have a good way out market as flavorings for the food and cosmetic industries and some application in the pharmaceutical industries. If the oxidative atmosphere is brought to a higher concentration respect to the inert carrier (especially if the fluidized bed configuration is used as reactoristic choice) the distribution of products loses the aromaticity to facilitate the formation of quinones or quinines (C6-cyclic dione structure). In both cases, the inert or oxidative atmosphere represents a limitation for making continuous and scalable processes. The huge amount of TAR formation during the condensation of the vapors coming out from the cracker reactor seems to be detrimental for the large scale development [61]. Due to this limit, in recent years, the catalytic cracking of lignin in a reducing atmosphere (presence of hydrogen) has been more and more studied, in order to avoid the formation of undesired products (char and tar), maximize the useful oil fractions, minimize its acidic content (in order to get it stable), and maximize its concentration in monomeric phenols [62,63]. If we refer to the char by-product, in the last three years many studies have been made for a possible upgrade of the char towards useful compounds like active carbon or as a simple soil amendment compound. From this perspective the valorization of the char could make the pyrolysis more attractive for its further development in bigger scale. Alternative depolymerization routes are also represented by microbial conversions for producing vanillic, ferulic and coumaric acids, the enzymatic oxidation for the production of pigments [64,65].

The controlled breaking of different linkage types in lignin needs detailed information on the stability of the bonds under different conditions and knowledge of the mechanisms of lignin decomposition. The etheric β-O-4 bond that crosslinks more than 50% of the phenolic monomers can be cleaved between 200 and 400 °C. At higher temperatures secondary reactions could bring radicalic carbons crossing over with formation of diphenilic C-C bonds much more stable. A very bad consequence might be the TAR formation as unsuitable effect. For this reason, lignin treatments typically yield oligomers, rarely a single monomer, so the cleavage can be promoted using a reducing atmosphere. In all these cases the use of functionalized catalysts that allow the simultaneous cleavage of etheric and diphenilic bonds, and the hydrogenation of the radicals generated by the thermal cracking seem to be a keyword in order to get the goal. In Figure 9 are shown the products that can be obtained with the various depolymerization techniques, depending from the atmosphere in which they take place.

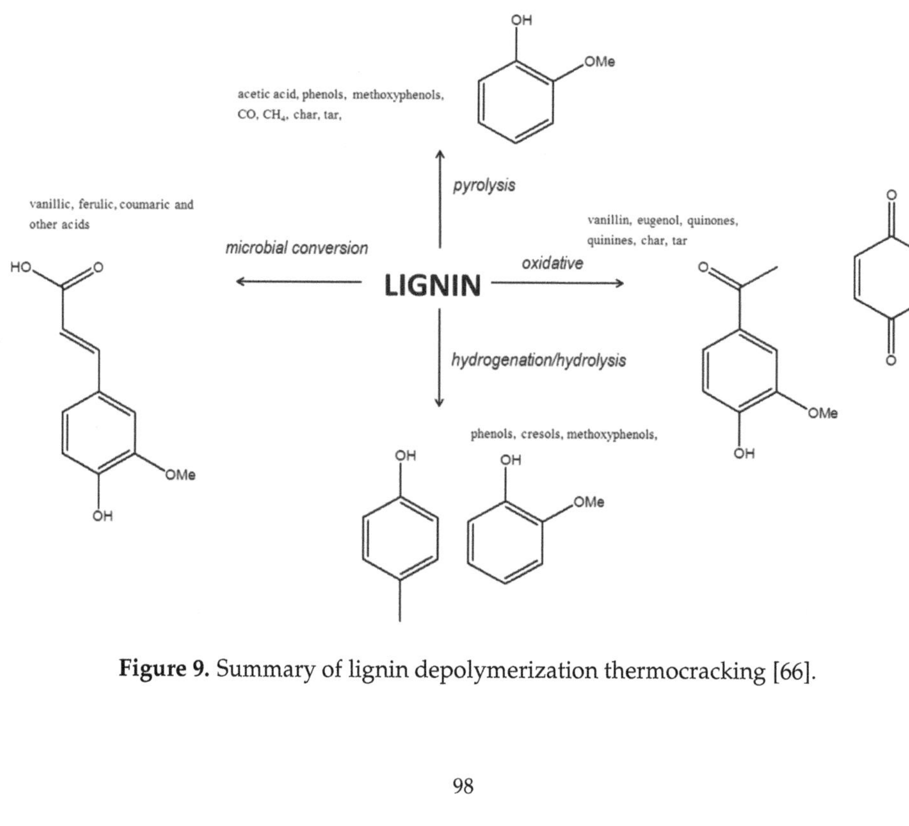

Figure 9. Summary of lignin depolymerization thermocracking [66].

If we refer to the sources of lignin available on the market, lignosulfonates still represents the most important product with a huge use as plasticizer in the concrete industries [67,68] as well as many others like emulsifiers but with lower impact in terms of tons sold to the tons produced [66]. In any case, the high concentration of sulfur and salts in its structures has strongly limited the valorization of this ligno-derivatives towards other applications, in particular its depolymerization towards valuable monomers [66]. Nowadays the development of second generation bioethanol plants has made possible to foresee in the near future alternative sources of very pure types of lignin products, sensibly less affected from sulfur or salts, so easier to transform in high valuable compounds and/or pure chemicals. From this perspective, the valorization of the lignin residues coupled to the second generation bioethanol plants makes it possible a new and more versatile scheme of a biorefinery platform in which the lignin transformation becomes more profitable than its energy valorization [69]. Many studies made in the past upon different kind of lignins (differing for the extraction process, because each extraction process can affect differently lignin native structure, its bond functionalities and its polymerization grade), have been revalued and deepened in order to better identify new and alternative routes for the production of emulsifiers [70], chelating agents for heavy metals removal from industrial effluent [71], active carbons (after activation by gasification) [72] as well as several depolymerization technologies via many catalytic, thermic or thermocatalytic cracking processes able to convert this feedstock to high value products having lower molecular weight [73].

8. Conclusions

The developments and the main results achieved during the scale-up at commercial scale of second-generation biorefineries is becoming more important with every year. If everything starts from the three components that constitute the feedstocks for this kind of concept plant—cellulose (in part and hemicellulose) and lignin—a plethora of processes have been separately developed for being later integrated among them in order to:

- Separate the main components
- Depolymerize or use as they are the components that are retained, which are a huge source of different chemicals, able to be entered in basic chemistry or other more complex applications

If the cellulose can open a route towards a plethora of chemicals, it is mandatory to operate in the inlet part of the biorefinery concept a series of operations able to depolymerize the starting matrix in order to get pure and low molecular weight sugars that can be then converted towards the desired chemicals, a route already widely known from the I generation biorefinery concept. If the conversion of cellulose and its derivatives can be considered a quite well-known pathway, lignin and its valorization is still undergoing. However numerous direct applications seem to be close to their first scale-up from small quantity to almost commercial amounts, as well as the scale-up of several depolymerization processes from pilot to semi-commercial scale.

Conflicts of Interest: The authors declare no conflict of interest.

References

1. Raschka, A.; Carus, M. Industrial Material Use of Biomass Basic Data for Germany, Europe and the World. Available online: http://www.nova-institut.de/bio/ (accessed on 19 February 2015).
2. Klemm, D.; Philipp, B.; Heinze, T.; Heinze, U.; Wagenknecht, W. Systematic of cellulose Fractionation. In *Comprehensive Cellulose Chemistry*, 1st ed.; Wiley & Sons: New York, NY, USA, 1998; Volume 2, pp. 221–223.
3. Ding, S.Y.; Himmel, M. The maize primary cell wall microfibril: A new model derived from direct visualization. *J. Agric. Food Chem.* **2006**, *54*, 597–606. [CrossRef] [PubMed]
4. Lee, D.K.; Owens, N.V.; Boer, A.; Jeranyama, P. *Composition of Herbaceous Biomass Feedstocks*; North Central Sun Grant Center, South Dakota State University: Brookings, SD, USA, 2007; Available online: ncsungrantsds.sdstate.org/uploads/publications/SGIN1-07.pdf (accessed on 25 June 2007).
5. De Wild, P. Biomass pyrolysis for chemicals. *Biofuel* **2011**, *2*, 185–208. [CrossRef]
6. Toledano, A.; Garcia, A.; Mondragon, I.; Labidi, J. Lignin separation and fractionation by ultrafiltration. *Sep. Purif. Technol.* **2010**, *71*, 38–43. [CrossRef]
7. Adler, E. Lignin chemistry—Past, present and future. *Wood Sci. Technol.* **1977**, *11*, 169–218. [CrossRef]
8. Dence, C.W.; Lin, S.Y. The determination of lignin. In *Methods in Lignin Chemistry*, 1st ed.; Springer: Berlin/Heidelberg, Germany, 1992; pp. 33–61.
9. Morohoshi, N.; Glasser, W.G. The structure of lignin in pulps. *Wood Sci. Technol.* **1979**, *13*, 165–178. [CrossRef]
10. Garcia Calvo, F.; Dobado, J.A. Lignin as Renewable raw material. *ChemSusChem* **2010**, *3*, 1227–1235. [CrossRef] [PubMed]
11. Li, C.; Knierim, B.; Manisseri, C.; Arora, R. Comparison of dilute acid and ionic liquid pretreatment of switchgrass: Biomass recalcitrance, delignification and enzymatic saccharification. *Bioresour. Technol.* **2010**, *101*, 4900–4906. [CrossRef] [PubMed]
12. Lynd, L.R.; Wyman, C.E.; Gerngross, T.U. Biocommodity engineering. *Biotechnol. Program* **1999**, *15*, 777–793. [CrossRef] [PubMed]
13. Yang, B.; Wyman, C. Pretreatment: The key to unlocking low-cost cellulosic ethanol. *Biofuels Bioprod. Biorefin.* **2008**, *2*, 26–40. [CrossRef]
14. Mosier, N.S.; Wyman, C.; Dale, B.; Elander, R. Features of promising technologies for pretreatment of lignocellulosic biomass. *Bioresour. Technol.* **2005**, *96*, 673–686. [CrossRef] [PubMed]

15. Brodeur, G.; Yau, E.; Badal, K.; Collier, J.; Ramachandran, K.B.; Ramakrishnan, S. Chemical and physicochemical pretreatment of lignocellulosic biomass: A review. *Enzyme Res.* **2011**, *2011*, 1–17. [CrossRef] [PubMed]
16. Menon, V.; Rao, M. Trends in bioconversion of lignocellulose: Biofuels, Platform Chemicals & Biorefinery Concept. *Prog. Energy Combust. Sci.* **2012**, *38*, 522–550.
17. Focher, B.; Marzetti, A.; Crescenzi, V. *Steam Explosion Techniques, Fundamentals and Industrial Applications*, 1st ed.; Gordon and Breach Publishers: Philadelphia, PA, USA, 1991; pp. 412–430.
18. Teymouri, F.; Perez, L.L.; Alizadeh, H.; Dale, B.E. Ammonia fiber explosion treatment of corn stover. *Appl. Biochem. Biotechnol.* **2004**, *116*, 951–963. [CrossRef]
19. Yu, G.; Yano, S.; Inoue, H.; Inoue, S.; Endo, T.; Sawayama, S. Pretreatment of rice straw by a hot-compressed water process for enzymatic hydrolysis. *Appl. Biochem. Biotechnol.* **2010**, *160*, 539–551. [CrossRef] [PubMed]
20. Kim, Y.; Mosier, N.S.; Ladisch, M.R. Enzymatic digestion of liquid hot water pretreated hybrid poplar. *Biotechnol. Prog.* **2009**, *25*, 340–348. [CrossRef] [PubMed]
21. Zhu, S.; Wu, Y.; Yu, Z.; Liao, J.; Zhang, Y. Pretreatment by microwave/alkali of rice straw and its enzymatic hydrolysis. *Process Biochem.* **2005**, *40*, 3082–3086. [CrossRef]
22. Zhao, H.; Cecil, L.J.; Gary, A.B.; Shu, Q.X.; Olarongbe, O.; Vernecia, N.P. Regenerating cellulose from ionic liquids for an accelerated enzymatic hydrolysis. *J. Biotechnol.* **2009**, *139*, 47–54. [CrossRef] [PubMed]
23. Hélène, O.B.; Lionel, M. Ionic liquids: Perspectives for organic and catalytic reactions. *J. Mol. Catal. A-Chem.* **2002**, *182*, 419–437.
24. Kurakake, M.; Ide, N.; Komaki, T. Biological pretreatment with two bacterial strains for enzymatic hydrolysis of office paper. *Curr. Microbiol.* **2007**, *54*, 424–428. [CrossRef] [PubMed]
25. Salvachúa, D.; Prieto, A.; López-Abelairas, M. Fungal pretreatment: An alternative in second-generation ethanol from wheat straw. *Bioresour. Technol.* **2011**, *102*, 7500–7506. [CrossRef] [PubMed]
26. Lau, M.W.; Dale, B.E. Cellulosic ethanol production from AFEX-treated cornstover using Saccharomyces cerevisiae 424A (LNH-ST). *Proc. Natl. Acad. Sci. USA* **2009**, *106*, 1368–1373. [CrossRef] [PubMed]
27. Philippidis, G.P. Cellulose bioconversion technology. In *Handbook on Bioethanol-Production and Utilization*, 1st ed.; Taylor and Francis: Washington, DC, USA, 1996; pp. 253–285.
28. Doran Peterson, J.; Jangid, A.; Brandon, S.K. Simultaneous Saccharification and Fermentation and Partial Saccharification and Co-Fermentation of Lignocellulosic Biomass for Ethanol Production. In *Biofuel*, 1st ed.; Humana Press: New York, NY, USA, 2009; pp. 263–280.
29. Lu, Y.F.; Wang, Y.H.; Xu, G.Q.; Chu, J.; Zhuang, Y.P.; Zhang, S.L. Influenece of high solid concentrations on enzymatic hydrolysis and fermentation of steamexploded cornstover biomass. *Appl. Biochem. Biotechnol.* **2010**, *160*, 360–369. [CrossRef] [PubMed]
30. Biermann, C.J. Pulping fundamentals. In *Handbook of Pulping and Papermaking*, 2nd ed.; Academic Press: San Diego, CA, USA, 1996; pp. 55–97.

31. Bajpai, P. *Biorefinery in the Pulp and Paper Industry*, 1st ed.; Academic Press: London, UK, 2013; pp. 52–54.
32. Koljonen, K.; Osterberg, M.; Kleen, M.; Furhmann, A.; Stenius, P. Precipitation of lignin and extractives on kraft pulp: Effect on surface chemistry, surface morphology and paper strength. *Cellulose* **2004**, *11*, 209–224. [CrossRef]
33. Elegir, G.; Daina, S.; Zia, L.; Bastet, G.; Orland, M. Laccase mediator system: Oxidation of recalcitrant lignin model structures present in residual kraft lignin. *Enzyme Microb. Technol.* **2005**, *37*, 340–346. [CrossRef]
34. Jongerius, A.L.; Jastrzebski, R.; Bruijnincx, C.A.; Weckhuysen, B.M. CoMO sulfide catalyst hydrodeoxygenation of lignin model compounds: An extended reaction network for the conversion of monomeric and dimeric substrates. *J. Catal.* **2012**, *285*, 315–323. [CrossRef]
35. Toledano, A.; Serrano, L.; Balu, A.M.; Luque, R.; Pineda, A.; Labidi, J. Fractionation of organosolv lignin from olive tree clippings and its valorization to simple phenolic compounds. *ChemSusChem* **2013**, *6*, 529–536. [CrossRef] [PubMed]
36. Pye, E.K.; Lora, J.H. The Alcell process: A proven alternative to kraft pulping. *Tappi* **1991**, *74*, 113–117.
37. Norbert, N.; Klaus, B. Acetocell—An innovative process for pulping, totally free from sulfur and chlorine. *Das Papier* **1989**, *47*, 16–24.
38. Parajo, J.C.; Alonso, J.L.; Vazquez, D.; Santos, V. Optimization of catalyzed acetosolv fractionation of pine wood. *Holzforschung* **1993**, *47*, 188–196. [CrossRef]
39. Black, N.P. ASAM alkaline sulfite process shows potential for large scale application. *Tappi* **1991**, *74*, 87–93.
40. Johansson, A.; Aaltonen, O.; Ylinen, P. Organosolv pulping—Methods and pulp properties. *Biomass* **1987**, *13*, 45–65. [CrossRef]
41. Saake, B.; Lummitsch, S.; Mormanee, R.; Lehnen, R.; Nimz, H. Production of pulp using Formacell method. *Das Papier* **1995**, *10*, 1–7.
42. Laamanen, L.; Sundquist, J.; Wartiovaara, S.; Kauliomki, S.; Poppius, S. Predefibering Lignocellulose with Peroxy (Formic or Acetic) Acid. U.S. Patent 4,793,898 A, 27 December 1988.
43. Lindner, A.; Wegener, G. Characterization of lignins from organosolv pulping according to the organocell process. *J. Wood Chem. Technol.* **1988**, *8*, 323–340. [CrossRef]
44. Pure Lignin. Available online: http://purelignin.com/lignin (accessed on 6 April 2015).
45. Mathiasson, A.; Kubt, D.G. Wood. *Wood Prod.* **1994**, *52*, 9–18. [CrossRef]
46. Lignoworks. Available online: www.lignoworks.ca (accessed on 6 April 2014).
47. Prolignin. Available online: www.woodwisdom.net/mm_files/do_902/ProLignin.pdf (accessed on 6 April 2015).
48. Wang, J.; Manley, R.S.J.; Feldman, D. Synthetic polymer-lignin copolymers and blends. *Prog. Polym. Sci.* **1992**, *17*, 611–646. [CrossRef]
49. Thielemans, W.; Wool, R.P. Novel Applications of Lignin in Composite Materials. *Biomacromolecules* **2005**, *6*, 1895–1905. [CrossRef] [PubMed]
50. Stewart, D. Lignin as a base material for materials applications: Chemistry, application and economics. *Ind. Crops Prod.* **2008**, *27*, 202–207. [CrossRef]

51. Cavdar, A.D.; Kalaycioglu, H.; Hiziroglu, S.; Mater, J. Some of the properties of oriented strand board manufactured using Kraft lignin phenolic resin. *Process Technol.* **2008**, *202*, 559–563. [CrossRef]
52. Park, Y.; Doherty, W.O.S.; Halleya, P.J. Developing lignin-based resin coatings and composites. *Ind. Crops Prod.* **2008**, *27*, 163–167. [CrossRef]
53. Tejado, A.; Kortaberria, G.; Peña, C.; Labidi, J.; Echeverría, J.M.; Mondragon, I. Lignins for phenol replacement in novolac-type phenolic formulations, Part I: Lignophenolic resins synthesis and characterization. *J. Appl. Polym. Sci.* **2007**, *106*, 2313–2319. [CrossRef]
54. Tejado, A.; Kortaberria, G.; Peña, C. Lignins for phenol replacement in novolac-type phenolic formulations. II. Flexural and compressive mechanical properties. *J. Appl. Polym. Sci.* **2008**, *107*, 159–165. [CrossRef]
55. Hu, T.Q. *Chemical Modification, Properties, and Usage of Lignin*, 1st ed.; Kluwer Academic: New York, NY, USA, 2002; pp. 43–44.
56. Gandini, A. Polymers from Renewable Resources: A Challenge for the Future of Macromolecular Materials. *Macromolecules* **2008**, *41*, 9491–9504. [CrossRef]
57. Arboform^{TM}. Available online: http://www.spin-project.eu/index.php?node_id=58.43&lang_id=1 (accessed on 6 April 2015).
58. Kadla, F.; Kubo, S.; Gilbert, R.D.; Venditti, R.A. *Lignin-Based Carbofibers, in Chemical Modification, Properties, and Usage of Lignin*, 1st ed.; Springer: Berlin, Germany, 2002; pp. 55–69.
59. Kersten, S.A.; Wang, X.; Prins, W.; Van Swaaij, W.P.M. Biomass Pyrolysis in a Fluidized Bed Reactor. Part 1: Literature Review and Model Simulations. *Ind. Eng. Chem. Res.* **2005**, *44*, 8773–8785. [CrossRef]
60. Bridgwater, A.V. Review of fast pyrolysis of biomass and product upgrading. *Biomass Bioenergy* **2012**, *38*, 68–94. [CrossRef]
61. Peacocke, G.V.C.; Bridgwater, A.V. Ablative plate pyrolysis of biomass for liquids. *Biomass Bioenergy* **1994**, *7*, 147–154. [CrossRef]
62. Nowakowski, D.J.; Bridgwater, A.V.; Elliott, D.C.; Meier, D.; De Wild, P. Lignin pyrolysis: Results from an international project. *J. Anal. Appl. Pyrolysis* **2010**, *88*, 53–72. [CrossRef]
63. Gooty, A.T.; Li, D.; Berruti, F.; Briens, C. Kraft-lignin pyrolysis and fractional condensation of its bio-oil vapors. *J. Anal. Appl. Pyrolysis* **2014**, *106*, 33–40. [CrossRef]
64. Tessini, C.; Müller, N.; Mardones, C.; Meier, D.; Berg, A.; Von Baer, D. Chromatographic approaches for determination of low-molecular mass aldehydes in bio-oil. *J. Chromatogr. A* **2012**, *1219*, 154–160. [CrossRef] [PubMed]
65. Pandey, M.P.; Kim, C.S. Lignin Depolymerization and Conversion: A Review of Thermochemical Methods. *Chem. Eng. Technol.* **2011**, *34*, 29–41. [CrossRef]
66. Bozel, J.J.; Holladay, J.E.; Johnson, D.; White, J.F. Top value-added chemicals from biomass. In *Results of Screening for Potential Candidates from Biorefinery Lignin*, 1st ed.; Pacific Northwest national Laboratory (PNNL): Richland, WA, USA; National Renewable Energy Laboratory: Golden, CO, USA, 2007.
67. Yang, H.; Yan, R.; Chen, H.; Ho Lee, D.; Zheng, C. Characteristics of hemicellulose, cellulose and lignin pyrolysis. *Fuel* **2007**, *86*, 1781–1788. [CrossRef]

68. Vishtal, A.; Kraslawski, A. Challenges in industrial application of technical lignin. *Bioresources* **2011**, *6*, 3547–3568.
69. Clark, J.H.; Deswarte, F.E.; Farmer, T.J. The integration of green chemistry into the future biorefinery. *Biofuels Bioprod. Biorefin.* **2009**, *3*, 72–90. [CrossRef]
70. Boeriu, C.G.; Bravo, D.; Gosselink, R.J.A.; Van Dam, J.A.G. Characterization of structure-dependent functional properties of lignin with infrared spectroscopy. *Ind. Crops Prod.* **2004**, *20*, 205–218. [CrossRef]
71. Sena-Martins, G.; Almeida-Vara, E.; Duarte, J.C. Eco-friendly new products from enzymatically modified industrial lignins. *Ind. Crops Prod.* **2008**, *27*, 189–195. [CrossRef]
72. Suhas, P.J.M.; Carrott, M.M.L. Lignin: From natural adsorbant to activated carbon. *Bioresour. Technol.* **2007**, *98*, 2301–2312. [CrossRef] [PubMed]
73. De Wild, P.; Der Laan, D.V.; Kloekhorst, A.; Heeres, E. Lignin valorization for chemicals and (transportation) fuels via (catalytic) pyrolysis and hydrodeoxygenation. *Environ. Prog. Sustain. Energy* **2009**, *28*, 461–469. [CrossRef]

Biodegradable and Biobased Polymers: Definitions, Standards, and Future Perspectives

Patrizia Cinelli, Maria-Beatrice Coltelli, Norma Mallegni and Andrea Lazzeri

Abstract: The growing concern of plastics disposal at the end of their life cycle has induced the addressing of specific standards and normatives to promote reuse, recycling and final disposal of the plastics with the aim to minimize the use of resources and select the best possible option for final disposal of the material, thus promoting recovering in terms of materials itself (mechanical or chemical recycling), energy, or biomass. Similar criteria has been considered also in materials production, thus promoting both the preferential use of renewable versus petro sources and a green chemistry approach for the synthesis of polymers and additives. The industrial production of commodity non biodegradable polymers from bio-based resources has induced some confusion in the definition of bio-based polymers versus biodegradable ones. In addition, the general assessment of biodegradability of a plastic is not sufficient for granting the fulfilment of the requirements of consolidated international standards for composting. Thus, it is important to have a clear idea of the definition of bio-based, biodegradable, and compostable as well as to be aware of the current legislation and of the sustainability concepts. Indeed, if not properly planned and addressed, the production of compostable materials might result as being less sustainable than that of petro derived, not biodegradable polymers. The present chapter reviews the current definition of bio-based, recyclable, biodegradable and compostable materials and refers to current standards and legislation. The definition of sustainable production for plastic materials and the methods currently adopted to evaluate the sustainability of products are also addressed.

---

1. Introduction

There is a growing awareness worldwide of the issue of plastic disposal since wastes produced by the use of plastics based on not degradable polymers are solid, visible, and usually quite persistent. This issue is particularly bound to plastic waste generated by packaging disposal. The management of plastic waste can be made by following different options. However, waste policy in most countries follows the so-called waste hierarchy concept, which also represents the European Commission's preferences towards different waste management options.

For this policy, the best options are:

- Reduction of waste by prevention of waste production,
- Recycling (organic or material recycling),
- Energy recovering,
- Disposal in landfill.

In Europe, this is expressed by the Waste Framework Directive 2008/98/EC, 19 November 2008 [1], of the European Parliament that gives specific attention to prevention, reuse and recycling. The option of recycling includes composting and anaerobic digestion as a type of organic recycling. For organic waste, the waste framework legislation encourages recycling in agriculture through composting, since this is considered the most environmentally friendly option for organic waste management. In addition, recycling as re-processing of production scraps in industries is highly recommended.

The Council Directive 1999/31 of 26 April 1999 on the landfill of waste [2], which considers the worst possible option for waste disposal, states that the European Member States are encouraged to plan a national strategy for the implementation of the reduction of biodegradable waste going to landfills. This can be achieved by means of organic recycling (composting and anaerobic digestion), material recycling and/or energy recycling (incineration with energy recovery).

Starting from 1999, the target of organic waste going to landfills at 75%, after five years was indicated, 50% after eight years, 35% after 15 years. This directive aims at reducing the amount of methane gas produced by anaerobic degradation of organic waste in landfills, and consequently supported the introduction of separate collection of biodegradable waste, by a proper sorting and consequent possibility of recovery and recycling.

More recently, the European Commission has adopted an ambitious Circular Economy Package, which includes revised legislative proposals on waste to stimulate Europe's transition towards a circular economy, which will boost global competitiveness, foster sustainable economic growth and generate new jobs. This revised legislative proposal on waste [3] sets clear targets for reduction of waste and establishes an ambitious and credible long-term path for waste management and recycling.

Key elements of the revised waste proposal include:

- A common EU target for recycling 65% of municipal waste by 2030;
- A common EU target for recycling 75% of packaging waste by 2030;
- A binding landfill target to reduce landfill to maximum of 10% of municipal waste by 2030;
- A ban on landfilling of separately collected waste;

- Promotion of economic instruments to discourage landfilling;
- Simplified and improved definitions and harmonised calculation methods for recycling rates throughout the EU;
- Concrete measures to promote re-use and stimulate industrial symbiosis, turning one industry's by-product into another industry's raw material;
- Economic incentives for producers to put greener products on the market and support recovery and recycling schemes (e.g., for packaging, batteries, electric and electronic equipment, vehicles).

In the field of materials recycling, the plastic materials represent the fraction most difficult to be managed. In fact, several different polymers are currently employed in packaging in order to respect the requirements necessary for different foods and beverages protection. Poly(ethylene), poly(propylene), poly(ethylene terephthalate) and poly(styrene) are the most employed, but also multilayer packaging are currently used. These plastic materials, usually derived from petrol, cannot be recycled without being separated from each other. Each polymer can be recycled by processing it at high temperature to obtain a viscous melt. However, if a preliminary separation is not carried out, polymer blends showing very low properties are obtained because of the immiscibility and incompatibility of the different polymers. Many efforts have been made to organize the recycling of the different types of plastics. Currently, poly(ethylene terephthalate) (PET) and rigid polyolefins are well recovered and recycled in active industrial chains, but some plastic fractions are still difficult to be recovered and recycled, especially those employed in multilayer packaging. The replacement of such complex fraction of packaging with a biodegradable one, managing its end life by composting, could be an interesting alternative.

Consequently, both in producers and consumers, there is a growing interest versus the use of "bio-polymers" for the production of "bio-plastic". Scientific efforts toward the design, synthesis, and production of sustainable or green materials have expanded tremendously in the last two decades [4]. At present, biopolymers share about 1% of the total market of polymers [5].

Governments are getting involved in promoting bio-based materials through initiatives such as the Lead Market Initiative (European Union, Brussels, Belgium) and BioPreferred (USA). The Lead Market Initiative actions range from improving the application of bio-based products to standardization, certification and labeling. The BioPreferred procurements refers to a preferential purchasing policy based on the public procurement program stating that bioplastic version is to be preferred if available versus a petro derived counterpart. As a consequence, federal agencies are obliged to use products from a stated BioPreferred list [6].

These initiatives are expected to facilitate the growth of innovative markets, such as that of bio-based materials, with high economic and social value and encourage its potential increase. Thus, a worldwide demand for replacing petroleum derived raw materials with renewable resource-based raw materials for the production of polymers has raised interest.

2. Definition

Even if bio-plastics already play an important role in the fields of packaging, agriculture, gastronomy, consumer electronics and automotive, to name a few, the industry for production and marketing of biodegradable and bio-based plastics is relatively young and growing fast. Also for this reason, in the area of bio-plastics, there is actually a lot of confusion on terminology. In Europe and worldwide, there are several industrial associations related to bioplastics.

We report some definition from the main associations in the sector.

SPI Plastic Industry Trade Association, Bioplastics Council [7]:

- Bioplastics: plastic that is biodegradable, has bio-based content, or both.
- Biodegradable Plastic: a plastic that undergoes biodegradation (process where the action of naturally-occurring micro-organisms such as bacteria, fungi, and algae induce the degradation of the material) accepted as industrial standards.
- From 2008, accepted industry standard specifications are: ASTM D6400, ASTM D6868, ASTM D7081 or EN 13432 [8–11].
- Bio-based content: fraction of the carbon content made up of biological materials or agricultural resources versus fossil carbon content, where the bio-based content is measured following the procedures set by ASTM D6866, where a product's bio-based carbon content is reported as a fraction of total organic carbon content (TOC) and not on its weight [12], while other analytical standard methods such as EN 16440 and ISO 16620-2 allow bio-based results to be reported as a fraction of total carbon (TC) that specifies a calculation method for the determination of the bio-based carbon content in monomers, polymers, and plastic materials and products, based on the ^{14}C content measurement [13,14].
- European Bioplastics [15]
- Bio-based: material or product (partly) derived from biomass (plants).
- Bioplastic: the term bio-plastics encompass a whole family of materials that are bio-based, biodegradable, or both.

Thus, it is clear that bio-based and biodegradable have not the same meaning. We can summarize that the term "bio-based" means just that the material or product is (partly) derived from biomass (plants), but its biodegradability has to be assessed. Some bio-based products can biodegrade in municipal or commercial composting facilities, home composting, and aquatic and roadside environments, while others will only biodegrade in very specific environments and some will not biodegrade at all. For example, poly(lactic acid) is compostable while bio-poly(ethylene) is not. Biodegradation refers to a chemical process carried on by micro-organisms that are present in the environment. These micro-organisms convert the organic carbon in the materials into natural substances such as water, carbon dioxide, and compost.

The process of biodegradation depends on the surrounding environmental conditions (e.g., location or temperature), on the material and on the application.

When defining a material to be biodegradable, we should also specify in which environment. Thus, some polymers can biodegrade in compost, but not in soil, some degrade in marine water and some not, etc.

Composting is considered as an organic recycling of material, where industrial composting is performed under controlled composting conditions at high temperature in large-scale composting plants, while home composting is performed at ambient temperature in a reduced scale.

3. Standards and Directives

The Packaging Directive 94/62/EC [16], and most recent amendments, set minimum recovery and recycling targets for each type of packaging product. This Directive address the amounts of biodegradable waste that can be land-filled or incinerated, and thus also compliance with the Landfill Directive, but does not affect recycling of bio-waste as defined in the Waste Framework Directive.

The following European standards provide a framework within which these standards can be used together to support the claim that packaging is in compliance with the essential requirements for packaging to be placed on the market as defined in the following directives:

- EN 13431:2000—Packaging. Requirements for Packaging Recoverable in the Form of Energy Recovery Including Specification of Minimum Inferior Calorific Value.
- EN 13432:2000—Packaging. Requirements for Packaging Recoverable Through Composting and Biodegradation. Test Scheme and Evaluation Criteria for the Final Acceptance of Packaging.
- CR 13695-1—Packaging. Requirements for Measuring and Verifying the Four Heavy Metals (Cr, Rd, Hg, Pb) and Their Release into the Environment, and Other Dangerous Substances Present in Packaging.
- EN 13427:2000—Packaging. Requirements for the Use of European Standards in the Field of Packaging Waste ("Umbrella Norm").
- EN 13428:2000—Packaging. Requirements Specific to Manufacturing and Composition. Prevention by Source Reduction.
- EN 13429:2000—Packaging. Reuse

- EN 13430:2000—Packaging. Requirements for Packaging Recoverable by Material Recycling.

The norm EN 13432 has been endorsed by the European Commission and therefore is a harmonized EU standard with a juridical value for defining a product as compostable.

A product can be defined as compostable as required by EN 13432 (2000) if it respects specific characteristics:

- The product must contain at least 50% organic matter and may not exceed the heavy metal limits specified in the standard.
- The products should mineralize for at least 90% within six months under controlled composting conditions, where mineralization is defined as the conversion of the organic C to Carbon dioxide (CO2) and biomass, and this characteristic is linked to the chemical composition of the sample.
- The product, in the form which enters the market, should, within a timeframe of 12 weeks, fragment in parts smaller than 2 mm under controlled composting conditions. It has to be outlined that this requirement refers to the physical form of the product instead of to the chemical composition. These characteristics are connected mostly to the thickness and the physical construction (e.g., laminate, coating, etc.) of the sample, and can result tricky to be met al.so for packaging based on biodegradable materials.
- The compost obtained at the end of the composting trial, which can also contain some no degraded residuals from the product, must not have any negative effects to the germination and growth of plants.

Thus, we can say that compostability comprises more than just biodegradability. A packaging that is compostable is always biodegradable, while a packaging which is biodegradable may not be compostable (since it might be too thick for disintegration or might release in the compost toxic substances). In order to avoid confusion on terminology and standards to be applied, several authorities have decide to promote (or even mandate) the definition "compostable" versus the definition "biodegradable".

In Europe, there are several certification logos used by institutes for industrial compostability. One of them is the 'Seedling' logo of European Bioplastics, issued by the certification institutes of Din Certco (Berlin, Germany) [17] and the Ok Compost of Vinçotte (Ghent, Belgium) [18] (Figure 1).

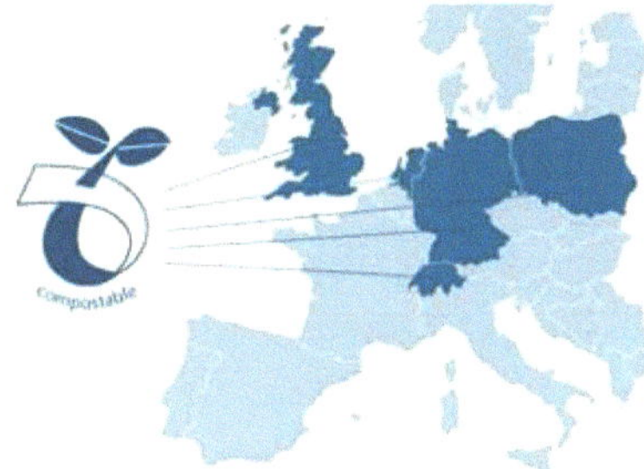

Figure 1. Logos for compostable materials.

Concerning home compostability, there is currently no international norm in place that defines the criteria to be adopted. Thus, contrast to industrial composting, which is a harmonised and controlled waste treatment option operated under specific and controlled conditions in specific plants with explicit permits and resulting compost as a regulated product, home composting is a gardening practise carried out by citizens on a voluntary basis, without permits and without obligations. The conditions in each garden compost pile and the actual treatment methods vary greatly, making it difficult to standardise this process [19].

Organizations that have defined criteria for home compostability include the certifier Vinçotte (Luxemburg, Belgium) with the Ok Compost Home programme (Figure 2) offers a certification scheme, while DIN CERTCO offers a certification for home compostability according to the Australian standard AS 5810 "Biodegradable plastics suitable for home composting". Italy has a national standard for composting at ambient temperature, UNI 11183:2006 "Plastic Materials Biodegradable At Room Temperature—Requirements And Test Methods". In November 2015, the French Standard NF T 51-800 "Plastics—Specifications for plastics suitable for home composting" was introduced.

Figure 2. Logo of Ok Compost in the Home program.

In the United Kingdom, we can find the Association for Organic Recycling with a home compostable certification scheme identical to the OK Compost Home programme, while there is a recent law in France that requires some single use items such as tableware, to be home compostable from 2017 with a minimum bio-sourced content of 30% (increasing progressively in subsequent years to 60% in 2025) [20].

Anyway, as recalled before, it must be considered that the biodegradability of a material is related to the environment and is frequently very different in compost, soil, anaerobic conditions, and particularly in water. This is related to the different microorganism present in the different substrates and also to differences in temperature, pHs, etc. Thus, specific standards have been made for the evaluation of the biodegradability in different environments. For example, to achieve the requirements for logos such as Ok Biodegradable Soil, a material should present more than 90% absolute or relative biodegradation within two years in soil, while for Ok Biodegradable Water, an aquatic biodegradability test, it is prescribed to reach 90% biodegradation within 56 days.

4. Bio-Based Materials

The use of materials that are bio-based is as well valuable since it allows saving petro resources and can valorize by-products or waste products of agriculture. Products on the market are made from a variety of natural feedstock including corn, potatoes, rice, tapioca, palm fiber, wood cellulose, wheat fiber and bagasse. Products are available for a wide range of applications in packaging such as cups, bottles, cutlery, plates, bags, film, etc.

The first generation of bio-based polymers was based on agricultural feedstock such as corn, potatoes, and other carbohydrate feedstock. Most recently, the focus has shifted versus by-products and waste product of agriculture with the aim to move away from food-based resources, this was supported also by advancement in biotechnology such as the use of bacterial fermentation processes.

Bio-based polymers can be produced from renewable resources in different ways:

- From natural bio-based polymers with eventually chemical modification to meet the requirements (starch, cellulose, chitosan, proteins, and their derivatives)
- Production of bio-based monomers by fermentation/conventional chemistry followed by polymerization (polylactic acid, polybutylene succinate, polyethylene)
- Production of bio-based polymers by bacteria (polyhydroxyalkanoates).
- Among bio-based polymers that are not biodegradable polyethylene (PE) produced from ethanol derived ethylene is the most important due to the large market sector involved. PE is produced by polymerization of ethylene under pressure, temperature, in the presence of a catalyst. Traditionally, ethylene is produced through steam cracking of naphtha or heavy oils or ethanol dehydration.

There are several products on the market that report the label of bio-based products. Braskem is producing polyetheylene from bio-based feedstock, sugarcane ethanol, in Brasil (Figure 3).

Figure 3. Biobased polyethylene (PE) Triunfo, Braskem, Brasil.

Braskem was able to produce bio-based PE and bio-based polyvinyl chloride (PVC) from bioethanol already in the 1980s, but, in that time, there was not such an issue in the use of petro sources for the production of polymers, and the limitation in biotechnology made the process not economically advantageous [21]. At present, bio-based green PE is produced by dehydration of ethanol that is derived by microbial fermentation of biomass. In this process, the juice of sugarcane is used that has a high content of sucrose. The juice is anaerobically fermented to produce ethanol. The ethanol produced is distilled to remove water, and it yields an azeotropic mixture of hydrous ethanol. A solid catalyst is used to dehydrated ethanol at high temperatures; in this way, it is produced ethylene and, subsequently, polyethylene [22,23].

On the market, we can also find bio-based poly(ethylene terephthalate), which is used for the production of bottles as publicized by the Coca-Cola Company (Figure 4).

Figure 4. Bottles produced with bio-based polyethylene terephthalate (PET).

Condensation polymerization of ethylene glycol (EG) and terephthalic acid (TA) is commonly used to produce PET. Petro derived EG can be replaced with a bio-based feedstock, as proposed by the suppliers of the Coca-Cola plant bottles. Anyway, terephthalic acid is still petro-derived, thus the resulting bio-based carbon content achieves a range of 28%.

In the context of the "Lead Market Initiative for Europe", the European Commission created the Mandate M/429 [24] addressed to the European Standardization bodies (CEN, CENELEC and ETSI) for the development of horizontal European standards for bio-based products. CEN initiated a new Technical Committee CEN/TC411 on "Bio-based products", which started working in the beginning of October 2011.

The main active institutes in this field are Vinçotte (Belgium) and Din Certco (Germany). They both have a ranking system based on the bio-based carbon content. Vinçotte has created a ranking system with stars, which are featured in the logo:

One star	20% < Biobased < 40%,	☆
Two star	40% < Biobased < 60%,	☆☆
Three star	60% < Biobased < 80%,	☆☆☆
Four stars	80% < Biobased.	☆☆☆☆

The corresponding logo is reported in Figure 5.

OK biobased VINÇOTTE	OK biobased VINÇOTTE	OK biobased VINÇOTTE	OK biobased VINÇOTTE
between 20 and 40 % Biobased	between 40 and 60 % Biobased	between 60 and 80 % Biobased	more than 80 % Biobased

Figure 5. Bio-based certification by Vincotte.

The system of Din Certco is based on a numbering of the bio-based carbon content (Figure 6).

Figure 6. Biobased Logo by Din Certco.

In the USA, the United States Department of Agriculture (USDA) sets the percentage of bio-based components required for a product to be defined to as bio-based, on a product-by-product basis. A minimum threshold of 50 percent of bio-based content for products to be considered bio-based has been recommended to USDA by the Institute for Local Self-Reliance (ILSR).

An image of the logo for bio-based products by USDA is reported in Figure 7.

Figure 7. Bio-based certification by United States Department of Agriculture (USDA).

Thus, summarizing, some materials can be compostable even if produced by petro sources, considering that also poly(lactic acid) (PLA) can be produced by petro sources. This group includes several synthetic polyesters produced by BASF, Mitsubishi.

While some materials can be both compostable and bio-based such as starch and some of its derivate, cellulose and some of its derivate, other polysaccharides such as chitosan, and polymers synthesized by micro-organisms such as the family of polyhydroxyalkanoates. Some other materials can be just bio-based but not compostable such as PE, PET and polyamides. Some examples are reported in Table 1.

Table 1. Some example of polymers being either: compostable petro sourced, compostable bio-based, bio-based not compostable.

Compostable Petro Sources	Compostable Bio-Based	Bio-Based Not Compostable
Poly(caprolactone) Poly(lactic acid) Poly(butylene succinate) Poly(butylene adipate—*co*-terephthalate)	Starch Based Poly(lactic acid) Polyhydroxyalkanoates Cellulose Based Chitosan/Chitin Animal and vegetal proteins	PE from bioethanol PET from bioethanol Polyamide 4, 10 or Polyamide 11

5. Sustainability

Being bio-based or compostable does not necessary mean that the material considered is more environmentally sustainable than the counterpart petro sourced or not compostable, but, for example, recyclable. For this reason, it is important to evaluate the sustainability and ecological benefits of a bio-based polymer in term of raw materials, production, application and end of life of the bio-based materials. This evaluation can be performed through a Life Cycle Assessment (LCA) study. The LCA represents an internationally standardised methodology that consists of four phases (ISO 14040, 2006 and ISO 14044, 2006) [25,26]:

1. Goal and scope definition,
2. Life cycle inventory analysis,
3. Life cycle impact assessment,
4. Life cycle interpretation.

The definition of the goal and scope includes decisions about the functional unit, which forms the basis of comparison, the product system to be studied, system boundaries, allocation procedures, assumptions made and limitations. The life cycle impact assessment (LCIA) converts the emissions and the raw material requirements from the life cycle inventory analysis into potential environmental impacts. Outputs with comparable effects (e.g., all acidifying components) are aggregated by use of so-called characterisation factors. This leads to a limited number of parameters, called impact categories (=Characterisation Step).

The collection, evaluation and discussion of the data are usually performed in line with the "International Reference Life Cycle Data System" (ILCD) Handbook and ISO standards (ISO 14040, 14044). The ILCD impact assessment method is applied to calculate the environmental impacts of the products under study. The method ILCD is a midpoint method, and includes 16 impact categories covering a broad range of environmental issues, not only the potential emissions of Green House Gases (GHG), those are: climate change; ozone depletion; human toxicity;

cancer effects; human toxicity, non-cancer effects; particulate matter; ionizing radiation HH (human health); ionizing radiation E (ecosystems); photochemical ozone formation; acidification; terrestrial eutrophication; freshwater eutrophication; marine eutrophication; freshwater ecotoxicity; land use; water resource depletion; mineral; fossil and renewable resource depletion.

Methods such as "ReCiPe" enables the normalisation and weighting of the outputs and the aggregation of the endpoint impact categories in three damage categories;

1. Human Health (unit: DALY = Disability adjusted life years; this means different disability caused by diseases are weighted)
2. Ecosystem Quality (unit: PDF $\times$ m^2 year; PDF = Potentially Disappeared Fraction of plant species)
3. Resources (unit: MJ surplus energy = Additional energy requirement to compensate lower future ore grade).

The environmental modelling is completed by the calculation of the total scores for the three damage categories. In several cases, further steps are not needed since these results provide sufficient information.

For example, "normalization" and "weighting" steps are referred to as "optional elements" in the ISO 14042 documents.

The weighting step is not used in some studies, in particular when comparative assertions are made, and planned to be disclosed to the public. The user of this methodology to decide if and how the damage assessment is applied, and if it is necessary to follow or deviate from the ISO standards.

The LCA completes the study on the environmental impact related to a bio-based material.

6. Future Perspectives

Actually, many bio-based polymers are produced on an industrial scale with good mechanical performance and stability. The large marketing of bio-based polymers in many applications is hindered by the performance-to-price ratio when compared with their conventional petro-derived counterparts. This remains a significant challenge to overcome for bio-based polymers.

The increasing in the demand for food and energy can induce a competition for renewable resources, since the global demand will increases over time. Thus, at present, the renewable feedstock used for manufacturing bio-based monomers and polymers often compete with the demand of resources for food-based products. The first-generation of bio-based fuel production is in expansion and consequently it will induce an unsustainable demand of biomass. Thus, will be a problem for the sustainability of biochemical and biopolymer production versus the use of biomass for food production [27].

Key Performance Indicators (KPIs) set out in the Strategic Innovation and Research Agenda (SIRA) Bio-Based Industries (BBI) roadmap [28].

By replacing fossil-based products with bio-based products, which tend to have a smaller carbon footprint, bio-based industries can make a critical contribution to Europe's climate goals.

Several initiatives are under consideration for using ligno cellulosic or cellulose-based feedstock for the production of usable sugars for biofuels, biochemicals, and biopolymers [29].

Thus, in the future, we envisage a growing production and use of bio-based products with a focus on those bio-based materials produced by renewable resources that are not in food competition such as by-products or waste products of agro-food production. In particular, the new trend of promoting production of bio-based polymers from no food resources is reducing the concern on the opportunity to utilize biomass for the production of plastic. For those polymers produced from edible sources, including starch and PLA, the favourite choice is using their production crops grown in lands that are not suitable for the production of edible crops. These lands can be used for the production of crops to be devoted to industrial manufacture, and the valorization of depressed areas is important where land is not used for agriculture anymore since it is not profitable.

Modern economies use a huge amount of plastics and, consequently, in Europe, the plastic consumption is very high, in the range of 50 million tons per year; in contrast, only 25% of plastic waste in Europe is currently recycled and the larger share is incinerated or even landfilled [30].

Considering the objectives of EC's policy of achieving a resource efficient recycling society, the landfilling of nearly 50% of plastic waste is a tremendous waste of resources [31], and an issue to be seriously approached, even considering that bioplastics still have a very limited share among plastics (approximately 1%), thus a large potential for growth [32,33]. Thus, bioplastics have the advantages of allowing a switch to renewable non-fossil feedstocks, preferably selected among waste, bio-products or over production biomasses, offering also the plus of and end of life modularity. This is important especially for those short-term single use applications, which, by nature (e.g., shoppers, mulching films, lightweight nets) or due to contamination (food, oils, fertilizers, ect), are hardly collectable and sometimes not recyclable (mechanically or organically) when non-biodegradable. To achieve these targets, more research and development have to be invested in further improving bioplastics' specific functionalities, and more cost-competitive solutions have to be found to ease the eventual large-scale superior performance and market-driven uptake.

7. Conclusions

In the last decade, there has been much interest towards bio-plastics and this brought about many rapid changes in the knowledge and classification of these materials. Their introduction on the market pushed the institutions to promulgate new regulations and a technical-scientific community to increase the knowledge and standardization of materials and tests. The replacement of commodity polymers currently employed in many sectors is still on-going and the necessity of projecting materials and goods as a function of both their origin and present end life is underlying this competition.

In some applications, bio-based polymers are close to replacing conventional polymers in the market. In particular, bio-based polymers can be employed in many applications, from commodity to hi-tech applications, due to advancement in biotechnologies, with consequent improvement of polymer quality as well as public awareness of the importance to use materials derived from renewable resources. Significant progress has been achieved in terms of production, processing, collection and bio recycling of bio-based compostable polymers.

Acknowledgments: All sources of funding of the study should be disclosed. Please clearly indicate grants that you have received in support of your research work. Clearly state if you received funds for covering the costs to publish in open access.

Conflicts of Interest: The authors declare no conflict of interest.

References

1. European Parliament and Council. Directive 2008/98/EC of 19 November 2008 on waste and repealing certain Directives. *Off. J. Eur. Union* **2008**, *L312/3*, 3–30.
2. European Parliament and Council. Directive 1999/31/EC of 26 April 1999 on the landfill of waste. *Off. J. Eur. Union* **1999**, *L182*, 1–19.
3. European Commission. Review of Waste Policy and Legislation. Available online: http://ec.europa.eu/environment/waste/target_review.htm (accessed on 16 October 2017).
4. Miller, S.A. Sustainable Polymers: Opportunities for the Next Decade. *ACS Macro Lett.* **2013**, *2*, 550–554. [CrossRef]
5. Babu, R.P.; O'Connor, K.; Seeram, R. Current progress on bio-based polymers and their future trends. *Prog. Biomater.* **2013**, *2*, 1–16. [CrossRef] [PubMed]
6. BioPreferred. Available online: www.biopreferred.gov (accessed on 25 July 2017).
7. Plastics Industry Association. Available online: http://www.plasticsindustry.org (accessed on 25 July 2017).
8. ASTM International. *Standard Specification for Labeling of Plastics Designed to Be Aerobically Composted in Municipal or Industrial Facilities*; ASTM D6400-12; ASTM International: West Conshohocken, PA, USA, 2012.

9. ASTM International. *Standard Specification for Labeling of End Items that Incorporate Plastics and Polymers as Coatings or Additives with Paper and Other Substrates Designed to be Aerobically Composted in Municipal or Industrial Facilities*; ASTM D6868-11; ASTM International: West Conshohocken, PA, USA, 2011.
10. ASTM International. *Standard Specification for Non-Floating Biodegradable Plastics in the Marine Environment*; ASTM D7081-05; ASTM International: West Conshohocken, PA, USA, 2005.
11. EN 13423:2000 Packaging - Requirements for packaging recoverable through composting and biodegradation - Test scheme and evaluation criteria for the final acceptance of packaging. Available online: https://www.en-standard.eu/ilnas-en-13432-packaging-requirements-for-packaging-recoverable-through-composting-and-biodegradation-test-scheme-and-evaluation-criteria-for-the-final-acceptance-of-packaging/?gclid=EAIaIQobChMIpazS55Lh3AIVidKyCh1q_AYIEAAYASAAEgKZhvD_BwE (accessed on 9 August 2018).
12. ASTM International. *Standard Test Methods for Determining the Biobased Content of Solid, Liquid, and Gaseous Samples Using Radiocarbon Analysis*; ASTM D6866-12; ASTM International: West Conshohocken, PA, USA, 2016.
13. European Committee for Standardization. *Bio-Based Products—Determination of the Bio-Based Carbon Content of Products Using the Radiocarbon Method*; CEN/TS 16640:2014; European Committee for Standardization: Brussels, Belgium, 2014.
14. International Organization for Standardisation. *Plastics—Biobased content—Part 2: Determination of Biobased Carbon Content*; ISO 16620-2; International Organization for Standardisation: Geneva, Switzerland, 2015.
15. European Bioplastics. Available online: www.european-bioplastics.org (accessed on 25 July 2017).
16. European Parliament and Council. Directive 94/62/EC of 20 December 1994 on packaging and packaging waste. *Off. J. Eur. Union* 1994, *L365*, 10–23.
17. DinDIN CERTCO. Available online: http://www.dincertco.de (accessed on 25 July 2017).
18. Ok Compost. Available online: http://www.okcompost.be (accessed on 25 July 2017).
19. Revised Mandate for Home Composting Standard, 2016. European Bioplastics. Available online: http://www.european-bioplastics.org/revised-mandate-for-home-composting-standard/ (accessed on 25 July 2017).
20. French Public Service Broadcasting Légifrance. Decree: Relatif Aux Modalités de Mise en Oeuvre de La Limitation Des Gobelets, Verres et Assiettes Jetables en Matière Plastique; 2016; Paris, France, 30 August 2016. Available online: https://www.legifrance.gouv.fr/affichTexte.do?cidTexte=JORFTEXT000033076240&categorieLien=id (accessed on 25 July 2017).
21. De Guzman, D. Bioplastic Development Increases with New Applications, 2010. Independent Chemical Information Service (ICIS). Available online: https://www.icis.com/resources/news/2010/10/25/9402443/bioplastic-development-increases-with-new-applications/ (accessed on 25 July 2017).

22. Guangwen, C.; Shulian, L.; Fengjun, J.; Quan, Y. Catalytic dehydration of bioethanol to ethylene over TiO2/γ-Al2O3 catalyst in microchannel reactors. *Catal. Today* **2017**, *125*, 111–119.
23. De Castro Morschbacker, A. A Method for the Production of One or More Olefins, an Olefin, and a Polymer. U.S. Patent 2010/0069691 A1, 18 March 2010.
24. European Commission. *Mandate Addressed to CEN, CENELEC and ETSI for the Elaboration of a Standardisation Programme for Bio-Based Products*; M/429; European Commission: Brussels, Belgium, 10 October 2008; Available online: http://ec.europa.eu/growth/tools-databases/mandates/index.cfm?fuseaction=refSearch.search&lang=DE# (accessed on 25 July 2017).
25. International Organisation for Standardization. *Environmental Management—Life Cycle Assessment—Principles and Framework*, 2nd ed.; ISO 14040:2006; International Organization for Standardisation: Geneva, Switzerland, 2006.
26. International Organisation for Standardization. *Environmental Management—Life Cycle Assessment—Requirements and Guidelines*; ISO 14044:2006; International Organization for Standardisation: Geneva, Switzerland, 2006.
27. Carus, M.; Carrez, D.; Kaeb, H.; Ravenstijn, J.; Venus, J. *Policy Paper on Bio-Based Economy in the EU. Level Playing Field for Bio-Based Chemistry and Material; Nova Paper 1 on Bio-Based Economy*; Nova Institute: Hürth, Germany, July 2011; Available online: www.bio-based.eu/policy/ (accessed on 16 October 2017).
28. Bio-based Industries Consortium. *Strategic Innovation and Research Agenda (SIRA)*; Brussels, Belgium, 2017. Available online: https://www.bbi-europe.eu/sites/default/files/sira-2017.pdf (accessed on 16 October 2017).
29. Jong, E.D.; Higson, A.; Walsh, P.; Maria, W. *Bio-based Chemicals: Value Added Products from Biorefineries*; Report on Behalf of the International Energy Agency (IEA) Bioenergy: Wageningen, The Netherlands, February 2010; Available online: http://www.ieabioenergy.com/publications/bio-based-chemicals-value-added-products-from-biorefineries/ (accessed on 25 July 2017).
30. PlasticsEurope. Plastics—the Facts 2014/2015. In *An Analysis of European Plastics Production, Demand and Waste Data*; PlasticEurope Report; PlasticsEurope: Brussels, Belgium, 2015; Available online: http://www.plasticseurope.org/documents/document/20150227150049-final_plastics_the_facts_2014_2015_260215.pdf (accessed on 16 October 2017).
31. European Commission. Plastic Waste—Strategy and Background. Available online: http://ec.europa.eu/environment/waste/plastic_waste.htm (accessed on 16 October 2017).
32. Persistence Market Research (PMR). Global Market Study on Foodservice Disposables: Rapid Growth Linked To Increasing Traction of Online and Home Delivery Food Services. Available online: https://www.persistencemarketresearch.com/market-research/foodservice-disposables-market.asp (accessed on 16 October 2017).

33. World Economic Forum (WEF). *The New Plastic Economy: Rethinking the Future of Plastics*; World Economic Forum: Geneva, Switzerland, January 2016; Available online: http://www3.weforum.org/docs/WEF_The_New_Plastics_Economy.pdf (accessed on 16 October 2017).

PART III
Fundamentals

Lignin: Isolation, Structure and Valorisation

Heiko Lange and Claudia Crestini

Abstract: Lignin is the second most abundant polymer in forest biomass after cellulose. Compared to the ubiquitous use of cellulose, lignin is currently simply being wasted. In fact, as a waste product in the cellulose production, it serves mainly for the generation of energy. The reason for this lies in the challenging structural features and the patchy understanding of the correlation between structural features and polymer characteristics displayed by various lignins. This chapter will introduce the general characteristics and peculiarities of lignin as biopolymer and will present techniques for investigating structural issues. Approaches toward a valorization of lignin will be highlighted, showing the possibilities for using lignin in chemistry and material sciences.

1. Introduction

The efficient use of all available biomass components is of utmost interest with respect to a sustainable use of renewable resources. Lignin, however, albeit being the second most-abundant component in forest biomass, is still not exploited to its full potential. Lignin is a rather complex natural polymer. Its straightforward use is severely impeded by several difficulties that reach back to the molecular level, and lignin is thus unfortunately still seen as a natural polymer with unpredictable properties.

Especially those areas of research that could eventually truly benefit from using and incorporating lignin as a functional groups-carrying polymeric compound seem to be paralyzed by the old paradigm that "lignin is not good for anything but burning it".

Prejudices and the admittedly tedious initial effort necessary to get an idea of the nature of the lignin at hand contribute to maintaining lignin in a niche. Methods to characterize a given lignin sample do exist, but these methods, although quite matured as such by now, still need further refinement, cross-correlation, and independent validation, in order to further improve and overall streamline the processes of acquiring structural data of lignin. Sound structural data are absolutely necessary for being able to analyse structural features with respect to reactivity of a given lignin, and for understanding which chemistries are suitable for derivatization, functionalization, and depolymerization of lignin samples isolated from different renewable sources.

The isolation of lignin, its structural characterization, and its utilization have been subject to research activities around the globe for decades. Quite some knowledge has been accumulated regarding lignin and its characteristics, and this knowledge is constantly refined with respect to technological advances in adjacent fields such as biorefinery, biotechnology, spectrometry, spectroscopy, etc. Achievements are regularly summarized in reviews and monographs [1–5]. Nevertheless, more fundamental studies on lignin are needed to arrive at a complete understanding; this is true for both lignin *in planta* and isolated lignins.

Lignin, derived from 'lignum', Latin for wood, was introduced as early as 1819 by de Candolle, but not until Payen scientifically described a wood component in 1838, which, from our current perspective must have been a lignin carbohydrate complex (LCC), the term lignin was connected closely to an isolable substance. Schulze, in 1865, chemically defined lignin as a polymer that was different from the cellulose components [5].

Lignin accounts for 15–35% of the dry mass of wood, depending on the type of wood; this makes this natural polyphenol being the second most abundant component in forest biomass after cellulose in absolute numbers [6].

The rather hydrophobic polymer, located mainly in the plant cell walls, is chemically and physically interacting with hemicelluloses and cellulose, respectively [7]: it is located in form of a mixture together with hemicelluloses, eventually forming the aforementioned LCCs, between the cellulose fibrils [1,8].

The tight interplay between these three major plant biopolymers renders the plant cell walls less permeable, leads to increased mechanical strength and rigidity, and thus serves to give not only stability to the plants, but also a better resistance to microbial attacks.

The different functions of lignin in the plant cause its distribution to vary significantly within the different parts of the plant, i.e., between stem, branching points, branches and leafs, and between the different walls of the plant cells themselves [9]. Middle lamella and primary cell walls show higher lignin concentrations than the secondary cell wall. 75–85% of all lignin in the plant, however, is located in the secondary cell walls, since these are significantly more voluminous.

The total amount of lignin present in the plant varies from *ca.* 20% in hardwoods, *ca.* 28% in softwoods and herbaceous angiosperms, and *ca.* 15% in monocots [2,7].

2. Biosynthesis, Isolation and Structural Features of Lignin

Unlike the structurally well understood cellulose and hemicelluloses, lignin does not display a regular repetition motif, as one might expect for a natural polymer closely interacting with other regularly structured polymers. Furthermore, its structure does not only change between different plant types, but also within the same plant depending on where it fulfils its function; structural changes as well as abundance of lignin thus govern and/or are governed by the need of the plant for stability, protection, etc.

The difficulty in understanding the complex structure of lignin does not only stem from the fact that the isolation processes themselves can give rise to structural modification of lignins, as discussed below, but also from the fact that the biosynthetic pathway is not yet fully understood. It is yet to be elucidated whether there are control mechanisms that determine the structure of lignin in plants, or whether it is allowed to form according to the conditions set by a microenvironment surrounding the site at which the chain propagation takes place. As a result of this presumably uncontrolled biosynthetic pathway that is roughly depicted in Figure 1, lignin displays random sequences of different interunit bonding motifs (Figure 2) [10–12].

Although lignin contains different asymmetric carbon atoms that are formed during its biosynthesis, a preferred absolute configuration, as one might expect from the close interplay with the stereogenic polymeric sugars, does not exist—another fact that suggests that the in vivo lignin formation is probably not controlled in detail. It has been shown, however, that the ratio of the different aromatic units does correlate with the ratio between *threo* and *erythro*-configured aliphatic parts in lignin [13]. Current efforts utilize genetically modified model plants such as, e.g., *Arabidopsis thalianato*, to improve understanding of the formation of lignin, its role in and for the plants, and its interaction with the other components in the plant cell walls [14–17].

The monomers consumed in the radical polymerisations are the three monolignols *p*-coumaryl alcohol (H-unit), coniferyl alcohol (G-unit) and sinapyl alcohol (S-unit), which are synthesised by the plants (Figure 1) [10,11,18]. Depending on the natural source from which the lignin is isolated, and also depending on the method with which the lignin has been extracted from the wood and separated from the cellulose and hemicellulose wood components, the abundances of the different monomer types, as well as the nature and the distribution of the interunit bonding motifs via which they are linked differ.

Three different general types of lignin are generally distinguishable on the basis of fundamental structural aspects: (i) softwood lignin, which is mainly comprised of G-type monomers and a small percentage of H-type monomers; (ii) hardwood lignin, which contains mainly S- and G-units; and (iii) grass lignin, which contains all three monomer types.

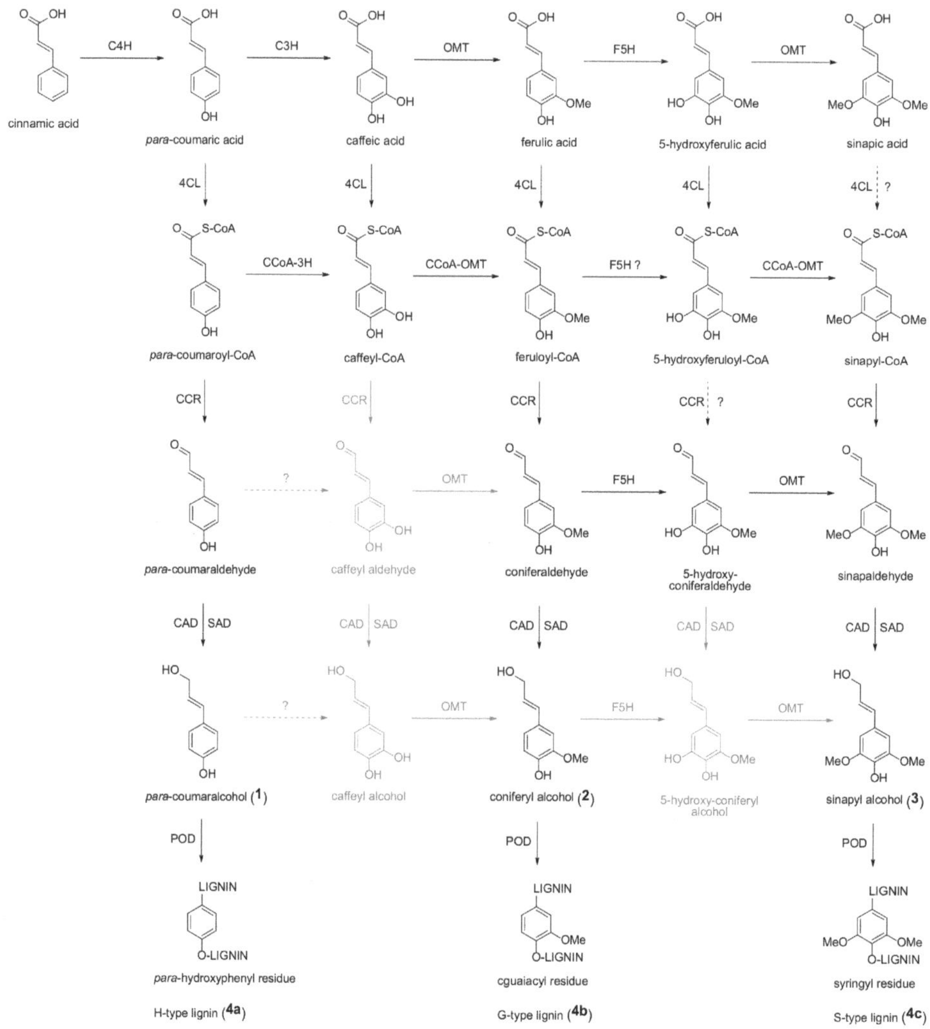

Figure 1. Main elements of the biosynthesis of lignin as described in Ref. [11]. (C4H—cinnamate-4-hydroxylase; C3H—cinnamate-3-hydroxylase; OMT—O-methyltransferase; F5H—ferulate-5-hydroxylase; 4CL—CoA-ligase; CCR—cinnamoyl co-enzyme A reductase; CCoA-3H—coumaroyl-co-enzyme A 3 hydroxylase; CCoA-OMT—coumaroyl-co-enzyme O-methyl transferase; CAD—cinnamyl alcohol dehydrogenase; SAD—short- chain alcohol dehydrogenase; POD—peroxidase.

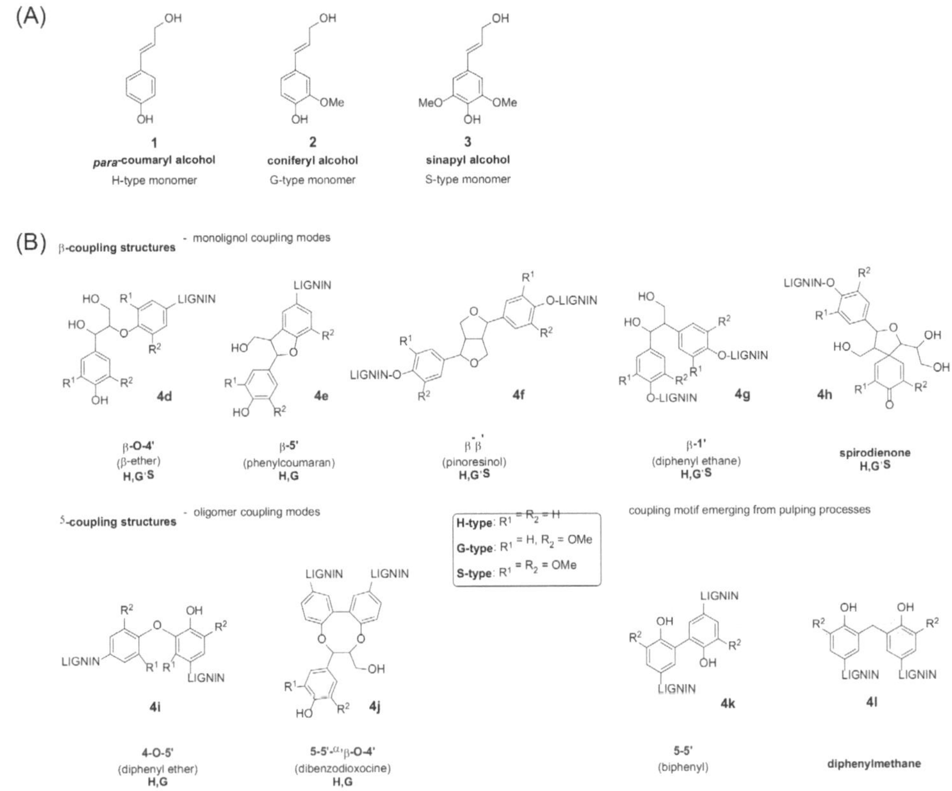

Figure 2. Characteristic interunit bonding motifs and functional groups found in different lignins.

Oxidative coupling of the monomers occurs, however, preferentially at only a limited number of positions, namely O-4, C-1, C-3, C-5 and C-β (IUPAC numbering), thus eventually reflecting some selectivity of enzymes putatively involved in the biosynthesis. Interesting recent findings suggest that the formation of lignin polymers does not end with the death of the plant cell, but that the major part of the lignin in a plant cell is produced after its death, as long as monomers are available [19]. This adds another source of uncertainty to the structural puzzle. Besides, based on the theoretically possible coupling modes, it is currently still controversially discussed in the community whether lignin *in planta* is a branched, and thus three-dimensional polymer, or whether it consists of linear chains that are rather oligomeric than polymeric (Figure 3) [20–22]. Newer studies have established that at least milled wood lignins and some organosolv lignins comprise linear oligomeric chains [22,23]. Newest studies revised the structure of kraft lignins, revealing the fact that it is generally composed of two main, structurally different parts: one consisting of truly polymeric chains, slightly brached, and one of small chains, brached and presumably exhibiting only a low amount of aliphatic moieties [24].

Different techniques and processes have been established for the isolation of lignin; the most important ones are those yielding: (i) milled wood lignin (MWL) (obtained via Björkman's procedure) (Figure 3B) [25]; (ii) acidolysis lignin [26,27]; (iii) enzymatic mild acidolysis lignin [28]; (iv) organosolv lignin (obtained in the Allcel process) [29,30]; (v) kraft lignin (KL) (Figure 4A) [31,32]; (vi) lignosulfonate (obtained in the Howard process) (Figure 4B) [33]; (vii) pyrolysis lignin [34–37]; and (viii) steam explosion lignin [38–40] (Table 1). Since any isolation process can affect the natural structure of lignin, i.e., isolation methods may introduce new functional groups, and/or cause partial degradation, and the isolated lignins are thus not necessarily representing the natural lignin polymers.

(A)

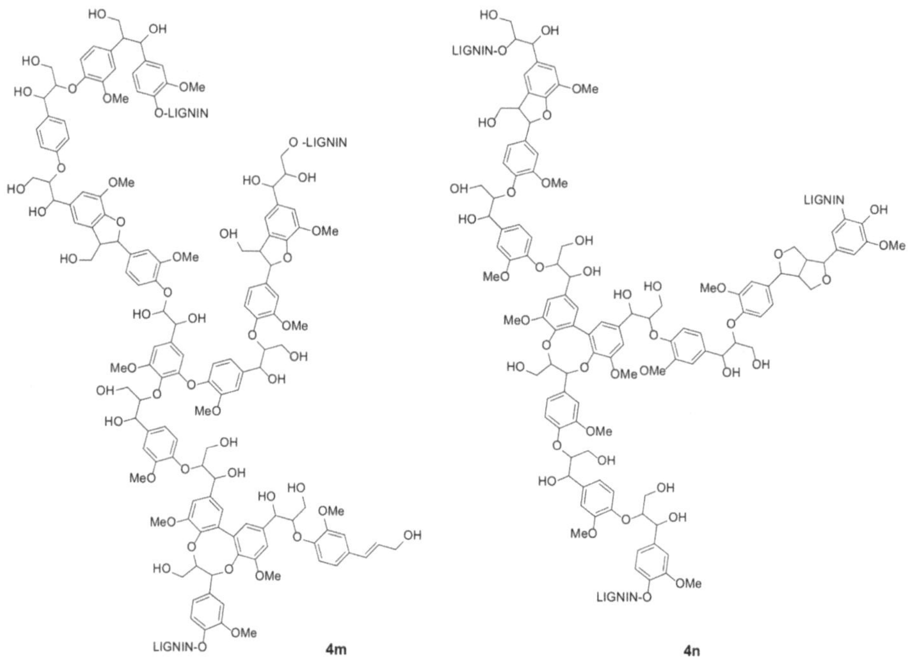

Figure 3. *Cont.*

(B)

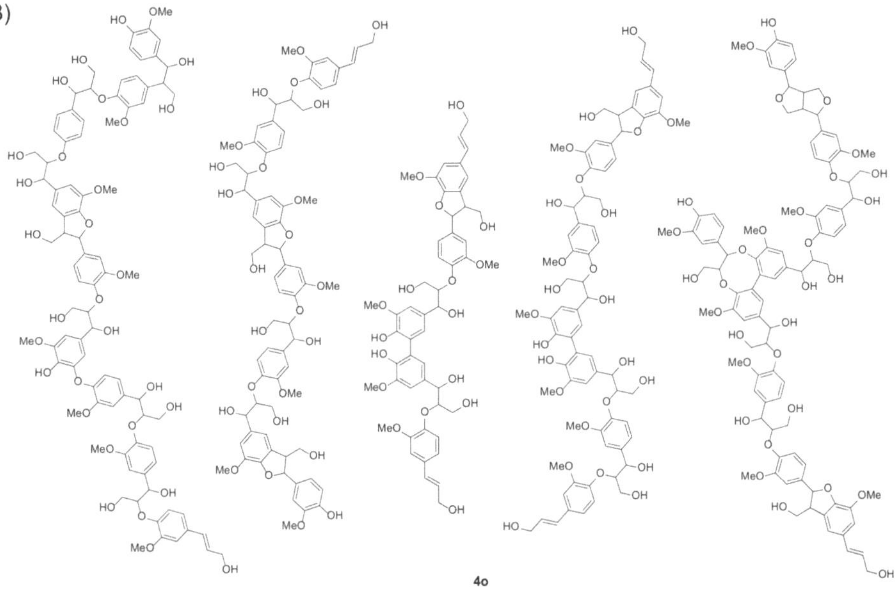

4o

Figure 3. Lignin structures showing characteristic interunit bonding motifs and functional groups for different types of lignin: (**A**) branched polymeric lignin (outdated view); (**B**) linear chains of oligomeric milled wood lignin (actual view).

More recent efforts aimed at circumventing this issue by (i) generating a 'stand-alone' 'synthetic lignin' in form of dehydrogenation polymers (DHP) by reacting monolignols in the presence of oxidative enzymes [41–43], or (ii) by evaluating and refining current standard analyses techniques as discussed below for the investigation of lignin inside undamaged woody tissues [44].

(A) (C)

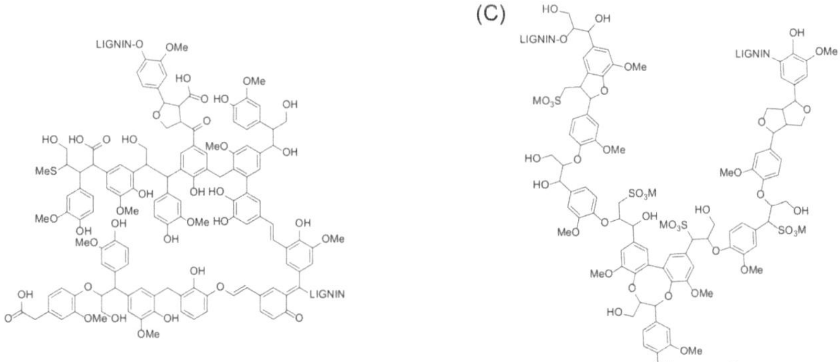

4p 4q

Figure 4. *Cont.*

(B)

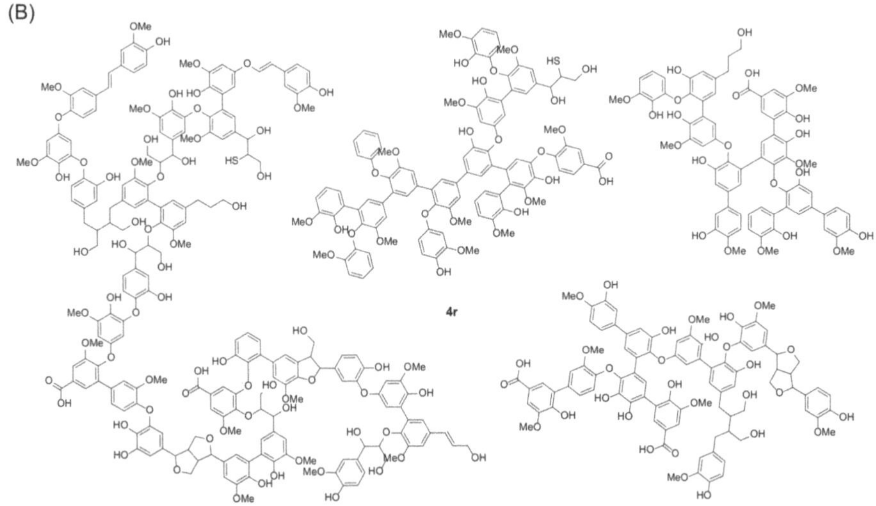

Figure 4. Lignin structures showing characteristic interunit bonding motifs and functional groups for different types of lignin (continued): (**A**) kraft lignin (outdated view); (**B**) kraft lignin (proposed revised structure composed of polymeric and oligomeric units); (**C**) lignosulfonate.

Table 1. Overview comparing the (average) characteristics of various lignins obtained through the different characteristic isolation techniques (for combined references on data shown here, refer to Ref. [45]).

Lignin Type	C9 Molecular Formula	Monomer Molecular Weight [u]	Number-Average Molecular Weight (M_n)	Poly-Dispersity
Milled wood lignin [a]	$C_9H_{7.80}O_{2.41}(OCH_3)_{0.95}$	198	2800–14,200	3.7–12.9
Cellulolytic enzyme lignin [b]	$C_9H_{8.02}O_{2.82}(OCH_3)_{0.90}$	187	~1900	5.7–6.7
Enzymatic mild acidolysis lignin (EMAL) [b]	$C_9H_{8.02}O_{2.82}(OCH_3)_{0.90}$	187	~2000	~3
Kraft lignin [c]	$C_9H_{8.5}O_{2.1}S_{0.1}(OCH_3)_{0.8}(CO_2H)_{0.2}$	180	1000–3000	2–4
Lignosulfonate (softwood) [d]	$C_9H_{8.5}O_{2.5}(OCH_3)_{0.85}(SO_3H)_{0.4}$	215–254	5000–20,000	4–9
Lignosulfonate (hardwood) [d]	$C_9H_{7.5}O_{2.5}(OCH_3)_{0.39}(SO_3H)_{0.6}$	188	5000–20,000	4–9
Organosolv lignin [e]	$C_9H_{8.53}O_{2.45}(OCH_3)_{1.04}$	188	>1000	2.4–6.4
Pyrolysis lignin [f]	$C_9H_{6.3\text{-}7.3}O_{0.6\text{-}1.4}(OCH_3)_{0.3\text{-}0.8}(OH)_{1\text{-}1.2}$	n.d.	300–600	2.0–2.2
Steam explosion lignin [g]	$C_9H_{8.53}O_{2.45}(OCH_3)_{1.04}$	188	1100–2300	1.5–2.8

[a] Norway spruce wood; [b] Isolated from milled Norway spruce wood; [c] Norway spruce wood; [d] Norway spruce wood; [e] Norway spruce wood; [f] Beech wood; [g] Japanese white birch wood and larch wood.

Total amounts of lignin in a sample, however, are still best and commonly determined as Klason lignin by the Klason method [46,47]. If necessary, the amount of Klason lignin can be corrected for the parts of lignin that are soluble under the conditions used in the Klason method, i.e., the so-called acid soluble lignin.

3. Biodegradation

As mentioned above, lignin plays a dual role for the plant: it gives stability to the plant cells and thus to the plant, and it serves to protect the plant and its cells against attacks by microbes or saprophytic organisms. On the molecular bases, the protection is effected by simply preventing the microbes to hydrolytically cleave the cellulosic polymers, by physically shielding the cellulose and presenting a polymer that does not give rise to facile (bio)degradation [48,49].

Most of the phenols are used for the construction of alkyl aryl ethers, and the redox potential needed for effective bond cleavages is too high, as that the plant-based oxidoreductases which are involved in the construction of lignin, could be strong enough to ('accidently') destroy it. Lignin degradation can thus only be effected by organisms outside the plant that host a special group of enzymes, known as lignocellulolytic enzymes [50–52]. Evolution of these organisms that serve to recycle woody biomass and to feed back the carbon sources in the carbon cycle of the planet started presumable in the first half of the Palaeozoic Era, when the first vascular plants grew on dry land mass [53,54]. The group of lignocellulolytic enzymes (cellulases [55], hemicellulases [56] and ligninases [57]), which is found widespread in different fungi, e.g., *Trichoderma reesei* (ascomycetes), *Phanerochaete chrysosporium* (white-rot) and *Fomitopsis palustris* (brown-rot) (both basidiomycetes phyla), comprises both cellulose and lignin degrading enzymes. Organisms such as, e.g., *Ceriporiopsis subvermispora* [58], *Phlebia spp.* [59,60], *Physisporinus rivulosus* [61] and *Dichomitus squalens* [60] do exist that are specialised only on lignin degradation, as are pathogenic species such as *Fusarium solani f. sp. glycines* [62] as well.

Only lignin degrading enzymes occasionally termed ligninases are laccases (phenol oxidases) and peroxidases (lignin peroxidase (LiP), manganese peroxidase (MnP) [53,63]. Noteworthy, oxidative degradation of lignin using laccases is effectuated by additional use of small molecules, so-called mediators, that serve as the actual substrates for the laccases, and then move in activated state into the lignocellulosic network where they attack the lignin structure as such.

In course of the emerging efforts to substitute fossil-based materials with more sustainable alternatives, both the biotechnological valorisation of lignin using some of the enzymes just mentioned as well as its biodegradation are intensively studied. The interested reader is encouraged to consult the recent literature on lignin degradation for further information, e.g., on mechanistic details [64–67].

4. Elucidating Structural Features in Lignin

Independent of the lignin-type as such, a full structural elucidation is only possible by using a combination of different types of analyses techniques. The traditional C9-formula of lignins, determined via elemental analysis, represents just one key-figure to characterise lignins. Systematic degradation and chemical modification using solvolytic methods such as acidolysis [68] and thioacidolysis [69], or oxidative methods such as ozonolysis [70,71] and the alkaline nitrobenzene oxidation [72] contributed to determine structural features and compositions of lignins in terms of the H/G/S-ratios etc. Acidolysis and thioacidolysis target interunit bonding motifs and yield different C9-fragments that allow analyses and thus inference to the structural motifs originally present in a lignin sample. While the parent method acidolysis uses aqueous hydrochloric acid for degrading polymeric structures, thioacidolysis uses thioethanol in the presence of Lewis-acidic boron trifluoride etherate. By now, thioacidolysis has become *the* solvolytic method used for the structural determination of lignins due to an overall superior performance in terms of yields [27,73], and due to the possibility to combine it with a preceding alkylation in order to deduce additional information regarding the degree of etherification. Typical structural motifs obtained in thioacidolyses of lignins are depicted in Figure 5A.

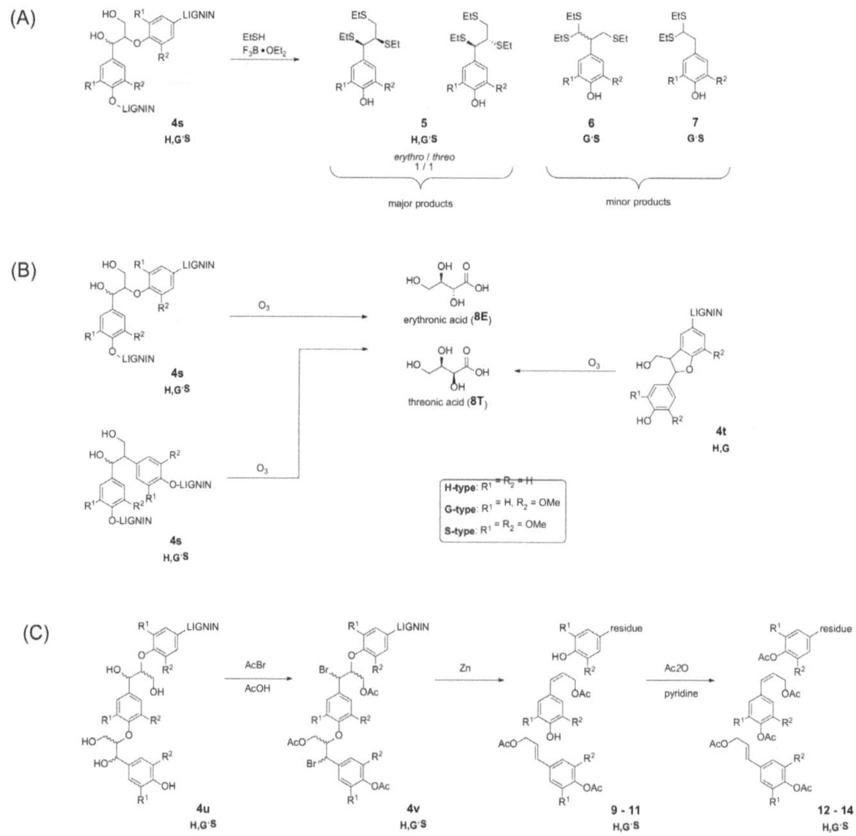

Figure 5. (**A**) Conversion of the most common interunit bonding motif arylglycerol-β-aryl ethers (β-*O*-4′) upon thioacidolysis. (**B**) Conversion of arylglycerol-β-aryl ethers (β-*O*-4′), diarylpropane (β-1) and phenylcoumaran (β-5′) motifs upon ozonolysis. (**C**) Conversion of arylglycerol-β-aryl ethers (β-*O*-4′) within the DFRC procedure.

Figure 5B shows the formation of erythronic acid and threonic acid as result of ozonolytically degrading the aromatic rings and other unsaturated moieties in lignin, e.g., stilbenes, while leaving the side-chains intact [13,74,75]. It has been shown that a skilful ozonolysis enables identification of interunit bonding motifs and, in addition, determination of the ratio between *threo*- and *erythro*-configured aliphatic moieties in lignins, which can give rise to stereochemical considerations.

In Figure 5C, the chemical transformations during a 'derivatisation followed by reductive cleavage' (DFRC) treatment of lignin are shown [76]; again, β-aryl ethers react preferentially to induce defined fragmentation for analysis. Important structural findings were possible using the DFRC method, particularly noteworthy is the fact that it is possible to proof the aforementioned racemic nature of lignin due to the mild chemistry that constitutes this method. This method also serves in combination with non-destructive advanced nuclear magnetic resonance (NMR) spectroscopy on lignins—excellent overviews have been written recently in Ref. [4] regarding NMR studies on lignins—and it can be used to determine whether a given lignin represents a linear or crosslinked polymer [22,28].

Combinations of refined degrading methods such as thioacidolysis and DFRC with non-destructive analysis tools such as NMR spectroscopy are used by default nowadays in lignin analysis. The most common, since most easily to be performed techniques comprise (i) size-exclusion chromatography for the determination of molecular mass key figures, such as the mean molecular weight, the weight-average molecular weight and the polydispersity index [77]; (ii) vibrational spectroscopy, especially Fourier-transform infrared (FT-IR) spectroscopy for delineating functional groups present in a lignin sample, albeit its drawback that it does not easily allow for quantification [78]; (iii) difference spectroscopy for the quantification of phenolic groups, ethylenic double bonds, non-condensed phenolic groups and phenylcoumarans, [79–81]; (iv) quantitative ^{13}C NMR for the elucidating the amount of methoxy groups, etc. [4,82,83]; (v) two-dimensional heterocorrelated NMR such as ^{1}H-^{13}C HSQC for determining interunit bonding motifs [4]; (vi) quantitative ^{31}P NMR on phosphitylated samples to determine both the nature and the abundance of the interunit bonding motifs [4,84]; (vii) electron paramagnetic resonance (EPR) or electron spin resonance (ESR) spectroscopy for qualitatively and quantitatively elucidating unpaired electrons [85]; and (viii) mass spectrometry, especially newer MALDI-TOF MS/MS methods, that seem to be capable of determining structural aspects and molecular mass key figures more accurately [36,44].

Although FT-IR spectroscopy is quick and informative as such, and albeit the UV-based determination of especially the phenolic end-groups is quick and often practically useful, overall superior results are obtained using NMR spectroscopy-based analysis methods, since NMR spectroscopy allows obtaining both detailed structural information and quantitative data. The main difficulty hampering the analyses using NMR methods is the low solubility of most lignins in the typical NMR solvents such as deuterated chloroform ($CDCl_3$) or deuterated dimethyl sulfoxide (DMSO-d_6). Solubility issues can be circumvented by carefully acetylating the lignin prior to analyses [22,86], accepting the fact that the observed shifts are affected by the presence of the introduced acetates. Figure 6A shows a quick-quantitative ^{1}H-^{13}C-HSQC (QQ-HSQC) spectrum of acetylated Norway spruce MWL in DMSO-d_6 [86].

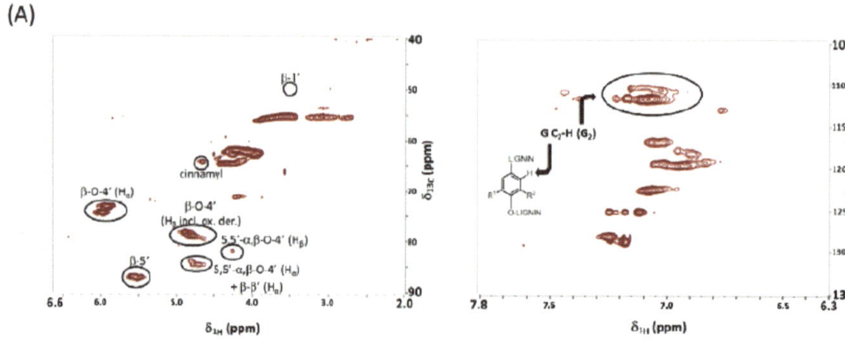

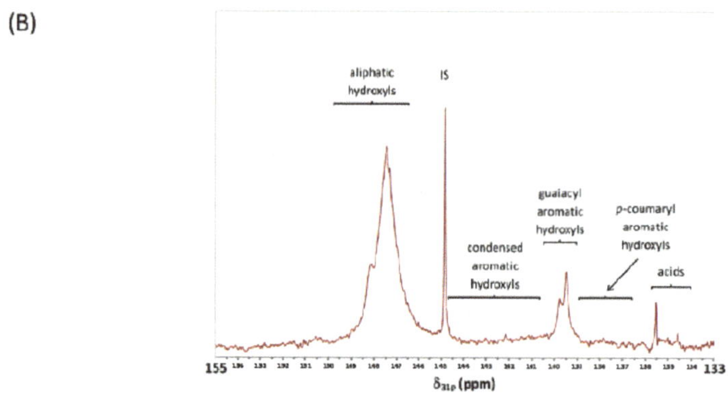

Figure 6. (**A**) Quantitative two-dimensional ^{1}H-^{13}C NMR ($HSQC_1$) of acetylated Norway spruce MWL in $CDCl_3$, aliphatic region (left) and aromatic region (right) [86]. (**B**) ^{31}P NMR of a phosphitylated sample of Norway spruce MWL in $CDCl_3$ in the presence of cholesterol as internal standard.

Different possibilities for the acquisition of quantitative HSQC spectra exist; often the available NMR equipment dictates the method that can be practically employed. In case a high-field (400 MHz or more) NMR machine equipped with a cryo-probe is available, either a quick-quantitative HSQC (QQ-HSQC) pulse sequence [87], or an approach consisting of a series of HSQC-measurements with incremented repetition times and mathematical backward-extrapolation, called $HSQC_0$ [88], can be adopted for the acquisition of lignin spectra [86]. While the results are comparable, the QQ-HSQC method has the drawback of being more difficult to be implemented, while the $HSQC_0$ method is significantly more time-consuming, and bears an additional error source due to the necessary extrapolation. In any case, quantification is achieved based on the intrinsic standard represented by the distinct shifts of the aromatic hydrogen atoms in 2-position, which can be easily detected, summed up, and used as reference. It is needless to say that the two-dimensional analyses should be accompanied by quantitative one-dimensional ^1H and ^{13}C NMR data. For obtaining these in case of lignins, again a careful adjustment of the pulse sequences and internal standards, especially in case of ^{13}C NMR, is necessary for obtaining viable data sets. Just as in case of the two-dimensional NMR studies, eventual solubility issues are conveniently evaded by working with acetylated samples [89].

Using an acetylated sample makes it impossible to determine the degree of natural esterification in the lignin sample. This characteristic number can be determined, however, by analysing the lignin sample before and after basic hydrolysis comparing the quantities of free hydroxyl groups. Both the number and the nature of free hydroxyl groups can be determined conveniently using ^{31}P NMR spectroscopy on phosphitylated lignins in the presence of an internal standard [84]. Figure 6B shows the phosphitylation reaction with 2-chloro-3,3,4,4-tetramethyl- 1,3,2-dioxophospholane [90] and a representative ^{31}P NMR spectrum of phosphitylated Norway spruce MWL in the presence of cholesterol as internal standard using a suitable pulse sequence for obtaining quantitative results. The different chemical environments around the phosphorous atoms lead to characteristic shifts, allowing for the differentiation and quantification. Typical ranges for the different groups distinguishable on the bases of the aforementioned phosphorous reagent are indicated in Figure 6B. During the phosphitylation, solubility improves significantly; for solubilising tenacious samples, ionic liquids can be used as solvents for ^{31}P NMR spectroscopy [91]. This quantitative analytical technique, perhaps the fastest, simplest and most reliable one for getting significant structural information about polyphenolic material, as withnessed by the huge amount of citations, is nowadays routinely used in many laboratories.

An alternative ^{31}P NMR-based protocol has been developed for the determination of quinones [92,93]. The determination of free radicals in lignin model studies was possible using a ^{31}P-containing spin trap reagent, such as, e.g., 5-diisopropoxyphosphoryl -5-methyl-1-pyrroline -*N*-oxide (DIPPMPO) [94,95]; these studies gave valuable insight in the mechanisms underlying enzymatic lignin degradation.

The phosphorous marker-based heteronuclear NMR spectroscopy is just one example for the use of non-lignin-inherent, NMR-active elements for the structural elucidation of lignins. While ^{31}P NMR is by far the most prominent one, ^{19}F NMR is used in order to identify functional groups via their interconversion into ^{19}F marker-containing derivatives; most often, ^{19}F NMR is used for the determination of carbonyl groups in lignin, via the in situ formation of trifluoromethyl ethers [96,97].

Much less commonly used heteronuclear NMR methods for lignin utilise ^{28}Si or ^{199}Hg NMR spectroscopy, in order to elucidate hydroxyl functionalities (^{28}Si) [98], and aromatic substitution patterns (^{199}Hg) [99]. ^{15}N NMR is less important for analysing lignins due to lack of nitrogen functionalities in standard lignins [100].

Analyses of the aforementioned mono- and two-dimensional NMR spectra is only possible based on extensive studies using monomeric, dimeric, and trimeric lignin models, which continuously help improving structural elucidation of lignins. In this respect it is important to identify natural impurities in lignins, such as tannins and lignans, which due to their chemical similarity cannot be always easily removed completely from isolated lignin.

Determining the key molecular weight figures of lignins can be as challenging as the determination of the structural motifs. Initial efforts used techniques such as vapor pressure osmometry [101] or cryoscopy [102], which exploit colligative properties of the polymers to determine average molecular weights of lignin. It was found, however, that these methods are too susceptible to influences emerging from lignin-solvent interactions.

A seemingly more promising way is the use of size exclusion chromatography (SEC) in form of gel permeation chromatography (GPC) [77,103–105]; however, this method is suffering from problems emerging from the diversity within the chemical structure of lignins: the structural differences between different lignins require theoretically a set of tailor-made standards for achieving a calibration of the SEC set-up that fits the characteristics of the lignin analyte. This is, however, simply not feasible. Efforts have been made in the past for estimating the error that is created when a standards made from a polymer such as polystyrene are used, in combination with only a number of synthetically derived lignin models, but this universal calibration did not lead to significantly superior results than a calibration only based on polystyrene standards, thus indicating that differences in the hydrodynamic volume of the two polymers cannot be the only issue to be considered [106]. Additional general problems arise from the low solubility of most lignin types in the solvents that are commonly used for SEC/GPC analyses, which makes it necessary to derivatise the sample; common derivatisations are acetobromination [107] or acetylation (and thus the same ones used for structural analysis via NMR), which cause ultimately the same structural changes discussed before [76]. Different detector-types are known to lead to different results: commonly used UV-based detectors and refractive index detectors were shown to perform less well compared to molecular weight sensitive detectors based on viscosimetry [108], or laser light scattering detectors [109–113]. Technical advances in the field of mass spectrometry made it possible to use MALDI-TOF mass analysis techniques for the determination of molar masses of lignin. Since this method should be independent from solubility issues and structural aspects, results are expected to be more accurate [114].

5. Mechanical and Thermal Properties Lignin

The interesting properties of lignin from a material's point of view are extensively investigated, and numerous efforts aim at exploiting mechanical and thermal characteristics in industrial applications [115]. With respect to such industrial applications, aspects such as flowability and processibility become important. The differences in structural characteristics among different lignins as discussed above naturally result in different macroscopic properties. As outlined in more detail above, isolated lignins display a rather broad distribution of chain lengths compared to other natural and synthetic polymers. The combination of these lignin-specific distribution of polymer-sizes and thus polymer characteristics, in addition to the presence of a large number of reactive functional groups and free radicals, makes it difficult to design general processes for lignin valorisations, especially with respect to the thermal behaviour of the lignin in the process. Several methods give insight into molecular motion and thermal stability of a polymer and to determine characteristic key figures such as the glass transition temperature (T_g) and the decomposition temperature. Thermal gravimetric analyses (TGA), differential scanning calorimetry (DSC), and dynamic mechanical analysis (DMA) / dynamic rheology are commonly used for determining these key figures, especially in studies aimed at tailoring lignins by fractionation as discussed below. Due to its structural and chemical characteristics, heating of lignin is accompanied by vaporisation of volatiles (including the water that might be more or less strongly bound to the hydroxyl groups), chemical reactions and thermal degradation, and thus results in a weight loss so, that TGA is most suitable for determining the physico–chemical properties of lignin with exception of T_g. Interpretation of TGA spectra of lignins, however, remains tedious, since the complexity of the natural polymer does not allow for quick conclusions. DSC represents the easiest way to get hold of T_g, but is often seen as being too simple as method, since non-negligible degrees of crosslinking—effectuated by radical combination reactions—have been observed during the initial DSC scan when temperatures are reached that are close to the T_g of the lignin, resulting in augmented molecular weights and polydispersity, so that the measured T_g is nothing but an artefact. These issues can be avoided using DMA analysis; T_g is determined here based on the analysis of the temperature-dependent evolutions of the storage (G′) and the loss (G″) moduli. As a general trend for a homogeneous series of lignin samples, i.e., a series of different batches of the same lignin, e.g., a Kraft lignin, it is found that T_g increases with increased molecular weight.

The successful correlation between molecular weight and key figures of the thermal stability seems to sustain efforts that aim at 'tailoring' a given lignin sample by simple fractionation in batches of narrower molecular weight distributions [116]. These fractionations can be done using consecutive rounds of extraction with organic solvents [117–119], fractionated precipitation [23,120–122], and ultrafiltration [123–128]. Most noteworthy is, however, the fact that it remains difficult to strictly correlate *structural* features with physico-chemcial characteristics of the fractionated lignin samples even in case of successful fractionations.

A series of studies investigated the processibility of kraft lignin in extruder-based applications: while the original kraft lignin used in the study proofed to be useless due to its thermal characteristics, a targeted functionalisation or a simple specific blocking of the phenolic hydroxyl groups, e.g., via formation of methyl ethers, led to a significant improvement of the thermal stability [129–132].

6. Chemical Valorisation of Lignin

Chemical valorisation of lignin is difficult. The structural features and the dense distribution of functional groups present a chemical challenge for 'synthesising' lignin-based materials. The situation is additionally complicated by the low solubility of the polymer and the tedious analytics that are necessary to validate any structural change beyond the simple conversion of the different hydroxyl groups. Nevertheless, the need for sustainable and biodegradable materials justifies the numerous efforts that were made and are still being made. In the context of this book chapter, the aim of the following paragraphs can thus only be to give a general overview and to point to the references that allow further insight.

Generally speaking, oxidative valorisation of lignin aims at the production of more complex aromatic compounds, and platform chemicals for the industrial production of fine chemicals. Reductive degradation, in contrast, aims at a general reduction of functionality and at the formation of simple bulk aromatic compounds or less defined small molecules for bio-oils or fuels.

7. Oxidative Modifications

Oxidative valorisation of lignin can be achieved by different means that range from purely chemical ones to biotechnology. These routes have recently been reviewed in detail [45,133], but shall be mentioned here as well in a wider context.

7.1. Metal-Free Oxidations

4-acetamido-TEMPO (5 mol %; TEMPO = 2,2,6,6-tetramethylpiperidine-*N*-oxyl) proofed to be a versatile catalyst when used in combination with HNO_3 and

HCl (10 mol % each) for the chemoselective aerobic oxidation of secondary benzylic alcohols within lignin model compounds and lignins [134]. The resulting aryl ketones can then serve as substrates for further catalytic cleavages; this step can be done photolytically, as has been demonstrated recently [135].

7.2. Organometallic Catalysts

One of the structurally least-complex metal-based catalysts for activating both molecular oxygen or hydrogen peroxide is methyltrioxo rhenium (MTO) [136,137]. Two peroxorhenium intermediates, a mono-peroxo η^2-complex ($[MeRe(O_2)O_2]$ (**15**)), and a *bis*-peroxo η^2-complex ($[MeRe(O_2)_2O]$ (**16**)) are involved in the activation of hydrogen peroxide (Figure 7A) [138]. In order to improve, and fine-tune catalyst performance, polystyrene or poly(4-vinylpyridine) beads have been used as support, or as material for encapsulation in order to immobilised MTO (Figure 7B). Oxygen transfer from these peroxo-complexes to the substrate occurs via a concerted mechanism that includes a butterfly-like transition state, **17**, which prevents the formation of intermediate radical species (Figure 7A). The concerted mechanism makes MTO a rather unique reagent for the oxidative valorisation of lignin, since it allows a more controlled chemistry by suppressing the presence of various species of very reactive oxygen-based radicals in the reaction mixture. MTO-activated hydrogen peroxide proofed to be a versatile catalyst for various lignin samples; its activity was explicitly shown for lignin model compounds, including challenging substrate such as activated phenols and methoxybenzenes [139–141]. The polymer-supported variant oxidised both phenolic and non-phenolic lignin model compounds, and due to the lowered Lewis acidity of the MTO catalyst upon immobilisation, also aliphatic groups can be oxidised while side-reactions targeting the aromatic rings and leading to over-oxidation can be suppressed.

Cobalt salen ([Co(salen)]) complexes (**21**) were shown to be stable in aqueous reaction media [142] and could thus be successfully exploited in oxidative lignin transformations. The oxidation proceeds mechanistically via the initial formation of a phenoxy-radical, which reacts with molecular oxygen to ultimately form oxidised lignin derivatives and models (Figure 7B) [143–146]. More recently, a bulky heterocyclic nitrogen base was incorporated in a new Co-containing Schiff base-based catalyst which was able to convert both S- and G-type lignin models in high yields; the conversion of S- and G-type structures is essential for an effective use of lignin as a chemical feedstock, and remained elusive before using other [Co(salen)] complexes [147].

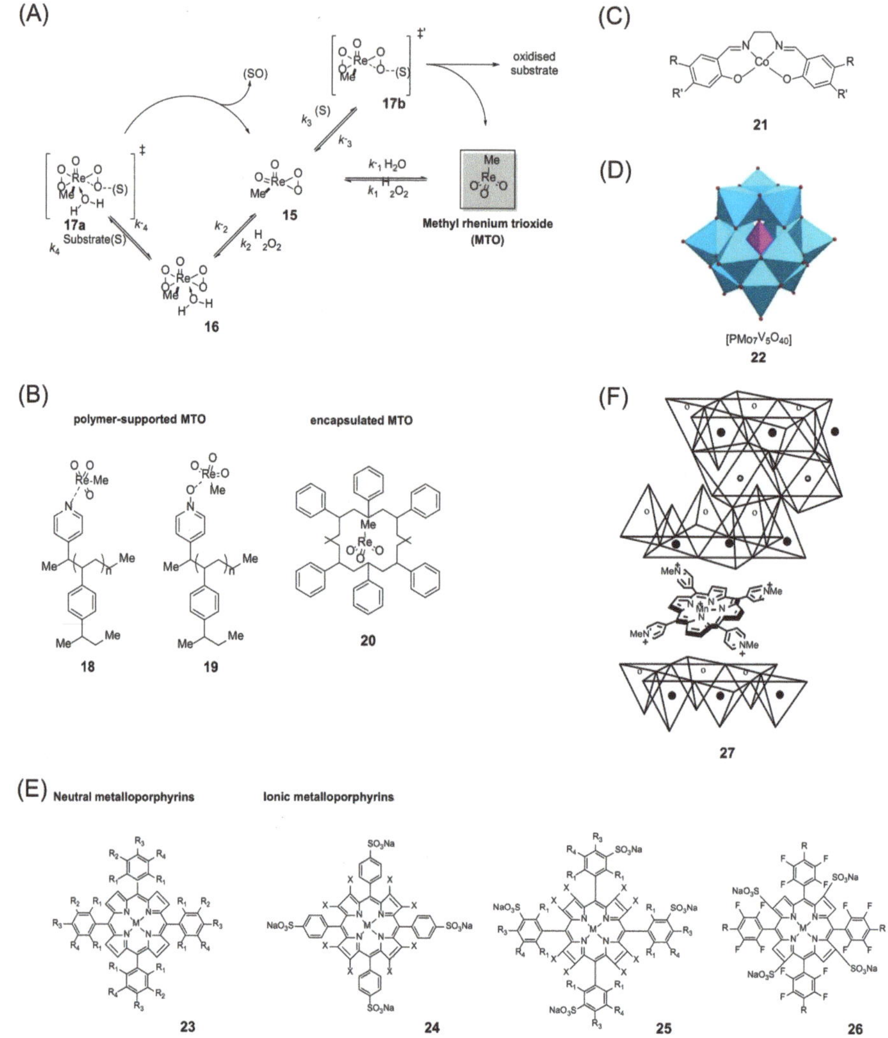

Figure 7. Organometallic catalysts for the oxidative valorisation of lignin: (**A**) Methyltrioxo rhenium (MTO) and mechanistic details of radical oxidation. (**B**) Polymer-supported versions of MTO. (**C**) Polyoxometalate (POM). (**D**) General structure of salen complexes. (**E**) Neutral and ionic metalloporphyrins. (**F**) Schematic illustrating a polymer-supported metalloporphyrin. For details and references, see main text.

Keggin-type polyoxometalates (POMs, **22**) (Figure 7C) with the general formula $[PM^1_xM^2_{12-x}O_{40}]$ can activate hydrogen peroxide and molecular oxygen for the oxidative valorisation of lignin and its various model compounds. Different protocols, including the catalytic use of POMs have been developed as "green" alternatives for the oxidative valorisation of lignin [148–151]. A couple of dedicated review articles exist in the archival literature regarding this topic [149,152]. Phenolic lignin model compounds were shown to be oxidised by C-C bond cleavage between the aryl ring and the adjacent hydroxy-substituted carbon atom using vanadium catalyst **45** (Figure 7) [153]. Labelling experiments indicated key mechanistic differences that serve to explain the differences in activity with respect to a different vanadium catalyst that favoured C-O bond cleavages [154].

The performance of the different catalysts with respect to the most common interunit bonding motif aryl glycerol-β-aryl ethers (β-*O*-4′) is shown in Figure 8.

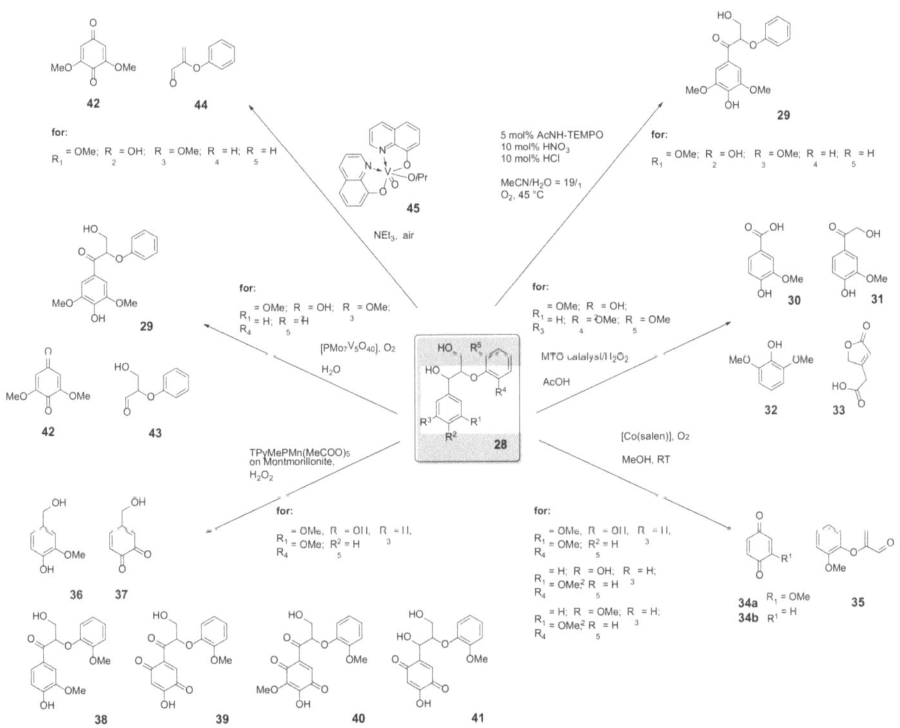

Figure 8. Conversion of the most common interunit bonding motif aryl glycerol-β-aryl ethers (β-*O*-4′) upon oxidative treatment with different organometallic catalysts and under metal-free conditions. References are given in discussion in the main text.

7.3. *Biomimetic Catalysts*

Synthetic metalloporphyrins represent biomimetic systems for both LiP- and MnP-containing highly oxidised metallo-oxo species similar to those found in, e.g., LiP I and LiP II [53]. Highly functionalized porphyrins such as those shown as structures **23–26** in Figure 7, bearing aryl substituents in the *meso*-positions of the heme motif, represent tuneable catalyst systems that oxidised both lignin model compounds and lignin. In contrast to the enzymes using the heme-type metal complexes, the metalloporphyrins can tolerate an excess of hydrogen peroxide, and are thus exhibiting a higher industrial potential. This feature is further improved by immobilising these catalysts on solid supports such as silica gel [155], naturally occurring clays [156], or artificial polymers [157], while maintaining effectiveness using mediators. Important findings have been reviewed before in dedicated articles [158,159].

7.4. Biotechnological Approaches

The industrial use of lignolytic enzymes that were mentioned above in course of the biodegradation of lignin has recently been discussed [160]. Nevertheless, for the sake of completeness, the most important ones shall be briefly discussed here as well.

The multicopper oxidase laccase (EC 1.10.3.2) oxidises suitable substrates under concomitant reduction of oxygen to water. Radical cationic species are generated starting from a phenolic substrate within an outer-sphere electron-transfer mechanism. Subsequent deprotonation then yields a phenoxy radical, which initiates the depolymerisation of lignin following an *exo*-depolymerisation mechanism [63,161–163]. Whether a given substrate is oxidised by laccases is not only a question of steric matching according to the oversimplified lock-and-key model, but also a question of oxidation potential. Although different laccases are known that exhibit very different redox potentials, ranging from $E^\circ = 0.43$ V (tree laccase from *Rhus vernicifera*) to 0.78 V (fungal laccase from *Polyporus versicolor*) [164], a laccase suitable to oxidise blocked phenolic substrates has not yet been identified. However, a combination of laccases and a radical mediator species, such as 1-hydroxybenzotriazole (HBT) [165], *N*-hydroxyacetanilide (NHA) [166], violuric acid or 2,2′-azinobis-3-ethyl-benzthiazoline-6-sulfonate (ABTS), respectively, can serve to oxidise blocked phenolic substrates in an 'indirect' way (Figure 9B) [167,168]. Noteworthy, initial studies indicated that the use of mediators cannot only boost activity as a whole, but also holds the possibility of actively influencing the mechanistic pathways, and thus controlling also product distributions. (Figure 9A): oxidation presumably does no longer occur via an electron transfer process in these cases, but via a hydrogen atom abstraction process [169].

Manganese peroxidases oxidatively depolymerise lignin under concomitant reduction of molecular oxygen or hydrogen peroxide to water [170]. Activation of molecular oxygen or hydrogen peroxide is achieved via a two-step process, in which an iron protoporphyrin IX activates the oxidant first for the oxidation of the manganese co-factor from Mn(II) to Mn(III) [171]. Carboxylic acid anions then chelate the Mn(III)-centre, forming a small, freely diffusible species functioning as oxidant for depolymerizing oxidatively lignin and lignin model compounds—comparable to the activated mediator in the laccase-mediator systems [172,173]. The reactive manganese species is rather selectively attacking methyl and methylene groups in *para*-position to the phenolic hydroxyl-groups in lignin. Analyses of the residual lignin suggest an *exo*-depolymerisation mechanism.

Large scale industrial use of the aforementioned enzymes, but not only of these enzymes, is, however, still a problem, with noteworthy exception of laccases. Drawbacks are related to the costs connected to enzyme production, the industrial use including necessary recovery steps, and eventual substitution of used enzymes due to inevitable loss of activity upon consecutive use [174]. Support of enzymes turned out to be a suitable mean to render them more applicable in industrial transformations. In this respect, both the immobilisation and the encapsulation of laccase and other enzymes, either alone or in combinations, have been reported [175–185]. The combination of enzymes in connection with the effect of the solid support on their mechanism is another versatile option for directing reaction outcomes [186].

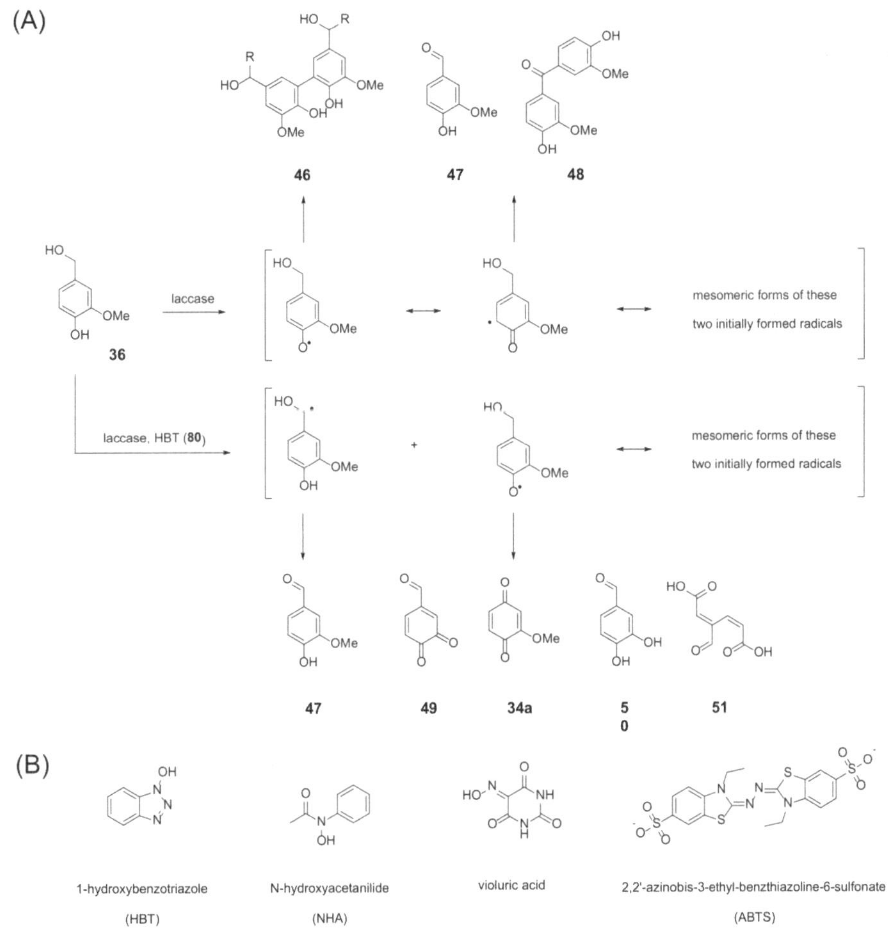

Figure 9. (**A**) Mechanistic implications of the use of mediators. (**B**) Common mediators.

7.5. Reductive Modifications

Reductive modification of lignin has been a research subject for decades [187], and due to the aforementioned rather destructive nature of this method, only a few recent examples are listed here, mainly for the sake of overall completeness of this chapter. It shall be noted explicitly, that a reductive valorisation of lignin is suitable only for the production of low-value applications, since the concomitant removal of versatile functional groups does not go together with the generation of functionalised materials that offer interaction points with other materials [154,188].

7.6. Metal-Free Reductions

Hydrosilanes were shown to convert aryl ethers to hydrolysable silyl ethers in the presence of $B(C_6F_5)_3$ in a general process that was used to convert model lignin compounds to both aryl silyl ethers and alkanes.

Softwood lignin itself could only be partially solubilized (30%) but hardwood lignin was efficiently reductively decomposed by the mono-functional hydrosilane $HMe_2SiOSiMe_3$ [189].

Lignin could be essentially entirely dissolved under moderate temperatures and pressures using ethanol/water mixtures. The molecular weights of the dissolved lignins were shown to be reduced (GPC and NMR studies) [190].

7.7. Heterogenic Transition Metal Catalysts

Cleavage and hydrodeoxygenation (HDO) of carbon-oxygen bonds in lignin model compounds were achieved using a Pd/Zn synergistic catalysis at high temperature (150 °C) and high hydrogen pressure (20 bar) (Figure 10A) [191]. The catalyst system showed a high selectivity and was recyclable. Tests on whole lignins, however, were not reported.

A heterogeneous nickel catalyst for the selective hydrogenolysis of aryl ethers to arenes and alcohols generated in situ from soluble nickel precursor $Ni(COD)_2$ or $Ni(CH_2TMS)_2(TMEDA)$ (COD—cyclooctadiene; TMS—trimethylsilyl; TMEDA—tetramethyl ethylendiamine) in the presence of a base additive (Figure 10A) [192]. The catalyst selectively cleaves C-O bonds in aryl ether models of lignin without hydrogenation of aromatic rings, and it operates at loadings down to 0.25 mol % at atmospheric hydrogen.

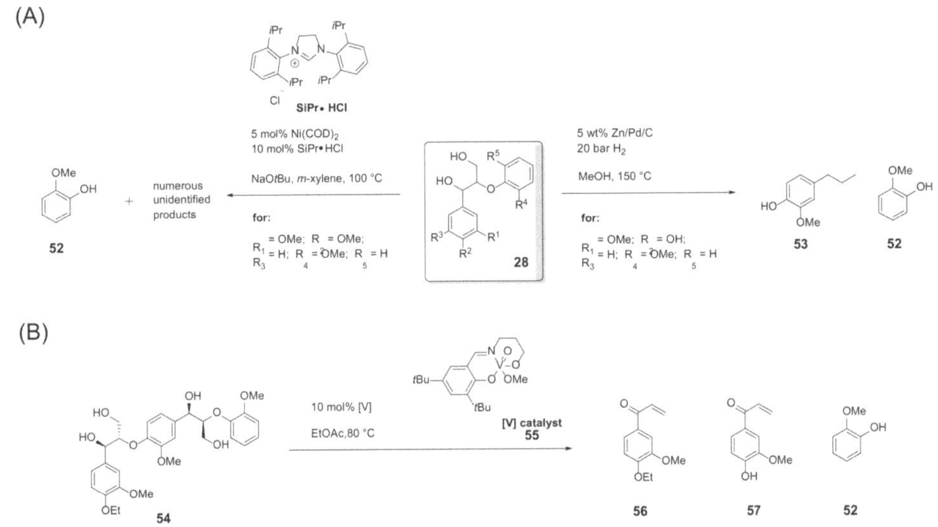

Figure 10. (**A**) Conversion of the most common interunit bonding motif aryl glycerol-β-aryl ethers (β-*O*-4′) upon reductive treatment with different organometallic catalysts. (**B**) Vanadium-catalysed reductive cleavage of a trimeric lignin model. References are given in the discussion in the main text.

7.8. Organometallic Catalysts

The conversion of lignin into low-boiling arenes instead of high-boiling phenols was realised by depolymerising lignin under simultaneous conversion of phenols into arenes using a tandem reaction catalysed by a tailored nickel-centred soluble organometallic catalyst [193]. Earlier studies using homogeneous Ni-catalyses for these purposes were not tried on lignin itself [194]. Another earlier study accomplished roughly 40 years ago on lignin model compounds used a homogeneous nickel–hydride complex [188]; selectivity in this system depended highly on the substrate specificities.

A vanadium species has also been reported to reductively degrade lignin by selectively attacking the C-O bonds of the aryl ethers forming the lignin backbone (Figure 10B) [154,195]. The studies were accompanied by a detailed structural analysis of the lignin that remained after catalysis: data confirm the selectivity of the catalyst. The experiments did reveal, however, that also the history of the lignin has an influence on the effectiveness of the transformation.

8. Applications of Lignin

A huge amount of applications of lignin exist and have been reported in both academic publications and patents, and recent reviews and monographs give excellent overviews [196]. Reported applications range from easy-to-be-realised uses of lignin as, e.g., filler, to more advanced applications in which the properties of lignin are used more distinctively. Industrial use of lignin, however, also includes its targeted destruction, as, e.g., for the production of bulk chemicals such as DMSO or vanillin [197,198]. Possible routes for the conversion of lignin to platform chemicals for the production of fine chemicals have been already outlined in the section on the chemical valorisation of lignin. In the following paragraphs, the use of lignin in low cost applications is just briefly touched, and more emphasis is placed on the use of lignin in nanomaterials and in pharmaceutical applications.

8.1. Lignin as Bulk Material in Co-Polymer Applications

Lignin has been used as low cost hydroxyl component for the formation of polyurethanes (PU) and polyesters (PE) [199–201]. Given the structural differences of various lignins, different qualities of PU are available for different applications [202–205]. As different lignins come with different distributions of aliphatic vs. aromatic hydroxyl groups, different chain lengths, etc., their use as component in polyurethane formation requires optimisation and careful PU characterisation. Generally, lignins represent the less elastic part of a PU.

Since lignin carries alcoholic and acidic functionalities at the same time, it has been used for the formation of PE [206], using application-depending stability-introducing acyl chlorides. It is obvious that the application will ultimately dictate which lignin is the most suitable one, but current research results underline generally the suitability of lignin as polymeric component for PEs [199].

Oxypropylation of lignin is used to adjust molecular weight characteristics of a lignin sample, and thus also to adjust the characteristics of a copolymer in which this modified lignins are incorporated, e.g., PU and PE formations [207,208].

Other polyblends incorporating lignin have been extensively studied, and were recently concisely reviewed [209]. Among them are phenol-formaldehyde adhesives [210]. The incorporation of lignin here allowed to reduce the amount of phenol by up to 60%, and to introduce new characteristics into the glue matrix, such as conductivity. Also lignin-epoxy adhesives were realised and studied extensively [211,212].

In order to increase their mechanical strength, light stability and partial biodegradability, polyolefins were blended with lignin [209,213,214]. As for other lignin applications, the quality of the lignin starting material had a great impact on the stability of the blends; different lignin types gave rise to different blends, whose characteristics additionally changed with varying amounts of lignin. UV-absorbing films have been reported, for example [215], as well as low-cost films with up to 50% lignosulfonate content [216].

Poly(vinyl chloride) (PVC) has been blended with lignin, driven by the same ideas of adjusting thermal strength and light stability [217–219]. Although the functional groups present in the two starting materials can easily undergo proton-donor/proton-acceptor interactions, complete miscibility is not automatically granted. Successfully grafted PVC-lignin blends showed an improvement of polymer characteristics in terms of yield and breaking strength.

Other lignin-containing blends have been realised using poly(vinyl acetate) (PVAc), poly(vinyl alcohol) (PVA), polysaccharides, e.g., starch and gums, etc. In all applications, the limiting factor is the solubility and the thermal stability of the lignin used [209].

Another possibility to make use of favourable thermal characteristics of lignin is its incorporation in the production of thermoplastics [129,220].

The mechanical strength of lignin is best exploited in the production of lignin-based carbon fibres, which represent a research object since the first example was reported in 1969 [221]. Lignin-based carbon fibres are obtained, in turn, upon carbonisation of lignin fibres that are formed upon melt-spinning [222,223] or electro-spinning; especially in the latter case, lignins are, however, usually still mixed with other polymeric materials such as polyethylene oxide (PEO or PEOX), [224,225] polyacrylonitrile (PAN) [226–228] or other plasticisers [229,230] to yield significant results. Only few reports exist detailing pure lignin fibers, that could eventually serve as carbon fibre precursors [231,232].

8.2. Applications in Nanomaterials

8.2.1. Capsules

Micro- and nanoparticles, micelles or capsules represent promising structures with respect to pharmaceutical applications. Recently, lignins were used to produce capsules capable of encapsulating either hydrophobic [233,234] or hydrophilic drugs [235]. Lignin capsules for hosting hydrophobic substances were produced by exploiting the amphiphilic character of kraft lignin: when dissolved in oil-in-water emulsions, the lignin polymers preferentially located at the oil-water interface, where they could be further intermolecularly crosslinked to form capsules [233]. This crosslinking was achieved using ultrasonification for radical activation, or by reacting diglycidyl ether-terminated poly(ethylene glycol) with the lignin polymers. The capsules could be filled with hydrophobic molecules, which in turn could be released upon addition of surfactants. The fundamental principle could be extended to other lignin types as well, realising a pH-dependent release mechanism for entrapped actives [234].

When treating lignin-containing aqueous solutions that were emulsified by the aid of a surfactant in an organic phase comprised of cyclohexane containing 2,4-toluene diisocyanate, a lignin-based polyurethane was formed at the interface between aqueous and organic phase [235]. Hydrophilic substances could be encapsulated upon formation, and subsequently released upon enzymatic degradation of the lignin shell.

8.2.2. Nanoparticles

Emulsion-based techniques can also be used to constitute lignin-based beads, simply by dispersing the lignin-containing phase: suspensions of black liquor in a mixture of oil and chlorobenzene in the presence of an emulsifier allowed the polymerisation of the lignin by crosslinking lignin oligomeric and polymeric chains using epichlorhydrin; spherical beads with diameters between 300 and 450 μm are obtained [236]. Less defined bead sizes were obtained when an alkaline lignin solution was dispersed in 1,2-dichloroethane and the lignin intermolecularly crosslinked in the aqueous droplets—either by mediation of epichlorhydrin [237] or by radical polymerisation processes involving acrylic acid moieties that were attached to the lignin before [238]. Attaching further functionalisation to these beads, e.g., sulfonhydrazine groups [237,238], made them attractive for applications in the wine industry, particularly for removing carbonyl-bearing compounds responsible for binding of sulphur dioxide in wines [239]. Lignin predissolved in tetrahydrofurane (THF)-water mixtures that were then poured in an excess of water precipitated in form of stable suspensions of nano-particles [240]; this way of lignin nano-particle generation was used in proof-of-concept studies as dug carriers [241].

Using a solution of lignin in ethylene glycol, the formation of lignin nanoparticles was achieved by gradually acidifying (hydrochloric acid) the solution [242].

A simple physical method based on ultrasonic irradiation can be used to build nanoparticles from lignins such as commercially available wheat straw lignin and Sarkanda grass lignin. It has been shown that the compositional and structural changes of the lignin in the produced nanoparticles were not significantly modified by the intensity of the ultrasound applied, but depended mainly on the nature of lignin itself [243].

When alkaline lignin solutions were added to cationic polyelectrolyte solutions of poly(diallyldimethylammonium chloride) (PDDAC), the cationic polymer was adsorbed on the lignin aggregates due to the display of overall negative surfaces stemming from the alkaline conditions chosen for solubilising the lignin. Similar to the process of creating LbL-protections for enzymes (*vide supra*), the now positively charged surface of the lignin-PDDAC hybrids interacted with negatively charged natural latex rubber particles to associate further in solution, resulting in natural rubber materials reinforced with finely dispersed nano-sized lignin particles. These new lignin-based hybrid-particles are characterised by improved mechanical and thermal properties [244].

Subjection of nanoparticles to crosslinking agents such as glutaraldehyde resulted in nanoparticles that showed useful stabilities over a wide pH-range when re-dispersed in water. Interestingly, the nanoparticles did not display cytotoxic activity when yeast or microalgae were exposed to them. Gradually adding water to a solution of acetylated lignin in THF resulted in the formation of nano-sized colloidal spheres that were water-dispersible and stable up to pH 12 [245]. Higher pH-values unblocked the phenols via hydrolysis of the acetic acid esters, causing higher degrees of electrostatic repulsion that effectively inhibited colloidisation.

8.2.3. Micro- and Nano-Porous Materials

Porous materials called aerogels are seen as promising materials for a range of applications, e.g., insulating material, adsorbents, catalysts, etc. [246,247]. Aerogels are obtainable via a sequence of sol-gel polymerisation of a resin, exchange of solvents and final drying. Most often, the resins used are of polyphenolic type, and as indicated during the discussion of the lower-value applications, lignin has been considered a potential source of phenols for this kind of resins.

Replacing part of resorcinol by lignin is enough for producing resorcinol-lignin-formaldehyde resins in aqueous solutions of sodium hydroxide [248].

Phenol-lignin-formaldehyde and tannin-lignin-formaldehyde resins have been produced using similar formulation approaches [249,250]. The maximum amount of lignin as replacement for the phenolic part could not exceed 50%. In any case, the replacement of simple phenols by lignin led to the formation of macropores (>50 nm) on the expense of mesopores (2–50 nm). Even if this caused an increase in the overall porosity and consequently a decrease in the bulk density, the total surface area was reduced, thus limiting the final material properties, such as the adsorption capacity or the insulating performances [248–250]. Aerogels based solely on lignin crosslinked by means of glycidyl-terminated oligo(ethylene glycol) or (propylene glycol) have been reported more recently [251]. More effective gel formations were observed in case of long-chain crosslinkers, and it was hypothesized that this is due to a more effective linking of lignin chains in this case.

Blocks of porous polymeric lignin-hybrids could be obtained using black liquor directly via the High Internal Phase Emulsion (HIPE) technique [252]. Epichlorhydrin proofed to be useful once more as crosslinker also in the presence of a surfactant: after the addition of oil, the mixture was heated to initiate epichlorhydrin-mediated crosslinking of lignin. Depending on the specifics of the oil and the surfactants used, monoliths of different porosity were obtained, exhibiting medium void sizes between 5 and 20 μm [252].

8.3. Pharmaceutical Applications

With respect to a use in pharmaceutical applications which normally come with significantly higher hurdles concerning safety and biocompatibility requirements, lignin is still under-utilised. New research is following some pioneering achievements, in which lignins have been used (i) as component for semi-interpenetrating networks (hydrogels) [253]; (ii) as anti-oxidative ingredients [254]; (iii) UV-absorbing component in films [255]; and for the generation of capsules for drug delivery approaches [233,256].

Anti-viral activities of lignin derivatives, also against HIV, have been investigated and reported [257,258].

Lignin is considered an interesting material with respect to anti-tumour treatments [259]: in several studies (i) polyphenols with structural motifs found in lignin, (ii) mixtures of the monomers lignin is composed of, or (iii) small aromatic molecules that could be the enzymatic degradation products of lignin were investigated, in order to evaluate the response of tumour development and growth and the effects on underlying and connected signalling pathways to these substances [260,261]. Often, the cytotoxic activity of lignin and lignin derivatives is the result of interactions of lignin with other substances such as ascorbic acid [262], or lignin containing natural complexes such as lignin–carbohydrate complexes (LCCs) [263].

9. Concluding Remarks

Despite great advances in the research on and with lignin, important questions are still unanswered, or have to be revisited in light of new findings. Possibilities do exist to convert lignin in valuable chemicals and materials, ranging from bulk applications to specialised nano-materials. Inclusion of these techniques into industrial processes will be one of the immediate next challenges, and more work is needed to compile a convincing process for lignin valorisation that starts with the biorefinery delivering a quality lignin suitable for further advanced transformations and ends with truly lignin-based, meaning mainly lignin-based high value products [264].

Conflicts of Interest: The authors declare no conflict of interest.

Abbreviations

The following abbreviations are used in this manuscript:

MWL	milled wood lignin
KL	kraft lignin
DHP	dehydrogenation polymers
LiP	lignin peroxidase
MnP	manganese peroxidase
DFRC	derivatisation followed by reductive cleavage
UV	ultra-violet
FT-IR	Fourier-transform infrared
NMR	nuclear magnetic resonance
HSQC	heteronuclear single quantum coherence
EPR	electron paramagnetic resonance
ESR	electron spin resonance

DMSO	dimethyl sulfoxide
DIPPMPO	5-diisopropoxyphosphoryl-5-methyl-1-pyrroline-*N*-oxide
TGA	thermal gravimetric analyses
DSC	differential scanning calorimetry
DMA	dynamic mechanical analysis
T_g	glass transition temperature
TEMPO	2,2,6,6-tetramethylpiperidine-*N*-oxyl
MTO	methyltrioxorhenium
POM	polyoxometalates
HBT	1-hydroxybenzotriazole
NHA	*N*-hydroxyacetanilide
ABTS	2,2′-azinobis-3-ethyl-benzthiazoline-6-sulfonate
COD	cyclooctadiene
TMS	trimethylsilyl
TMEDA	tetramethylethylendiamine
PU	polyurethanes
PE	polyesters
PVC	poly(vinyl chloride)
PVAc	poly(vinyl acetate)
PVA	poly(vinyl alcohol)
PDDAC	poly(diallyldimethylammonium chloride)
HIPE	high internal phase emulsion

References

1. Adler, E. Lignin chemistry—Past, present and future. *Wood Sci. Technol.* **1977**, *11*, 169–218. [CrossRef]
2. Sarkanen, K.V.; Ludwig, C.H. *Lignins: Occurrence, Formation, Structure and Reactions*; Wiley-Interscience: New York, NY, USA, 1971.
3. Hu, T.Q. *Chemical Modification, Properties, and Usage of Lignin*; Springer Science & Business Media: New York, NY, USA, 2002.
4. Heitner, C.; Dimmel, D.; Schmidt, J. (Eds.) *Lignin and Lignans: Advances in Chemistry*, 1st ed.; CRC Press: Boca Raton, FL, USA, 2010.
5. Sjöström, E. *Wood Chemistry: Fundamentals and Applications*, 2nd ed.; Academic Press Inc.: San Diego, CA, USA, 1993.
6. Lewis, N.G.; Yamamoto, E. Lignin: Occurrence, Biogenesis and Biodegradation. *Annu. Rev. Plant Physiol. Plant Mol. Biol.* **1990**, *41*, 455–496. [CrossRef] [PubMed]
7. Lewis, N.G.; Davin, L.B.; Sarkanen, S. Lignin and Lignan Biosynthesis: Distinctions and Reconciliations. In *Lignin and Lignan Biosynthesis*; Lewis, N.G., Sarkanen, S., Eds.; ACS Symposium Series; American Chemical Society: Washington, DC, USA, 1998; Volume 697, pp. 1–27.
8. Glasser, W.G.; Sarkanen, S. *Lignin: Properties and Materials*; American Chemical Society: Washington, DC, USA, 1989.

9. European Liaison Committee for Pulp and Paper. Proceedings of the 8th International Symposium on Wood and Pulping Chemistry, Congrex, Blue & White Conferences, Helsinki, Finland, 6–9 June 1995; Jyväskylä Gummerus Kirjapaino Oy: Jyväskylä, Finnland, 1995.
10. Vanholme, R.; Demedts, B.; Morreel, K.; Ralph, J.; Boerjan, W. Lignin Biosynthesis and Structure. *Plant Physiol.* **2010**, *153*, 895–905. [CrossRef] [PubMed]
11. Boerjan, W.; Ralph, J.; Baucher, M. Lignin Biosynthesis. *Annu. Rev. Plant Biol.* **2003**, *54*, 519–546. [CrossRef] [PubMed]
12. Donaldson, L.A. Lignification and lignin topochemistry—An ultrastructural view. *Phytochemistry* **2001**, *57*, 859–873. [CrossRef]
13. Akiyama, T.; Goto, H.; Nawawi, D.S.; Syafii, W.; Matsumoto, Y.; Meshitsuka, G. Erythro/threo ratio of β-*O*-4-5 structures as an important structural characteristic of lignin. Part 4: Variation in the erythro/threo ratio in softwood and hardwood lignins and its relation to syringyl/guaiacyl ratio. *Holzforschung* **2005**, *59*, 276–281. [CrossRef]
14. Romani, A.; Lattanzio, V.; Quideau, S. *Recent Advances in Polyphenol Research*; John Wiley & Sons: Chichester, UK, 2014.
15. Rogers, L.A.; Campbell, M.M. The genetic control of lignin deposition during plant growth and development. *New Phytol.* **2004**, *164*, 17–30. [CrossRef]
16. Anterola, A.M.; Lewis, N.G. Trends in lignin modification: A comprehensive analysis of the effects of genetic manipulations/mutations on lignification and vascular integrity. *Phytochemistry* **2002**, *61*, 221–294. [CrossRef]
17. Grima-Pettenati, J.; Goffner, D. Lignin genetic engineering revisited. *Plant Sci.* **1999**, *145*, 51–65. [CrossRef]
18. Whetten, R.; Sederoff, R. Lignin Biosynthesis. *Plant Cell* **1995**, *7*, 1001–1013. [CrossRef] [PubMed]
19. Pesquet, E.; Zhang, B.; Gorzsás, A.; Puhakainen, T.; Serk, H.; Escamez, S.; Barbier, O.; Gerber, L.; Courtois-Moreau, C.; Alatalo, E.; et al. Non-Cell-Autonomous Postmortem Lignification of Tracheary Elements in Zinnia elegans. *Plant Cell Online* **2013**, *25*, 1314–1328. [CrossRef] [PubMed]
20. Adler, E. Structural Elements of Lignin. *Ind. Eng. Chem.* **1957**, *49*, 1377–1383. [CrossRef]
21. Holtman, K.M.; Chang, H.-M.; Jameel, H.; Kadla, J.F. Elucidation of Lignin Structure through Degradative Methods: Comparison of Modified DFRC and Thioacidolysis. *J. Agric. Food Chem.* **2003**, *51*, 3535–3540. [CrossRef] [PubMed]
22. Crestini, C.; Melone, F.; Sette, M.; Saladino, R. Milled Wood Lignin: A Linear Oligomer. *Biomacromolecules* **2011**, *12*, 3928–3935. [CrossRef] [PubMed]
23. Lange, H.; Schiffels, P.; Sevastyanova, O.; Sette, M.; Crestini, C. Fractionated Wheat Straw Organosolv Lignin—High Quality Polymeric and Oligomeric Polyphenolic Building Blocks. *ACS Sustain. Chem. Eng.* **2016**, *4*, 5136–5151. [CrossRef]
24. Crestini, C.; Lange, H.; Sette, M.; Argyropoulos, D.S. On the structure of softwood kraft lignin. *Green Chem.* **2017**, *19*, 4104–4121. [CrossRef]
25. Björkman, A. Studies on finely divided wood. Part 1. Extraction of lignin with neutral solvents. *Sven. Papperstidn.* **1956**, *59*, 477–485.

26. Lundquist, K. Acidolysis. In *Methods in Lignin Chemistry*; Lin, S.Y., Dence, C.W., Eds.; Springer Series in Wood Science; Springer: Berlin/Heidelberg, Germany, 1992; pp. 289–300.
27. Lapierre, C.; Monties, B.; Rolando, C.; de Chirale, L. Thioacidolysis of Lignin: Comparison with Acidolysis. *J. Wood Chem. Technol.* **1985**, *5*, 277–292. [CrossRef]
28. Guerra, A.; Filpponen, I.; Lucia, L.A.; Argyropoulos, D.S. Comparative Evaluation of Three Lignin Isolation Protocols for Various Wood Species. *J. Agric. Food Chem.* **2006**, *54*, 9696–9705. [CrossRef] [PubMed]
29. McDonough, T.J. The chemistry of organosolv delignification. *TAPPI J.* **1993**, *76*, 186–193.
30. Pye, E.K.; Lora, J.H. The Alcell process. A proven alternative to kraft pulping. *TAPPI J.* **1991**, *74*, 113–118.
31. Gierer, J. Chemical aspects of kraft pulping. *Wood Sci. Technol.* **1980**, *14*, 241–266. [CrossRef]
32. Chakar, F.S.; Ragauskas, A.J. Review of current and future softwood kraft lignin process chemistry. *Ind. Crops Prod.* **2004**, *20*, 131–141. [CrossRef]
33. Howard, G.C. Utilization of Sulfite Liquor. *Ind. Eng. Chem.* **1934**, *26*, 614–617. [CrossRef]
34. Scholze, B.; Meier, D. Characterization of the water-insoluble fraction from pyrolysis oil (pyrolytic lignin). Part I. PY–GC/MS, FTIR, and functional groups. *J. Anal. Appl. Pyrolysis* **2001**, *60*, 41–54. [CrossRef]
35. Scholze, B.; Hanser, C.; Meier, D. Characterization of the water-insoluble fraction from fast pyrolysis liquids (pyrolytic lignin): Part II. GPC, carbonyl goups, and 13C-NMR. *J. Anal. Appl. Pyrolysis* **2001**, *58–59*, 387–400. [CrossRef]
36. Bayerbach, R.; Nguyen, V.D.; Schurr, U.; Meier, D. Characterization of the water-insoluble fraction from fast pyrolysis liquids (pyrolytic lignin): Part III. Molar mass characteristics by SEC, MALDI-TOF-MS, LDI-TOF-MS, and Py-FIMS. *J. Anal. Appl. Pyrolysis* **2006**, *77*, 95–101. [CrossRef]
37. Bayerbach, R.; Meier, D. Characterization of the water-insoluble fraction from fast pyrolysis liquids (pyrolytic lignin). Part IV: Structure elucidation of oligomeric molecules. *J. Anal. Appl. Pyrolysis* **2009**, *85*, 98–107. [CrossRef]
38. Li, J.; Gellerstedt, G.; Toven, K. Steam explosion lignins; their extraction, structure and potential as feedstock for biodiesel and chemicals. *Bioresour. Technol.* **2009**, *100*, 2556–2561. [CrossRef] [PubMed]
39. Tanahashi, M.; Tamabuchi, K.; Goto, T.; Aoki, T.; Karina, M.; Higuchi, T. Characterization of steam-exploded wood, 2: Chemical changes of wood components by steam explosion. *Wood Res. Jpn.* **1988**, *75*, 1–12.
40. Hemmingson, J.A. Steam-Explosion Lignins: Fractionation, Composition, Structure and Extractives. *J. Wood Chem. Technol.* **1987**, *7*, 527–553. [CrossRef]
41. Van Parijs, F.R.D.; Morreel, K.; Ralph, J.; Boerjan, W.; Merks, R.M.H. Modeling Lignin Polymerization. I. Simulation Model of Dehydrogenation Polymers. *Plant Physiol.* **2010**, *153*, 1332–1344. [CrossRef] [PubMed]

42. Tuomela, M.; Hatakka, A.; Raiskila, S.; Vikman, M.; Itävaara, M. Biodegradation of radiolabelled synthetic lignin (14C-DHP) and mechanical pulp in a compost environment. *Appl. Microbiol. Biotechnol.* **2001**, *55*, 492–499. [CrossRef] [PubMed]
43. Hatcher, P.G.; Minard, R.D. Comparison of dehydrogenase polymer (DHP) lignin with native lignin from gymnosperm wood by thermochemolysis using tetramethylammonium hydroxide (TMAH). *Org. Geochem.* **1996**, *24*, 593–600. [CrossRef]
44. Araújo, P.; Ferreira, M.S.; de Oliveira, D.N.; Pereira, L.; Sawaya, A.C.H.F.; Catharino, R.R.; Mazzafera, P. Mass Spectrometry Imaging: An Expeditious and Powerful Technique for Fast in Situ Lignin Assessment in Eucalyptus. *Anal. Chem.* **2014**, *86*, 3415–3419. [CrossRef] [PubMed]
45. Lange, H.; Decina, S.; Crestini, C. Oxidative upgrade of lignin—Recent routes reviewed. *Eur. Polym. J.* **2013**, *49*, 1151–1173. [CrossRef]
46. Jung, H.-J.G.; Varel, V.H.; Weimer, P.J.; Ralph, J. Accuracy of Klason Lignin and Acid Detergent Lignin Methods As Assessed by Bomb Calorimetry. *J. Agric. Food Chem.* **1999**, *47*, 2005–2008. [CrossRef] [PubMed]
47. Hatfield, R.D.; Jung, H.-J.G.; Ralph, J.; Buxton, D.R.; Weimer, P.J. A comparison of the insoluble residues produced by the Klason lignin and acid detergent lignin procedures. *J. Sci. Food Agric.* **1994**, *65*, 51–58. [CrossRef]
48. Dashtban, M.; Schraft, H.; Syed, T.A.; Qin, W. Fungal biodegradation and enzymatic modification of lignin. *Int. J. Biochem. Mol. Biol.* **2010**, *1*, 36–50. [PubMed]
49. Ruiz-Dueñas, F.J.; Martínez, Á.T. Microbial degradation of lignin: How a bulky recalcitrant polymer is efficiently recycled in nature and how we can take advantage of this. *Microb. Biotechnol.* **2009**, *2*, 164–177. [CrossRef] [PubMed]
50. Kubicek, C.P. (Ed.) *Fungi and Lignocellulosic Biomass*; John Wiley & Sons: Ames, IA, USA; Chichester, UK; Oxford, UK, 2012.
51. Sánchez, C. Lignocellulosic residues: Biodegradation and bioconversion by fungi. *Biotechnol. Adv.* **2009**, *27*, 185–194. [CrossRef] [PubMed]
52. Dashtban, M.; Schraft, H.; Qin, W. Fungal Bioconversion of Lignocellulosic Residues; Opportunities & Perspectives. *Int. J. Biol. Sci.* **2009**, *5*, 578–595. [PubMed]
53. Tien, M.; Kirk, T.K. Lignin-Degrading Enzyme from the Hymenomycete Phanerochaete chrysosporium Burds. *Science* **1983**, *221*, 661–663. [CrossRef] [PubMed]
54. Taylor, T.N.; Osborn, J.M. The importance of fungi in shaping the paleoecosystem. *Rev. Palaeobot. Palynol.* **1996**, *90*, 249–262. [CrossRef]
55. Bayer, E.A.; Chanzy, H.; Lamed, R.; Shoham, Y. Cellulose, cellulases and cellulosomes. *Curr. Opin. Struct. Biol.* **1998**, *8*, 548–557. [CrossRef]
56. Ljungdahl, L.G. The Cellulase/Hemicellulase System of the Anaerobic FungusOrpinomycesPC-2 and Aspects of Its Applied Use. *Ann. N. Y. Acad. Sci.* **2008**, *1125*, 308–321. [CrossRef] [PubMed]
57. Weng, J.-K.; Li, X.; Bonawitz, N.D.; Chapple, C. Emerging strategies of lignin engineering and degradation for cellulosic biofuel production. *Curr. Opin. Biotechnol.* **2008**, *19*, 166–172. [CrossRef] [PubMed]

58. Guerra, A.; Mendonça, R.; Ferraz, A.; Lu, F.; Ralph, J. Structural Characterization of Lignin during Pinus taeda Wood Treatment with Ceriporiopsis subvermispora. *Appl. Environ. Microbiol.* **2004**, *70*, 4073–4078. [CrossRef] [PubMed]
59. Arora, D.S.; Sharma, R.K. Enhancement in in vitro digestibility of wheat straw obtained from different geographical regions during solid state fermentation by white rot fungi. *BioResources* **2009**, *4*, 909–920.
60. Fackler, K.; Gradinger, C.; Hinterstoisser, B.; Messner, K.; Schwanninger, M. Lignin degradation by white rot fungi on spruce wood shavings during short-time solid-state fermentations monitored by near infrared spectroscopy. *Enzyme Microb. Technol.* **2006**, *39*, 1476–1483. [CrossRef]
61. Hildén, K.; Hakala, T.K.; Maijala, P.; Lundell, T.K.; Hatakka, A. Novel thermotolerant laccases produced by the white-rot fungus Physisporinus rivulosus. *Appl. Microbiol. Biotechnol.* **2007**, *77*, 301–309. [CrossRef] [PubMed]
62. Lozovaya, V.V.; Lygin, A.V.; Zernova, O.V.; Ulanov, A.V.; Li, S.; Hartman, G.L.; Widholm, J.M. Modification of phenolic metabolism in soybean hairy roots through down regulation of chalcone synthase or isoflavone synthase. *Planta* **2007**, *225*, 665–679. [CrossRef] [PubMed]
63. Ferm, R.; Kringstad, K.P.; Cowling, E.B. Formation of free radicals in milled wood lignin and syringaldehyde by phenol-oxidizing enzymes. *Sven. Papperstidn.* **1972**, *75*, 859–865.
64. Aresta, M.; Dibenedetto, A.; Dumeignil, F. *Biorefinery: From Biomass to Chemicals and Fuels*; Walter de Gruyter: Berlin, Germany; Boston, MA, USA, 2012.
65. Sannigrahi, P.; Pu, Y.; Ragauskas, A. Cellulosic biorefineries—Unleashing lignin opportunities. *Curr. Opin. Environ. Sustain.* **2010**, *2*, 383–393. [CrossRef]
66. Argyropoulos, D.S. (Ed.) *Materials, Chemicals and Energy from Forest Biomass*; ACS Symposium Series 954; American Chemical Society: Washington, DC, USA, 2007.
67. Scultz, T.P. *Lignin: Historical, Biological, and Materials Perspectives*; American Chemical Society: Washington, DC, USA; Cary, NC, USA, 2000.
68. Pometto, A.L.; Crawford, D.L. Simplified Procedure for Recovery of Lignin Acidolysis Products for Determining the Lignin-Degrading Abilities of Microorganisms. *Appl. Environ. Microbiol.* **1985**, *49*, 879–881. [PubMed]
69. Lapierre, C.; Monties, B.; Rolando, C. Preparative Thioacidolysis of Spruce Lignin: Isolation and Identification of Main Monomeric Products. *Holzforsch. Int. J. Biol. Chem. Phys. Technol. Wood* **2009**, *40*, 47–50. [CrossRef]
70. Khudoshin, A.G.; Mitrofanova, A.N.; Lunin, V.V. Lignin transformations and reactivity upon ozonation in aqueous media. *Russ. J. Phys. Chem. A* **2012**, *86*, 360–365. [CrossRef]
71. Sarkanen, K.V.; Islam, A.; Anderson, C.D. Ozonation. In *Methods in Lignin Chemistry*; Lin, S.Y., Dence, C.W., Eds.; Springer Series in Wood Science; Springer: Berlin/Heidelberg, Germany, 1992; pp. 387–406.
72. Chen, C.-L. Nitrobenzene and Cupric Oxide Oxidations. In *Methods in Lignin Chemistry*; Lin, S.Y., Dence, C.W., Eds.; Springer Series in Wood Science; Springer: Berlin/Heidelberg, Germany, 1992; pp. 301–321.

73. Rolando, C.; Monties, B.; Lapierre, C. Thioacidolysis. In *Methods in Lignin Chemistry*; Lin, S.Y., Dence, C.W., Eds.; Springer Series in Wood Science; Springer: Berlin/Heidelberg, Germany, 1992; pp. 334–349.
74. Aimi, H.; Matsumoto, Y.; Meshitsuka, G. Lignin fragments rich in detached side-chain structures found in water-soluble LCC. *J. Wood Sci.* **2005**, *51*, 252–255. [CrossRef]
75. Akiyama, T.; Sugimoto, T.; Matsumoto, Y.; Meshitsuka, G. Erythro/threo ratio of β-*O*-4 structures as an important structural characteristic of lignin. I: Improvement of ozonation method for the quantitative analysis of lignin side-chain structure. *J. Wood Sci.* **2002**, *48*, 210–215. [CrossRef]
76. Lu, F.; Ralph, J. DFRC Method for Lignin Analysis. 1. New Method for β-Aryl Ether Cleavage: Lignin Model Studies. *J. Agric. Food Chem.* **1997**, *45*, 4655–4660. [CrossRef]
77. Gellerstedt, G. Gellerstedt, G. Gel Permeation Chromatography. In *Methods in Lignin Chemistry*; Lin, S.Y., Dence, C.W., Eds.; Springer Series in Wood Science; Springer: Berlin/Heidelberg, Germany, 1992; pp. 487–497.
78. Faix, O. Fourier Transform Infrared Spectroscopy. In *Methods in Lignin Chemistry*; Lin, S.Y., Dence, C.W., Eds.; Springer Series in Wood Science; Springer: Berlin/Heidelberg, Germany, 1992; pp. 83–109.
79. Aulin-Erdtman, G. Spectrographic Contributions to Lignin Chemistry V. Phenolic groups in Spruce Lignin. *Sven. Papperstidn.* **1954**, *57*, 754–760.
80. Aulin-Erdtman, G.; Hegbom, L. Spectrographic Contributions to Lignin Chemistry VIII: Δε-Studies on Braun's Native Lignins from Coniferous Woods. *Sven. Papperstidn.* **1958**, *61*, 187–210.
81. El Mansouri, N.-E.; Salvadó, J. Analytical methods for determining functional groups in various technical lignins. *Ind. Crops Prod.* **2007**, *26*, 116–124. [CrossRef]
82. Landucci, L.L.; Ralph, S.A.; Hammel, K.E. 13C NMR Characterization of Guaiacyl, Guaiacyl/Syringyl and Syringyl Dehydrogenation Polymers. *Holzforsch. Int. J. Biol. Chem. Phys. Technol. Wood* **2009**, *52*, 160–170. [CrossRef]
83. Nimz, H.H.; Robert, D.; Faix, O.; Nemr, M. Carbon-13 NMR Spectra of Lignins, 8. Structural Differences between Lignins of Hardwoods, Softwoods, Grasses and Compression Wood. *Holzforsch. Int. J. Biol. Chem. Phys. Technol. Wood* **2009**, *35*, 16–26. [CrossRef]
84. Saake, B.; Argyropoulos, D.S.; Beinhoff, O.; Faix, O. A comparison of lignin polymer models (DHPs) and lignins by 31P NMR spectroscopy. *Phytochemistry* **1996**, *43*, 499–507. [CrossRef]
85. Hon, D.N.-S. Electron Spin Resonance (ESR) Spectroscopy. In *Methods in Lignin Chemistry*; Lin, S.Y., Dence, C.W., Eds.; Springer Series in Wood Science; Springer: Berlin/Heidelberg, Germany, 1992; pp. 274–286.
86. Sette, M.; Lange, H.; Crestini, C. Quantitative HSQC Analyses of Lignin: A Practical Comparison. *Comput. Struct. Biotechnol. J.* **2013**, *6*, 1–7. [CrossRef] [PubMed]
87. Peterson, D.J.; Loening, N.M. QQ-HSQC: A quick, quantitative heteronuclear correlation experiment for NMR spectroscopy. *Magn. Reson. Chem.* **2007**, *45*, 937–941. [CrossRef] [PubMed]

88. Hu, K.; Westler, W.M.; Markley, J.L. Simultaneous Quantification and Identification of Individual Chemicals in Metabolite Mixtures by Two-Dimensional Extrapolated Time-Zero 1H−13C HSQC (HSQC0). *J. Am. Chem. Soc.* **2011**, *133*, 1662–1665. [CrossRef] [PubMed]
89. Ralph, J.; Landucci, L. NMR of Lignins. In *Lignin Lignans*; Heitner, C., Dimmel, D., Schmidt, J., Eds.; CRC Press: Boca Raton, FL, USA; London, UK; New York, NY, USA, 2010; pp. 137–243.
90. Granata, A.; Argyropoulos, D.S. 2-Chloro-4,4,5,5-tetramethyl-1,3,2-dioxaphospholane, a Reagent for the Accurate Determination of the Uncondensed and Condensed Phenolic Moieties in Lignins. *J. Agric. Food Chem.* **1995**, *43*, 1538–1544. [CrossRef]
91. King, A.W.T.; Zoia, L.; Filpponen, I.; Olszewska, A.; Xie, H.; Kilpeläinen, I.; Argyropoulos, D.S. In Situ Determination of Lignin Phenolics and Wood Solubility in Imidazolium Chlorides Using 31P NMR. *J. Agric. Food Chem.* **2009**, *57*, 8236–8243. [CrossRef] [PubMed]
92. Argyropoulos, D.S.; Heitner, C. 31P NMR Spectroscopy in Wood Chemistry. Part VI. Solid State 31P NMR of Trimethyl Phosphite Derivatives of Chromophores and Carboxylic Acids Present in Mechanical Pulps; a Method for the Quantitative Determination of ortho-Quinones. *Holzforsch. Int. J. Biol. Chem. Phys. Technol. Wood* **2009**, *48*, 112–116.
93. Argyropoulos, D.S.; Zhang, L. Semiquantitative Determination of Quinonoid Structures in Isolated Lignins by 31P Nuclear Magnetic Resonance. *J. Agric. Food Chem.* **1998**, *46*, 4628–4634. [CrossRef]
94. Argyropoulos, D.S.; Li, H.; Gaspar, A.R.; Smith, K.; Lucia, L.A.; Rojas, O.J. Quantitative 31P NMR detection of oxygen-centered and carbon-centered radical species. *Bioorg. Med. Chem.* **2006**, *14*, 4017–4028. [CrossRef] [PubMed]
95. Zoia, L.; Argyropoulos, D.S. Characterization of free radical spin adducts of the DIPPMPO using mass spectrometry and (31)P NMR. *Eur. J. Mass Spectrom. Chichester Engl.* **2010**, *16*, 175–185. [CrossRef] [PubMed]
96. Ahvazi, B.C.; Crestini, C.; Argyropoulos, D.S. 19F Nuclear Magnetic Resonance Spectroscopy for the Quantitative Detection and Classification of Carbonyl Groups in Lignins. *J. Agric. Food Chem.* **1999**, *47*, 190–201. [CrossRef] [PubMed]
97. Ahvazi, B.C.; Argyropoulos, D.S. 19F Nuclear Magnetic Resonance Spectroscopy for the Elucidation of Carbonyl Groups in Lignins. 1. Model Compounds. *J. Agric. Food Chem.* **1996**, *44*, 2167–2175. [CrossRef]
98. Herzog, H.; Burba, P.; Buddrus, J. Quantification of hydroxylic groups in a river humic substance by ^{29}Si-NMR. *Fresenius J. Anal. Chem.* **1996**, *354*, 375–377.
99. Neirinck, V.; Robert, D.; Nardin, R. Use of indirect ^{199}Hg NMR detection for aromatic mercury compounds of biological interest. *Magn. Reson. Chem.* **1993**, *31*, 815–822. [CrossRef]
100. Potthast, A.; Schiene, R.; Fischer, K. Structural Investigations of *N*-modified Lignins by ^{15}N-NMR Spectroscopy and Possible Pathways for Formation of Nitrogen Containing Compounds Related to Lignin. *Holzforsch. Int. J. Biol. Chem. Phys. Technol. Wood* **2009**, *50*, 554–562.

101. Pla, F. Vapor Pressure Osmometry. In *Methods in Lignin Chemistry*; Lin, S.Y., Dence, C.W., Eds.; Springer Series in Wood Science; Springer: Berlin/Heidelberg, Germany, 1992; pp. 509–517.
102. Gross, S.K.; Sarkanen, K.; Schuerch, C. Determinations of Molecular Weight of Lignin Degradation Products by Three Methods. *Anal. Chem.* **1958**, *30*, 518–521. [CrossRef]
103. Obiaga, T.I.; Wayman, M. Improved calibration procedure for gel permeation chromatography of lignins. *J. Appl. Polym. Sci.* **1974**, *18*, 1943–1952. [CrossRef]
104. Chen, F.; Li, J. Aqueous Gel Permeation Chromatographic Methods for Technical Lignins. *J. Wood Chem. Technol.* **2000**, *20*, 265–276. [CrossRef]
105. Salanti, A.; Zoia, L.; Tolppa, E.-L.; Orlandi, M. Chromatographic Detection of Lignin–Carbohydrate Complexes in Annual Plants by Derivatization in Ionic Liquid. *Biomacromolecules* **2012**, *13*, 445–454. [CrossRef] [PubMed]
106. Lange, H.; Rulli, F.; Crestini, C. Gel Permeation Chromatography in Determining Molecular Weights of Lignins: Critical Aspects Revisited for Improved Utility in the Development of Novel Materials. *ACS Sustain. Chem. Eng.* **2016**, *4*, 5167–5180. [CrossRef]
107. Asikkala, J.; Tamminen, T.; Argyropoulos, D.S. Accurate and Reproducible Determination of Lignin Molar Mass by Acetobromination. *J. Agric. Food Chem.* **2012**, *60*, 8968–8973. [CrossRef] [PubMed]
108. Ringena, O.; Lebioda, S.; Lehnen, R.; Saake, B. Size-exclusion chromatography of technical lignins in dimethyl sulfoxide/water and dimethylacetamide. *J. Chromatogr. A* **2006**, *1102*, 154–163. [CrossRef] [PubMed]
109. Gidh, A.V.; Decker, S.R.; Vinzant, T.B.; Himmel, M.E.; Williford, C. Determination of lignin by size exclusion chromatography using multi angle laser light scattering. *J. Chromatogr. A* **2006**, *1114*, 102–110. [CrossRef] [PubMed]
110. Gidh, A.V.; Decker, S.R.; See, C.H.; Himmel, M.E.; Williford, C.W. Characterization of lignin using multi-angle laser light scattering and atomic force microscopy. *Anal. Chim. Acta* **2006**, *555*, 250–258. [CrossRef]
111. Fredheim, G.E.; Braaten, S.M.; Christensen, B.E. Molecular weight determination of lignosulfonates by size-exclusion chromatography and multi-angle laser light scattering. *J. Chromatogr. A* **2002**, *942*, 191–199. [CrossRef]
112. Braaten, S.M.; Christensen, B.E.; Fredheim, G.E. Comparison of Molecular Weight and Molecular Weight Distributions of Softwood and Hardwood Lignosulfonates. *J. Wood Chem. Technol.* **2003**, *23*, 197–215. [CrossRef]
113. Cathala, B.; Saake, B.; Faix, O.; Monties, B. Association behaviour of lignins and lignin model compounds studied by multidetector size-exclusion chromatography. *J. Chromatogr. A* **2003**, *1020*, 229–239. [CrossRef] [PubMed]
114. Richel, A.; Vanderghem, C.; Simon, M.; Wathelet, B.; Paquot, M. Evaluation of Matrix-Assisted Laser Desorption/Ionization Mass Spectrometry for Second-Generation Lignin Analysis. *Anal. Chem. Insights* **2012**, *7*, 79. [CrossRef] [PubMed]
115. Lora, J.H.; Glasser, W.G. Recent Industrial Applications of Lignin: A Sustainable Alternative to Nonrenewable Materials. *J. Polym. Environ.* **2002**, *10*, 39–48. [CrossRef]

116. Li, H.; McDonald, A.G. Fractionation and characterization of industrial lignins. *Ind. Crops Prod.* **2014**, *62*, 67–76. [CrossRef]
117. Thring, R.W.; Vanderlaan, M.N.; Griffin, S.L. Fractionation of Alcell lignin by sequential solvent extraction. *J. Wood Chem. Technol.* **1996**, *16*, 139–154. [CrossRef]
118. Mörck, R.; Reimann, A.; Kringstad, K.P. Fractionation of Kraft Lignin by Successive Extraction with Organic Solvents. III. Fractionation of Kraft Lignin from Birch. *Holzforsch. Int. J. Biol. Chem. Phys. Technol. Wood* **2009**, *42*, 111–116. [CrossRef]
119. Wang, K.; Xu, F.; Sun, R. Molecular Characteristics of Kraft-AQ Pulping Lignin Fractionated by Sequential Organic Solvent Extraction. *Int. J. Mol. Sci.* **2010**, *11*, 2988–3001. [CrossRef] [PubMed]
120. Sun, R.; Tomkinson, J.; Bolton, J. Effects of precipitation pH on the physico-chemical properties of the lignins isolated from the black liquor of oil palm empty fruit bunch fibre pulping. *Polym. Degrad. Stab.* **1999**, *63*, 195–200. [CrossRef]
121. García, A.; Toledano, A.; Serrano, L.; Egüés, I.; González, M.; Marín, F.; Labidi, J. Characterization of lignins obtained by selective precipitation. *Sep. Purif. Technol.* **2009**, *68*, 193–198. [CrossRef]
122. Cui, C.; Sun, R.; Argyropoulos, D.S. Fractional Precipitation of Softwood Kraft Lignin: Isolation of Narrow Fractions Common to a Variety of Lignins. *ACS Sustain. Chem. Eng.* **2014**, *2*, 959–968. [CrossRef]
123. Wallberg, O.; Jönsson, A.-S. Influence of the Membrane Cut-off during Ultrafiltration of Kraft Black Liquor with Ceramic Membranes. *Chem. Eng. Res. Des.* **2003**, *81*, 1379–1384. [CrossRef]
124. Keyoumu, A.; Sjödahl, R.; Henriksson, G.; Ek, M.; Gellerstedt, G.; Lindström, M.E. Continuous nano- and ultra-filtration of kraft pulping black liquor with ceramic filters: A method for lowering the load on the recovery boiler while generating valuable side-products. *Ind. Crops Prod.* **2004**, *20*, 143–150. [CrossRef]
125. Jönsson, A.-S.; Nordin, A.-K.; Wallberg, O. Concentration and purification of lignin in hardwood kraft pulping liquor by ultrafiltration and nanofiltration. *Chem. Eng. Res. Des.* **2008**, *86*, 1271–1280. [CrossRef]
126. Toledano, A.; Serrano, L.; Garcia, A.; Mondragon, I.; Labidi, J. Comparative study of lignin fractionation by ultrafiltration and selective precipitation. *Chem. Eng. J.* **2010**, *157*, 93–99. [CrossRef]
127. Helander, M.; Theliander, H.; Lawoko, M.; Henriksson, G.; Zhang, L.; Lindstrom, M.E. Fractionation of technical lignin: Molecular mass and pH effects. *BioResources* **2013**, *8*, 2270–2282. [CrossRef]
128. Sevastyanova, O.; Helander, M.; Chowdhury, S.; Lange, H.; Wedin, H.; Zhang, L.; Ek, M.; Kadla, J.F.; Crestini, C.; Lindström, M.E. Tailoring the molecular and thermo–mechanical properties of kraft lignin by ultrafiltration. *J. Appl. Polym. Sci.* **2014**, *131*, e40799. [CrossRef]
129. Hilburg, S.L.; Elder, A.N.; Chung, H.; Ferebee, R.L.; Bockstaller, M.R.; Washburn, N.R. A universal route towards thermoplastic lignin composites with improved mechanical properties. *Polymer* **2014**, *55*, 995–1003. [CrossRef]

130. Sadeghifar, H.; Cui, C.; Argyropoulos, D.S. Toward Thermoplastic Lignin Polymers. Part 1. Selective Masking of Phenolic Hydroxyl Groups in Kraft Lignins via Methylation and Oxypropylation Chemistries. *Ind. Eng. Chem. Res.* **2012**, *51*, 16713–16720. [CrossRef]
131. Cui, C.; Sadeghifar, H.; Sen, S.; Argyropoulos, D.S. Toward Thermoplastic Lignin Polymers; Part II: Thermal & Polymer Characteristics of Kraft Lignin & Derivatives. *BioResources* **2013**, *8*, 864–886.
132. Argyropoulos, D.S. Towards Thermoplastic Lignin Polymers: Progress in the Utilization of Kraft Lignin for the Synthesis of Heat Stable Polymer Melts. *J. Biotechnol. Biomater.* **2013**, *3*, 123. [CrossRef]
133. Crestini, C.; Crucianelli, M.; Orlandi, M.; Saladino, R. Oxidative strategies in lignin chemistry: A new environmental friendly approach for the functionalisation of lignin and lignocellulosic fibers. *Catal. Today* **2010**, *156*, 8–22. [CrossRef]
134. Rahimi, A.; Azarpira, A.; Kim, H.; Ralph, J.; Stahl, S.S. Chemoselective Metal-Free Aerobic Alcohol Oxidation in Lignin. *J. Am. Chem. Soc.* **2013**, *135*, 6415–6418. [CrossRef] [PubMed]
135. Nguyen, J.D.; Matsuura, B.S.; Stephenson, C.R.J. A Photochemical Strategy for Lignin Degradation at Room Temperature. *J. Am. Chem. Soc.* **2014**, *136*, 1218–1221. [CrossRef] [PubMed]
136. Genin, H.S.; Lawler, K.A.; Hoffmann, R.; Herrmann, W.A.; Fischer, R.W.; Scherer, W. Multiple Bonds between Main-Group Elements and Transition Metals. 138. Polymeric Methyltrioxorhenium: Some Models for Its Electronic Structure. *J. Am. Chem. Soc.* **1995**, *117*, 3244–3252. [CrossRef]
137. Hermann, W.A.; Fischer, R.W. Multiple Bonds between Main-Group Elements and Transition Metals. 136. "Polymerization" of an Organometal Oxide: The Unusual Behavior of Methyltrioxorhenium (VII) in Water. *J. Am. Chem. Soc.* **1995**, *117*, 3223–3230. [CrossRef] [PubMed]
138. Herrmann, W.A.; Fischer, R.W.; Scherer, W.; Rauch, M.U. Methyltrioxorhenium (VII) as Catalyst for Epoxidations: Structure of the Active Species and Mechanism of Catalysis. *Angew. Chem. Int. Ed. Engl.* **1993**, *32*, 1157–1160. [CrossRef]
139. Crestini, C.; Pro, P.; Neri, V.; Saladino, R. Methyltrioxorhenium: A new catalyst for the activation of hydrogen peroxide to the oxidation of lignin and lignin model compounds. *Bioorg. Med. Chem.* **2005**, *13*, 2569–2578. [CrossRef] [PubMed]
140. Argyropoulos, D.S.; Sun, Y.; Berry, R.M.; Bouchard, J. Reaction of dimethyldioxirane with lignin model compounds. *J. Pulp Pap. Sci.* **1996**, *22*, 84–90.
141. Bianchini, G.; Crucianelli, M.; Crestini, C.; Saladino, R. Catalytic MTO-based C–H insertion reactions of hydrogen peroxide: An investigation on the polymeric support role in heterogeneous conditions. *Top. Catal.* **2006**, *40*, 221–227. [CrossRef]
142. Haikarainen, A.; Sipilä, J.; Pietikäinen, P.; Pajunen, A.; Mutikainen, I. Synthesis and characterization of bulky salen-type complexes of Co, Cu, Fe, Mn and Ni with amphiphilic solubility properties. *J. Chem. Soc. Dalton Trans.* **2001**, 991–995. [CrossRef]

143. Zanini, S.; Canevali, C.; Orlandi, M.; Tolppa, E.-L.; Zoia, L.; Riccardi, C.; Morazzoni, F. Radical formation on CTMP fibers by argon plasma treatments and related lignin chemical changes. *BioResources* **2008**, *3*, 995–1006.
144. Bolzacchini, E.; Chiavetto, L.B.; Canevali, C.; Morazzoni, F.; Orlandi, M.; Rindone, B. Oxidation of propenoidic phenols catalysed by *N,N′*-ethylenebis(salicylideneiminato)cobalt(II) [Cosalen]: Reactivity and spectroscopic studies. *J. Mol. Catal. Chem.* **1996**, *112*, 347–351. [CrossRef]
145. Bolzacchini, E.; Canevali, C.; Morazzoni, F.; Orlandi, M.; Rindone, B.; Scotti, R. Spectromagnetic investigation of the active species in the oxidation of propenoidic phenols catalysed by [*N,N′*-bis(salicylidene)ethane-1,2-diaminato]cobalt(II). *J. Chem. Soc. Dalton Trans.* **1997**, *24*, 4695–4700. [CrossRef]
146. Canevali, C.; Orlandi, M.; Pardi, L.; Rindone, B.; Scotti, R.; Sipila, J.; Morazzoni, F. Oxidative degradation of monomeric and dimeric phenylpropanoids: Reactivity and mechanistic investigation. *J. Chem. Soc. Dalton Trans.* **2002**, *15*, 3007–3014. [CrossRef]
147. Biannic, B.; Bozell, J.J. Efficient Cobalt-Catalyzed Oxidative Conversion of Lignin Models to Benzoquinones. *Org. Lett.* **2013**, *15*, 2730–2733. [CrossRef] [PubMed]
148. Bujanovic, B.; Reiner, R.S.; Ralph, S.A.; Atalla, R.H. Polyoxometalate Delignification of Birch Kraft Pulp and Effect on Residual Lignin. *J. Wood Chem. Technol.* **2011**, *31*, 121–141. [CrossRef]
149. Bujanovic, B.; Ralph, S.; Reiner, R.; Hirth, K.; Atalla, R. Polyoxometalates in Oxidative Delignification of Chemical Pulps: Effect on Lignin. *Materials* **2010**, *3*, 1888–1903. [CrossRef]
150. Tonucci, L.; Coccia, F.; Bressan, M.; d'Alessandro, N. Mild Photocatalysed and Catalysed Green Oxidation of Lignin: A Useful Pathway to Low-Molecular-Weight Derivatives. *Waste Biomass Valorization* **2012**, *3*, 165–174. [CrossRef]
151. Lv, H.; Geletii, Y.V.; Zhao, C.; Vickers, J.W.; Zhu, G.; Luo, Z.; Song, J.; Lian, T.; Musaev, D.G.; Hill, C.L. Polyoxometalate water oxidation catalysts and the production of green fuel. *Chem. Soc. Rev.* **2012**, *41*, 7572–7589. [CrossRef] [PubMed]
152. Grigoriev, V.A.; Hill, C.L.; Weinstock, I.A. Polyoxometalate Oxidation of Phenolic Lignin Models. In *Oxidative Delignification Chemistry*; Argyropoulos, D.S., Ed.; ACS Symposium Series; American Chemical Society: Washington, DC, USA, 2001; Volume 785, pp. 297–312.
153. Hanson, S.K.; Wu, R.; Silks, L.A. C-C or C-O Bond Cleavage in a Phenolic Lignin Model Compound: Selectivity Depends on Vanadium Catalyst. *Angew. Chem. Int. Ed.* **2012**, *51*, 3410–3413. [CrossRef] [PubMed]
154. Son, S.; Toste, F.D. Non-Oxidative Vanadium-Catalyzed C-O Bond Cleavage: Application to Degradation of Lignin Model Compounds. *Angew. Chem. Int. Ed.* **2010**, *49*, 3791–3794. [CrossRef] [PubMed]
155. Ghiaci, M.; Molaie, F.; Sedaghat, M.E.; Dorostkar, N. Metalloporphyrin covalently bound to silica. Preparation, characterization and catalytic activity in oxidation of ethyl benzene. *Catal. Commun.* **2010**, *11*, 694–699. [CrossRef]
156. Shimada, M.; Habe, T.; Higuchi, T.; Okamoto, T.; Dorostkar, N. Biomimetic approach to lignin degradation. *Holzforschung* **1977**, *41*, 277–285. [CrossRef]

157. Pattou, D.; Labat, G.; Defrance, S.; Seris, J.L.; Meunier, B. Preparation, catalytic activity of supported metalloporphyrin catalysts on poly(vinylpyridine) polymers and related molecular modeling studies (case of a proximal ligand provided by the polymer itself). *Bull. Soc. Chim. Fr.* **1994**, *131*, 78–88.
158. Crestini, C.; Saladino, R.; Tagliatesta, P.; Boschi, T. Biomimetic degradation of lignin and lignin model compounds by synthetic anionic and cationic water soluble manganese and iron porphyrins. *Bioorg. Med. Chem.* **1999**, *7*, 1897–1905. [CrossRef]
159. Crestini, C.; Pastorini, A.; Tagliatesta, P. Metalloporphyrins immobilized on motmorillonite as biomimetic catalysts in the oxidation of lignin model compounds. *J. Mol. Catal. Chem.* **2004**, *208*, 195–202. [CrossRef]
160. Mendonça Maciel, M.J.; Castro e Silva, A.; Telles Ribeiro, H.C. Industrial and biotechnological applications of ligninolytic enzymes of the basidiomycota: A review. *Electron. J. Biotechnol.* **2010**, *13*, 1–13. [CrossRef]
161. Caldwell, E.S.; Steelink, C. Phenoxy radical intermediate in the enzymatic degradation of lignin model compounds. *Biochim. Biophys. Acta BBA Gen. Subj.* **1969**, *184*, 420–431. [CrossRef]
162. Lundquist, K.; Kristersson, P. Exhaustive laccase-catalysed oxidation of a lignin model compound (vanillyl glycol) produces methanol and polymeric quinoid products. *Biochem. J.* **1985**, *229*, 277–279. [CrossRef] [PubMed]
163. Higuchi, P.T. Lignin biochemistry: Biosynthesis and biodegradation. *Wood Sci. Technol.* **1990**, *24*, 23–63. [CrossRef]
164. Shleev, S.; Christenson, A.; Serezhenkov, V.; Burbaev, D.; Yaropolov, A.; Gorton, L.; Ruzgas, T. Electrochemical redox transformations of T1 and T2 copper sites in native Trametes hirsuta laccase at gold electrode. *Biochem. J.* **2005**, *385*, 745–754. [CrossRef] [PubMed]
165. Call, H.P.; Mücke, I. History, overview and applications of mediated lignolytic systems, especially laccase-mediator-systems (Lignozym®-process). *J. Biotechnol.* **1997**, *53*, 163–202. [CrossRef]
166. Poppius-Levlin, K.; Tamminen, T.; Kalliola, A.; Ohra-aho, T. Characterization of Residual Lignins in Pulps Delignified by Laccase/N-Hydroxyacetanilide. In *Oxidative Delignification Chemistry*; Argyropoulos, D.S., Ed.; ACS Symposium Series; American Chemical Society: Washington, DC, USA, 2001; Volume 785, pp. 358–372.
167. Messerschmidt, A.; Huber, R. The blue oxidases, ascorbate oxidase, laccase and ceruloplasmin Modelling and structural relationships. *Eur. J. Biochem.* **1990**, *187*, 341–352. [CrossRef] [PubMed]
168. Johannes, C.; Majcherczyk, A. Natural Mediators in the Oxidation of Polycyclic Aromatic Hydrocarbons by Laccase Mediator Systems. *Appl. Environ. Microbiol.* **2000**, *66*, 524–528. [CrossRef] [PubMed]
169. Fabbrini, M.; Galli, C.; Gentili, P. Comparing the catalytic efficiency of some mediators of laccase. *J. Mol. Catal. B Enzym.* **2002**, *16*, 231–240. [CrossRef]
170. Paice, M.G.; Bourbonnaise, R.; Reid, I.D.; Archibald, F.S.; Jurasek, L. Oxidative bleaching enzymes: A review. *J. Pulp Pap. Sci.* **1995**, *21*, J280–J284.

171. Wariishi, H.; Akileswaran, L.; Gold, M.H. Manganese peroxidase from the basidiomycete Phanerochaete chrysosporium: Spectral characterization of the oxidized states and the catalytic cycle. *Biochemistry* **1988**, *27*, 5365–5370. [CrossRef] [PubMed]
172. Wariishi, H.; Valli, K.; Gold, M.H. Oxidative cleavage of a phenolic diarylpropane lignin model dimer by manganese peroxidase from Phanerochaete chrysosporium. *Biochemistry* **1989**, *28*, 6017–6023. [CrossRef]
173. Tuor, U.; Wariishi, H.; Schoemaker, H.E.; Gold, M.H. Oxidation of phenolic arylglycerol.beta.-aryl ether lignin model compounds by manganese peroxidase from Phanerochaete chrysosporium: Oxidative cleavage of an.alpha.-carbonyl model compound. *Biochemistry* **1992**, *31*, 4986–4995. [CrossRef] [PubMed]
174. Rodríguez Couto, S.; Osma, J.F.; Saravia, V.; Gübitz, G.M.; Toca Herrera, J.L. Coating of immobilised laccase for stability enhancement: A novel approach. *Appl. Catal. Gen.* **2007**, *329*, 156–160. [CrossRef]
175. Krajewska, B. Application of chitin- and chitosan-based materials for enzyme immobilizations: A review. *Enzyme Microb. Technol.* **2004**, *35*, 126–139. [CrossRef]
176. Cho, Y.K.; Bailey, J.E. Immobilization of enzymes on activated carbon: Selection and preparation of the carbon support. *Biotechnol. Bioeng.* **1979**, *21*, 461–476. [CrossRef] [PubMed]
177. Abadulla, E.; Tzanov, T.; Costa, S.; Robra, K.-H.; Cavaco-Paulo, A.; Gübitz, G.M. Decolorization and Detoxification of Textile Dyes with a Laccase from Trametes hirsuta. *Appl. Environ. Microbiol.* **2000**, *66*, 3357–3362. [CrossRef] [PubMed]
178. Ryan, S.; Schnitzhofer, W.; Tzanov, T.; Cavaco-Paulo, A.; Gübitz, G.M. An acid-stable laccase from Sclerotium rolfsii with potential for wool dye decolourization. *Enzyme Microb. Technol.* **2003**, *33*, 766–774. [CrossRef]
179. Kandelbauer, A.; Maute, O.; Kessler, R.W.; Erlacher, A.; Gübitz, G.M. Study of dye decolorization in an immobilized laccase enzyme-reactor using online spectroscopy. *Biotechnol. Bioeng.* **2004**, *87*, 552–563. [CrossRef] [PubMed]
180. Held, C.; Kandelbauer, A.; Schroeder, M.; Cavaco-Paulo, A.; Guebitz, G.M. Biotransformation of phenolics with laccase containing bacterial spores. *Environ. Chem. Lett.* **2005**, *3*, 74–77. [CrossRef]
181. Rochefort, D.; Kouisni, L.; Gendron, K. Physical immobilization of laccase on an electrode by means of poly(ethyleneimine) microcapsules. *J. Electroanal. Chem.* **2008**, *617*, 53–63. [CrossRef]
182. Hébert, M.; Rochefort, D. Investigation of Microencapsulated Laccase as Enzyme Immobilization Template for Application in Bio-fuel Cell. *ECS Trans.* **2008**, *16*, 85–97.
183. Crestini, C.; Perazzini, R.; Saladino, R. Oxidative functionalisation of lignin by layer-by-layer immobilised laccases and laccase microcapsules. *Appl. Catal. Gen.* **2010**, *372*, 115–123. [CrossRef]
184. Guazzaroni, M.; Crestini, C.; Saladino, R. Layer-by-Layer coated tyrosinase: An efficient and selective synthesis of catechols. *Bioorg. Med. Chem.* **2012**, *20*, 157–166. [CrossRef] [PubMed]

185. Perazzini, R.; Saladino, R.; Guazzaroni, M.; Crestini, C. A novel and efficient oxidative functionalization of lignin by layer-by-layer immobilised Horseradish peroxidase. *Bioorg. Med. Chem.* **2011**, *19*, 440–447. [CrossRef] [PubMed]
186. Crestini, C.; Melone, F.; Saladino, R. Novel multienzyme oxidative biocatalyst for lignin bioprocessing. *Bioorg. Med. Chem.* **2011**, *19*, 5071–5078. [CrossRef] [PubMed]
187. Lloyd, J.A.; Ralph, J. *Hydrogenation and Hydrogenolysis of Wood Lignin: A Review*; Forest Products Division, Forest Research Institute: Rotorua, New Zealand, 1977.
188. Eachus, S.W.; Dence, C.W. Hydrogenation of Lignin Model Compounds in the Presence of a Homogeneous Catalyst. *Holzforsch. Int. J. Biol. Chem. Phys. Technol. Wood* **2009**, *29*, 41–48. [CrossRef]
189. Zhang, J.; Chen, Y.; Brook, M.A. Reductive Degradation of Lignin and Model Compounds by Hydrosilanes. *ACS Sustain. Chem. Eng.* **2014**, *2*, 1983–1991. [CrossRef]
190. Zakzeski, J.; Jongerius, A.L.; Bruijnincx, P.C.A.; Weckhuysen, B.M. Catalytic Lignin Valorization Process for the Production of Aromatic Chemicals and Hydrogen. *ChemSusChem* **2012**, *5*, 1602–1609. [CrossRef] [PubMed]
191. Parsell, T.H.; Owen, B.C.; Klein, I.; Jarrell, T.M.; Marcum, C.L.; Haupert, L.J.; Amundson, L.M.; Kenttämaa, H.I.; Ribeiro, F.; Miller, J.T.; et al. Cleavage and hydrodeoxygenation (HDO) of C–O bonds relevant to lignin conversion using Pd/Zn synergistic catalysis. *Chem. Sci.* **2013**, *4*, 806–813. [CrossRef]
192. Sergeev, A.G.; Webb, J.D.; Hartwig, J.F. A Heterogeneous Nickel Catalyst for the Hydrogenolysis of Aryl Ethers without Arene Hydrogenation. *J. Am. Chem. Soc.* **2012**, *134*, 20226–20229. [CrossRef] [PubMed]
193. Wang, X.; Rinaldi, R. A Route for Lignin and Bio-Oil Conversion: Dehydroxylation of Phenols into Arenes by Catalytic Tandem Reactions. *Angew. Chem. Int. Ed.* **2013**, *52*, 11499–11503. [CrossRef] [PubMed]
194. Sergeev, A.G.; Hartwig, J.F. Selective, Nickel-Catalyzed Hydrogenolysis of Aryl Ethers. *Science* **2011**, *332*, 439–443. [CrossRef] [PubMed]
195. Chan, J.M.W.; Bauer, S.; Sorek, H.; Sreekumar, S.; Wang, K.; Toste, F.D. Studies on the Vanadium-Catalyzed Nonoxidative Depolymerization of Miscanthus giganteus-Derived Lignin. *ACS Catal.* **2013**, *3*, 1369–1377. [CrossRef]
196. Duval, A.; Lawoko, M. A review on lignin-based polymeric, micro- and nano-structured materials. *React. Funct. Polym.* **2014**, *85*, 78–96. [CrossRef]
197. Lebo, S.E.; Gargulak, J.D.; McNally, T.J. Lignin. In *Kirk-Othmer Encyclopedia of Chemical Technology*; John Wiley & Sons, Inc.: New York, NY, USA, 2000.
198. Calvo-Flores, F.G.; Dobado, J.A. Lignin as Renewable Raw Material. *ChemSusChem* **2010**, *3*, 1227–1235. [CrossRef] [PubMed]
199. Bonini, C.; D'Auria, M.; Emanuele, L.; Ferri, R.; Pucciariello, R.; Sabia, A.R. Polyurethanes and polyesters from lignin. *J. Appl. Polym. Sci.* **2005**, *98*, 1451–1456. [CrossRef]
200. Gandini, A.; Belgacem, M.N.; Guo, Z.-X.; Montanari, S. Lignins as Macromonomers for Polyesters and Polyurethanes. In *Chemical Modification, Properties, and Usage of Lignin*; Hu, T.Q., Ed.; Springer: New York, NY, USA, 2002; pp. 57–80.

201. Thielemans, W.; Can, E.; Morye, S.S.; Wool, R.P. Novel applications of lignin in composite materials. *J. Appl. Polym. Sci.* **2002**, *83*, 323–331. [CrossRef]
202. Saito, T.; Perkins, J.H.; Jackson, D.C.; Trammel, N.E.; Hunt, M.A.; Naskar, A.K. Development of lignin-based polyurethane thermoplastics. *RSC Adv.* **2013**, *3*, 21832–21840. [CrossRef]
203. Cateto, C.A.; Barreiro, M.F.; Ottati, C.; Lopretti, M.; Rodrigues, A.E.; Belgacem, M.N. Lignin-based rigid polyurethane foams with improved biodegradation. *J. Cell. Plast.* **2014**, *50*, 81–95. [CrossRef]
204. Cateto, C.A.; Barreiro, M.F.; Rodrigues, A.E.; Belgacem, M.N. Kinetic study of the formation of lignin-based polyurethanes in bulk. *React. Funct. Polym.* **2011**, *71*, 863–869. [CrossRef]
205. Cheradame, H.; Detoisien, M.; Gandini, A.; Pla, F.; Roux, G. Polyurethane from kraft lignin. *Br. Polym. J.* **1989**, *21*, 269–275. [CrossRef]
206. Vilela, C.; Sousa, A.F.; Fonseca, A.C.; Serra, A.C.; Coelho, J.F.J.; Freire, C.S.R.; Silvestre, A.J.D. The quest for sustainable polyesters—Insights into the future. *Polym. Chem.* **2014**, *5*, 3119–3141. [CrossRef]
207. Cateto, C.A.; Barreiro, M.F.; Rodrigues, A.E.; Belgacem, M.N. Optimization Study of Lignin Oxypropylation in View of the Preparation of Polyurethane Rigid Foams. *Ind. Eng. Chem. Res.* **2009**, *48*, 2583–2589. [CrossRef]
208. Nadji, H.; Bruzzèse, C.; Belgacem, M.N.; Benaboura, A.; Gandini, A. Oxypropylation of Lignins and Preparation of Rigid Polyurethane Foams from the Ensuing Polyols. *Macromol. Mater. Eng.* **2005**, *290*, 1009–1016. [CrossRef]
209. Feldman, D. Lignin and Its Polyblends—A Review. In *Chemical Modification, Properties, and Usage of Lignin*; Hu, T.Q., Ed.; Springer: New York, NY, USA, 2002; pp. 81–99.
210. Abdelwahab, N.A.; Nassar, M.A. Preparation, optimisation and characterisation of lignin phenol formaldehyde resin as wood adhesive. *Pigment Resin Technol.* **2011**, *40*, 169–174. [CrossRef]
211. Feldman, D.; Banu, D.; Luchian, C.; Wang, J. Epoxy–lignin polyblends: Correlation between polymer interaction and curing temperature. *J. Appl. Polym. Sci.* **1991**, *42*, 1307–1318. [CrossRef]
212. Feldman, D.; Banu, D.; Khoury, M. Epoxy–lignin polyblends. III. Thermal properties and infrared analysis. *J. Appl. Polym. Sci.* **1989**, *37*, 877–887. [CrossRef]
213. Cazacu, G.; Pascu, M.C.; Profire, L.; Kowarski, A.I.; Mihaes, M.; Vasile, C. Lignin role in a complex polyolefin blend. *Ind. Crops Prod.* **2004**, *20*, 261–273. [CrossRef]
214. Kharade, A.Y.; Kale, D.D. Lignin-filled polyolefins. *J. Appl. Polym. Sci.* **1999**, *72*, 1321–1326. [CrossRef]
215. Košíková, B.; Kačuráková, M.; Demianová, V. Photooxidation of the Composite Lignin/Polypropylene Films. *Chem. Pap.* **1993**, *47*, 132–136.
216. Casenave, S.; Aït-Kadi, A.; Riedl, B. Mechanical behaviour of highly filled lignin/polyethylene composites made by catalytic grafting. *Can. J. Chem. Eng.* **1996**, *74*, 308–315. [CrossRef]

217. Hofmann, G.H. Polymer Blend Modification of PVC. In *Polymer Blends and Mixtures*; Walsh, D.J., Higgins, J.S., Maconnachie, A., Eds.; NATO ASI Series 89; Springer: Dordrecht, The Netherlands, 1985; pp. 117–148.
218. Feldman, D.; Banu, D.; El-Raghi, S. Poly(Vinyl Chloride)-Lignin Blends for Outdoor Applicationin Building. *J. Macromol. Sci. Part A* **1994**, *31*, 555–571. [CrossRef]
219. Feldman, D.; Banu, D. Contribution to the study of rigid PVC polyblends with different lignins. *J. Appl. Polym. Sci.* **1997**, *66*, 1731–1744. [CrossRef]
220. Nägele, H.; Pfitzer, J.; Ziegler, L.; Inone-Kauffmann, E.R.; Eckl, W.; Eisenreich, N. Lignin Matrix Composites from Natural Resources—ARBOFORM®. In *Bio-Based Plastics*; Kabasci, S:, Ed.; John Wiley & Sons Ltd.: Chichester, UK, 2013; pp. 89–115.Resources—ARBOFORM®. In *Bio-Based Plastics*; Kabasci, S., Ed.; John Wiley & Sons Ltd.: Chichester, UK, 2013; pp. 89–115.
221. Otani, S.; Fukuoka, Y.; Igarashi, B.; Sasaki, K. Method for Producing Carbonized Lignin Fiber. U.S. Patent 3,461,082 A, 12 August 1969.
222. Chatterjee, S.; Jones, E.B.; Clingenpeel, A.C.; McKenna, A.M.; Rios, O.; McNutt, N.W.; Keffer, D.J.; Johs, A. Conversion of Lignin Precursors to Carbon Fibers with Nanoscale Graphitic Domains. *ACS Sustain. Chem. Eng.* **2014**, *2*, 2002–2010. [CrossRef]
223. Kadla, J.F.; Kubo, S.; Gilbert, R.D.; Venditti, R.A. Lignin-Based Carbon Fibers. In *Chemical Modification, Properties, and Usage of Lignin*; Hu, T.Q., Ed.; Springer: New York, NY, USA, 2002; pp. 121–137.
224. Dallmeyer, I.; Ko, F.; Kadla, J.F. Electrospinning of Technical Lignins for the Production of Fibrous Networks. *J. Wood Chem. Technol.* **2010**, *30*, 315–329. [CrossRef]
225. Poursorkhabi, V.; Mohanty, A.K.; Misra, M. Electrospinning of aqueous lignin/poly(ethylene oxide) complexes. *J. Appl. Polym. Sci.* **2015**, *132*, 41260. [CrossRef]
226. Li, Q.; Xie, S.; Serem, W.K.; Naik, M.T.; Liu, L.; Yuan, J.S. Quality carbon fibers from fractionated lignin. *Green Chem.* **2017**, *19*, 1628–1634. [CrossRef]
227. Ding, R.; Wu, H.; Thunga, M.; Bowler, N.; Kessler, M.R. Processing and Characterization of low-cost electrospun carbon fibers from organosolv lignin/polyacrylonitrile blends. *Carbon* **2016**, *100*, 121–131. [CrossRef]
228. He, J.-H.; Wan, Y.-Q.; Yu, J.-Y. Effect of Concentration on Electrospun Polyacrylonitrile (PAN) Nanofibers. *Fibers Polym.* **2008**, *9*, 140–142. [CrossRef]
229. Lai, C.; Zhou, Z.; Zhang, L.; Wang, X.; Zhou, Q.; Zhao, Y.; Wang, Y.; Wu, X.-F.; Zhu, Z.; Fong, H. Free-standing and mechanically flexible mats consisting of electrospun carbon nanofibers made from a natural product of alkali lignin as binder-free electrodes for high-performance supercapacitors. *J. Power Sources* **2014**, *247*, 134–141. [CrossRef]
230. Ago, M.; Okajima, K.; Jakes, J.E.; Park, S.; Rojas, O.J. Lignin-Based Electrospun Nanofibers Reinforced with Cellulose Nanocrystals. *Biomacromolecules* **2012**, *13*, 918–926. [CrossRef] [PubMed]
231. Dallmeyer, I.; Lin, L.T.; Li, Y.; Ko, F.; Kadla, J.F. Preparation and Characterization of Interconnected, Kraft Lignin-Based Carbon Fibrous Materials by Electrospinning. *Macromol. Mater. Eng.* **2014**, *299*, 540–551. [CrossRef]

232. Dallmeyer, I.; Ko, F.; Kadla, J.F. Correlation of Elongational Fluid Properties to Fiber Diameter in Electrospinning of Softwood Kraft Lignin Solutions. *Ind. Eng. Chem. Res.* **2014**, *53*, 2697–2705. [CrossRef]
233. Tortora, M.; Cavalieri, F.; Mosesso, P.; Ciaffardini, F.; Melone, F.; Crestini, C. Ultrasound Driven Assembly of Lignin into Microcapsules for Storage and Delivery of Hydrophobic Molecules. *Biomacromolecules* **2014**, *15*, 1634–1643. [CrossRef] [PubMed]
234. Bartzoka, E.D.; Lange, H.; Tiehl, K.; Crestini, C. Coordination Complexes and one Step Assembly of Lignin for Versatile Nano-Capsule Engineering. *ACS Sustain. Chem. Eng.* **2016**, *4*, 5194–5203. [CrossRef]
235. Yiamsawas, D.; Baier, G.; Thines, E.; Landfester, K.; Wurm, F.R. Biodegradable lignin nanocontainers. *RSC Adv.* **2014**, *4*, 11661–11663. [CrossRef]
236. Chen, G.-F.; Liu, M.-H. Adsorption of L-lysine from aqueous solution by spherical lignin beads: Kinetics and equilibrium studies. *BioResources* **2012**, *7*, 298–314. [CrossRef] [PubMed]
237. Saidane, D.; Barbe, J.-C.; Birot, M.; Deleuze, H. Preparation of functionalized Kraft lignin beads. *J. Appl. Polym. Sci.* **2010**, *116*, 1184–1189. [CrossRef]
238. Saidane, D.; Barbe, J.-C.; Birot, M.; Deleuze, H. Preparation of functionalized lignin beads from oak wood alkaline lignin. *J. Appl. Polym. Sci.* **2013**, *128*, 424–429. [CrossRef]
239. Blasi, M.; Barbe, J.-C.; Maillard, B.; Dubourdieu, D.; Deleuze, H. New Methodology for Removing Carbonyl Compounds from Sweet Wines. *J. Agric. Food Chem.* **2007**, *55*, 10382–10387. [CrossRef] [PubMed]
240. Lievonen, M.; Valle-Delgado, J.J.; Mattinen, M.-L.; Hult, E.-L.; Lintinen, K.; Kostiainen, M.A.; Paananen, A.; Szilvay, G.R.; Setälä, H.; Österberg, M. A simple process for lignin nanoparticle Preparation. *Green Chem.* **2016**, *18*, 1416–1422. [CrossRef]
241. Figueiredo, P.; Lintinen, K.; Kiriazis, A.; Hynninen, V.; Liu, Z.; Bauleth-Ramos, T.; Rahikkala, A.; Correia, A.; Kohout, T.; Sarmento, B.; et al. In vitro evaluation of biodegradable lignin-based nanoparticles for drug delivery and enhanced antiproliferation effect in cancer cells. *Biomaterials* **2017**, *121*, 97–108. [CrossRef] [PubMed]
242. Frangville, C.; Rutkevičius, M.; Richter, A.P.; Velev, O.D.; Stoyanov, S.D.; Paunov, V.N. Fabrication of Environmentally Biodegradable Lignin Nanoparticles. *ChemPhysChem* **2012**, *13*, 4235–4243. [CrossRef] [PubMed]
243. Gilca, I.A.; Popa, V.I.; Crestini, C. Obtaining lignin nanoparticles by sonication. *Ultrason. Sonochem.* **2015**, *23*, 369–375. [CrossRef] [PubMed]
244. Jiang, C.; He, H.; Jiang, H.; Ma, L.; Jia, D.M. Nano-lignin filled natural rubber composites: Preparation and characterization. *Express Polym. Lett.* **2013**, *7*, 480–493. [CrossRef]
245. Qian, Y.; Deng, Y.; Qiu, X.; Li, H.; Yang, D. Formation of uniform colloidal spheres from lignin, a renewable resource recovered from pulping spent liquor. *Green Chem.* **2014**, *16*, 2156–2163. [CrossRef]
246. Hüsing, N.; Schubert, U. Aerogels—Airy Materials: Chemistry, Structure, and Properties. *Angew. Chem. Int. Ed.* **1998**, *37*, 22–45. [CrossRef]
247. Gesser, H.D.; Goswami, P.C. Aerogels and related porous materials. *Chem. Rev.* **1989**, *89*, 765–788. [CrossRef]

248. Chen, F.; Xu, M.; Wang, L.; Li, J. Preparationand and characterization of organic aerogels by the lignin-resorcinol-formaldehyde copolymer. *BioResources* **2011**, *6*, 1262–1272.
249. Grishechko, L.I.; Amaral-Labat, G.; Szczurek, A.; Fierro, V.; Kuznetsov, B.N.; Celzard, A. Lignin–phenol–formaldehyde aerogels and cryogels. *Microporous Mesoporous Mater.* **2013**, *168*, 19–29. [CrossRef]
250. Grishechko, L.I.; Amaral-Labat, G.; Szczurek, A.; Fierro, V.; Kuznetsov, B.N.; Pizzi, A.; Celzard, A. New tannin–lignin aerogels. *Ind. Crops Prod.* **2013**, *41*, 347–355. [CrossRef]
251. Perez-Cantu, L.; Liebner, F.; Smirnova, I. Preparation of aerogels from wheat straw lignin by cross-linking with oligo(alkylene glycol)-α,ω-diglycidyl ethers. *Microporous Mesoporous Mater.* **2014**, *195*, 303–310. [CrossRef]
252. Forgacz, C.; Birot, M.; Deleuze, H. Synthesis of porous emulsion-templated monoliths from a pulp mill by-product. *J. Appl. Polym. Sci.* **2013**, *129*, 2606–2613. [CrossRef]
253. Raschip, I.E.; Vasile, C.; Ciolacu, D.; Cazacu, G. Semi-interpenetrating Polymer Networks Containing Polysaccharides. I Xanthan/Lignin Networks. *High Perform. Polym.* **2007**, *19*, 603–620. [CrossRef]
254. Ugartondo, V.; Mitjans, M.; Vinardell, M.P. Applicability of lignins from different sources as antioxidants based on the protective effects on lipid peroxidation induced by oxygen radicals. *Ind. Crops Prod.* **2009**, *30*, 184–187. [CrossRef]
255. Liu, X.; Wang, J.; Li, S.; Zhuang, X.; Xu, Y.; Wang, C.; Chu, F. Preparation and properties of UV-absorbent lignin graft copolymer films from lignocellulosic butanol residue. *Ind. Crops Prod.* **2014**, *52*, 633–641. [CrossRef]
256. Ten, E.; Ling, C.; Wang, Y.; Srivastava, A.; Dempere, L.A.; Vermerris, W. Lignin Nanotubes As Vehicles for Gene Delivery into Human Cells. *Biomacromolecules* **2014**, *15*, 327–338. [CrossRef] [PubMed]
257. Ichimura, T.; Otake, T.; Mori, H.; Maruyama, S. HIV-1 Protease Inhibition and Anti-HIV Effect of Natural and Synthetic Water-soluble Lignin-like Substances. *Biosci. Biotechnol. Biochem.* **1999**, *63*, 2202–2204. [CrossRef] [PubMed]
258. Raghuraman, A.; Tiwari, V.; Zhao, Q.; Shukla, D.; Debnath, A.K.; Desai, U.R. Viral Inhibition Studies on Sulfated Lignin, a Chemically Modified Biopolymer and a Potential Mimic of Heparan Sulfate. *Biomacromolecules* **2007**, *8*, 1759–1763. [CrossRef] [PubMed]
259. Ugartondo, V.; Mitjans, M.; Vinardell, M.P. Comparative antioxidant and cytotoxic effects of lignins from different sources. *Bioresour. Technol.* **2008**, *99*, 6683–6687. [CrossRef] [PubMed]
260. Mikulášová, M.; Košíková, B. Modulation of mutagenicity of various mutagens by lignin derivatives. *Mutat. Res. Toxicol. Environ. Mutagen.* **2003**, *535*, 171–180. [CrossRef]
261. Petersen, M.; Alfermann, W.A. The production of cytotoxic lignans by plant cell cultures. *Appl. Microbiol. Biotechnol.* **2001**, *55*, 135–142. [CrossRef]
262. Sakagami, H.; Satoh, K.; Aiuchi, T.; Nakaya, K.; Takeda, M. Stimulation of ascorbate-induced hypoxia by lignin. *Anticancer Res.* **1997**, *17*, 1213–1216. [PubMed]

263. Kawano, M.; Thet, M.M.; Makino, T.; Kushida, T.; Sakagami, H. DNA Microarray Analysis of Signaling Pathway in Macrophages Stimulated by Lignin–Carbohydrate Complex from Lentinus edodes Mycelia (LEM) Extract. *Anticancer Res.* **2010**, *30*, 2567–2576. [PubMed]
264. Strassberger, Z.; Tanase, S.; Rothenberg, G. The pros and cons of lignin valorisation in an integrated biorefinery. *RSC Adv.* **2014**, *4*, 25310–25318. [CrossRef]

Cellulose Nanocrystals: Multifunctional Systems for Biomedical Applications

Elena Fortunati and José M. Kenny

Abstract: Nanocrystalline cellulose, micro- and nano-fibrillated cellulose and bacterial cellulose nanofibres, have become fascinating structures to design new biomaterials. Derived from the most abundant and renewable biopolymer, they are drawing an incredible level of attention. This growing interest is related to their important and useful physical and chemical properties. Their morphological and topographical likenesses allow the exploitation of cellulosic materials in the development of nanostructured composites targeting tissue engineering and regeneration. In this chapter, cellulosic materials, and in particular cellulose nanocrystals, are highlighted as promising alternatives in the production of nanocomposite dense and porous structures as well as hydrogels. Special focus will be given to recent developments of cellulose-based nanocomposite substrates that induce cell attachment and proliferation. Challenges and future opportunities of cellulose nanocrystal-based systems in the biomedical field will be discussed as well as obstacles remaining for their large use.

---

1. Introduction

Cellulose is a sustainable and abundant raw material, that represents a key source on an industrial scale. For millennia, cellulose has been used in the form of wood fibers as an energy source, for building materials, and for clothing [1]. The production of nanoscaled cellulose fibers and their application in composite and nanocomposite approach have expanded the attention due to their high strength and stiffness combined with low weight, biodegradability, renewability and sustainability [2].

A number of possibilities have been considered to obtain and extract cellulose-based micro- and/or nano-metric fillers. In particular, microfibrillated cellulose (MFC) is formed by fibrous cellulose structures from wood or plant and it is characterized by several tens of microns in length and few tens of nanometers thick [3]. In other words, MFC is formed by long, flexible and entangled cellulose nanofibres, where both amorphous and crystalline phases are present [4]. Microfibrillated cellulose can be obtained from wood, agricultural by-products and waste, such as crops aimed at textile production (e.g., sisal) or aimed at food production (e.g., fruit crops, like pineapple, and cereal crops, such as wheat, sorghum, etc.). MFC can be extracted using different mechanical procedures as high-pressure homogenization step, whilst in some cases the process has been optimized by applying chemical and enzymatic pre-treatments of the cellulose raw material [5].

Nanocelluloses are categorized into three groups: (1) Cellulose nanocrystals (CNC) or nanowhiskers (CNW), (2) Cellulose nanofibrils (CNF) and (3) Bacterial cellulose (BC). The yield of the extraction process (i.e., the quantity of nanocellulose obtained from a given weight of microfiber), depends on both the crystallinity of the specific plant fiber and the procedure adopted for the extraction [6,7]. Due to their unique features, cellulose and, in particular, cellulose nanocrystals have garnered a tremendous level of attention in the materials community, which can be confirmed by the increasing number of scientific publications in the field over the past decades and they found particular attention in the nanocomposite approach.

Nanocomposites are usually defined as a combination of two or more materials or phases in which one of the phases has at least one dimension in the nanometer range (1–100 nm) [8–10]. In biodegradable nanocomposites, also called bionanocomposites, matrices may be a biodegradable material derived from renewable resources while the reinforcement phase can include plant fibers and renewable resources or synthetic inorganic materials. Plant-based nanocellulose and bacterial cellulose are included in this definition [11]. Moreover, in a nanocomposite approach, the idea to combine different reinforcement phases with diverse specific and active properties to obtain multifunctional systems, gives the possibility to respond to practical needs for the final application [12–14]. The main reason to utilize cellulose at nanoscale in composite materials is because one can potentially exploit the high stiffness of the cellulose crystal for reinforcement [15]. Because of its unique structural properties combined with hydrophilic nature, biocompatibility, transparency and non-toxicity cellulose structures are attractive candidates for a wide range of applications related to biomedical and biotechnology applications [16]. The term biocompatibility is defined as the ability of a material to perform with an appropriated host response in a specific biomedical application [17]. These properties are the most relevant to consider these materials as good candidates for tissue engineering and regenerative medicine applications, both as tri-dimensional and porous scaffolds or as drug-delivery device. Nanotechnology can be applied across different application areas: the ability to control the material at the nanoscale and the evaluation of their influence in the micro- and macroscopic properties provides the opportunity to develop new bionanocomposite systems in a previously dimensions. Biodegradable nanocomposites are attractive materials for novel biomedical applications due to the matching of the length scales of their structure and the components of the extracellular matrix (ECM). Under these conditions, the development of nanostructured biomaterials for medical applications implies a multidisciplinary research approach. In this sense, much work has already focused on designing ideal biomedical devices from cellulose and especially bacterial cellulose, such as artificial skin, artificial blood vessels, artificial cornea, heart valve prosthesis, artificial urethra, artificial bone, artificial cartilage, artificial porcine knee menisci, and deliveries of drug, hormone and protein [18–20]. As an introduction, the prospects for the various biomedical applications of cellulose-based materials are shown in Figure 1.

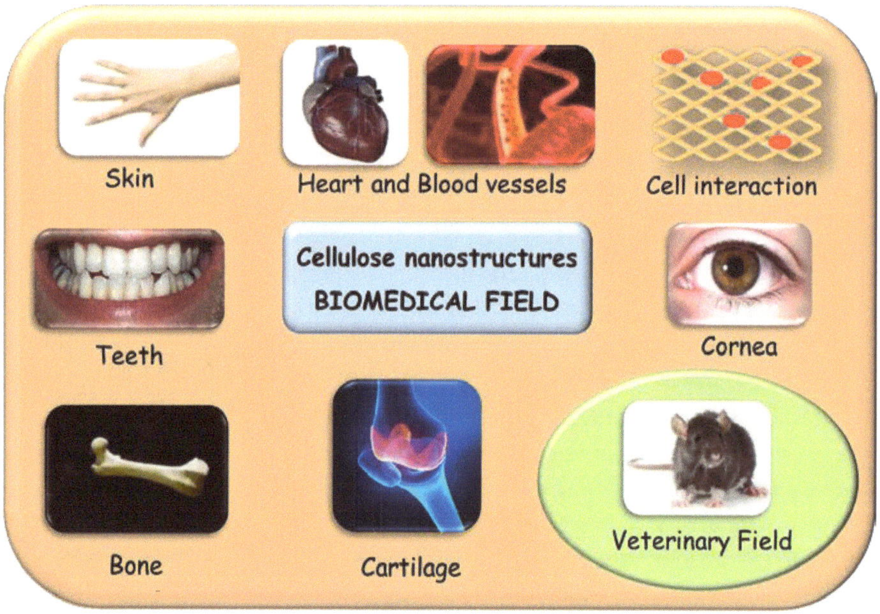

Figure 1. Perspective biomedical applications of cellulose based nanomaterials.

In this chapter we introduced an overview about the research executed with cellulosic materials in the biomedical field, with particular emphasis on tissue engineering and regenerative applications. The main focus will be on the development of nanocomposite systems and hydrogels based on cellulose components, in particular cellulose nanocrystal, within the previously described fields of application. The addition of plant-based and bacterial-based cellulose at nanoscale to enhance thermal and mechanical properties, as well as the biodegradability and sustainability of polymer matrices are also analyzed. This chapter intends to show that biomaterials from cellulosic sources have the potential to play an important role in human health and current research trends and a number of recent advances developed in our laboratories are reported and discussed.

2. Lignocellulosic Materials

Cellulose is one of the most important natural and abundant renewable polymers on earth and it has been used as a chemical raw material for about 150 years. Its annual production is estimated to be over 7.5×10^{10} tons [21]. Cellulose is widely distributed in higher plants; wood, consisting of up to 50% cellulose, is the most important raw material source for cellulose. Furthermore, it is also distributed in annual crops, and even in several marine animals (for example, tunicates), and to a lesser degree in algae, fungi, bacteria, invertebrates, and even amoeba [1]. Cellulose consists of a linear homopolysaccharide composed of β-D-glucopyranose units linked together by β-1-4-linkages. The repeat unit is a dimer of glucose, known as cellobiose. Each monomer has three hydroxyl groups, and it is, therefore, obvious that these hydroxyl groups and their ability to form hydrogen bonds play a major role in directing the crystalline packing and also governing the physical properties of cellulose [22]. The number of glucose units or the degree of polymerization is up to 20,000, but its value can vary among a wide range, and the value is around 10,000 in wood [22]. In nature, cellulose does not occur as an isolated individual molecule, but it is found as assemblies of individual cellulose chain-forming fibers. The morphological hierarchy is defined by elementary fibrils, which pack into larger units called microfibrils, which are in turn assembled into fibers [21,23]. Within the cellulose fibrils there are regions where the cellulose chains are arranged in a highly ordered structure (crystallites), and regions that are disordered (amorphous-like) [24].

Currently, studies on isolation, characterization and selection of prospective applications of novel forms of cellulose structures, variously termed crystallites, nanocrystals, whiskers, nanofibrils, and nanofibres, are generating interest. Novel methods for their production range from top-down methods involving enzymatic/chemical/physical methodologies for their isolation from wood and forest/agricultural residues, to the bottom-up approach of cellulose nanofibrils from glucose by bacteria [1].

2.1. Plant-Based Materials

Wood is a natural composite material with a hierarchical architecture in which cellulose, hemicellulose and lignin form a highly porous anisotropic microstructure [11]. Cellulose fibers can be classified according to their origin and grouped into leaf: abaca, cantala, curaua, date palm, henequen, pineapple, sisal, banana; seed: cotton; bast: flax, hemp, jute, ramie; fruit: coir, kapok, oil palm; grass: alfa, bagasse, bamboo; stalk: straw (cereal). The bast and leaf (the hard fibers) types are the most commonly used in composite applications [25]. Commonly used plant fibers are cotton, jute, hemp, flax, ramie, sisal, coir, henequen, and kapok [26]. Plant fibers are constitutes of cellulose fibers, consisting of helically wound cellulose microfibrils, bound together by an amorphous lignin matrix. Lignin keeps the water in fibers, acts as a protection against biological attack and as a stiffener to give stem its resistance against gravity forces and wind. Hemicelluloses, that represent acompatibilizer between cellulose and lignin [26], are located mainly in the secondary cell walls, and together with cellulose and lignin, build up the structure of the plants.

Cellulose microfibrils extracted by a mechanical disintegration process from wood cell were first obtained by Herrick et al. [27] and Tubark et al. [28], in 1983 and named microfibrillated cellulose. MFC, composed of expanded high-volume cellulose, is produced by different mechanical treatment, consist of high-pressure homogenization and/or grinding [27,28]. However, this production route is normally associated with high energy consumption for fiber defibrillation [29] and some different alternative pre-treatments have been proposed, such as mechanical cutting, acid hydrolysis, enzymatic pre-treatment, and the introduction of charged groups [30–32]. Cellulose microfibrils are long and flexible nanoparticles presenting lateral dimensions in the order of 3–100 nm, and length generally in the micrometre scale depending on the source of cellulose, defibrillation process and pre-treatment. The Young's modulus of cellulose fibrils has been reported to be 138 GPa and the extremely good strength properties and good thermal stability of the fibrils make them suitable for use as reinforcement in nanocomposites and paper products [33].

CNC, characterized by their typical acicular structure and dimension ranged from 3 to 10 nm in width and 100–300 nm in length has been isolated from a wide variety of cellulosic sources, including plants, [34] microcrystalline cellulose, [13] animals, [35] bacteria, and algae [1] by hydrolysis with mineral acids. The isolation of CNC from cellulose source materials occurs in two stages: a pre-treatment of the raw precursor material and a second step that could be a controlled chemical or enzymatic treatment (generally hydrolysis) to remove the amorphous regions of the cellulose polymer [6]. Siqueira et al. [36] investigated various combinations of mechanical shearing, acid and enzymatic hydrolysis in order to produce nanoparticles from sisal pulp showing as CNC could be obtained by enzymatic hydrolysis.

Cellulose nanocrystals have better mechanical properties than the majority of the commonly used reinforcing materials, and for these reasons, they are perfect candidates to be used as reinforcements phase in thermoplastic and/or thermosetting matrices in a nanocomposite approach with new opportunities in biomedical applications.

2.2. Bacterial Cellulose

Bacterial cellulose (BC), also called bacterial nanocellulose (BNC), microbial cellulose, or biocellulose, is formed by aerobic bacteria, such as acetic acid bacteria of the genus Gluconacetobacterxylinum, as a pure component of their biofilms. These bacteria are wide-spread in nature where the fermentation of sugars and plant carbohydrates takes place. In contrast to MFC and CNC materials isolated from plant based cellulose sources, BC is formed as a polymer and nanomaterial by biotechnological assembly processes from low-molecular weight carbon sources, such as d-glucose. The bacteria are cultivated in aqueous nutrient media, and the BNC is excreted as exopolysaccharide at the interface to the air. The resulting BNC hydrogel is composed of a nanofibre network (fibre diameter: 20–100 nm) enclosing up to 99% water. Cellulose derived from bacteria species has the advantage of being free from wax, lignin, hemicellulose and pectin, which are present in plant-based cellulosic materials. This BNC proved to be very pure cellulose with a high weight-average molecular weight (Mw), high crystallinity, and good mechanical stability. BC possesses a Young's modulus of about 114 GPa with theoretical values between 130 GPa and 145 GPa, depending on its crystallinity [37]. These values exceed that of synthetic glass fibres (about 70 GPa) and aramid fibres (about 67 GPa), considering that BC has a lower density (1.25 g cm^{-3}) than glass fibres (2.5 g cm^{-3}). BC exists naturally as a nanosized material (between 24 and 86 nm in diameter and several micrometres in length) [38] and it has a surface area of about 37 m^2 g^{-1} [39]. Such properties are highly advantageous for the production of composite materials as this implies that for the same amount of material, the interface between the matrix and the reinforcement will be larger for BC compared to micrometre-scale natural fibres. However, BC is extremely hydrophilic in nature and it will often have poor interfacial adhesion to hydrophobic polymer matrices.

The potential applications of BC are quite vast; it can be used in the preparation of materials targeting decellularised devices for wound healing or as scaffold for skin engineering after seeded with epithelial cells or more recently in vascular grafts [40,41]. Bodin et al. [42] have also proposed BC as meniscus implant since BC can be produced in a meniscus shape. The nanoscaled structured and topographical similarities with collagen make BC very attractive for cell immobilization, cell migration, and the production of ECMs. Finally, Klemm et al. [43], proved that celluloses with different functionalities open novel fields of application, specifically as nanoporous hydrophilic membranes, nanoscaffold-based composite materials, and medical implants for repair of hard and soft tissue, as well as cardiovascular substitutes.

3. Tissue Regeneration

Tissue engineering (TE) is a multidisciplinary field concentrated on the development and application of knowledge in chemistry, physics, engineering, life and clinical sciences to the solution of critical medical problems, such as tissue loss and organ failure [8].

Biomaterials play important roles in regenerative medicine and tissue engineering controlling cellular activities and functions [44]. Although remarkable advances have been made, much work remains in the development of biomaterials that integrate biological identification and signaling properties of the extracellular environment for practical applications in tissue engineering and regenerative medicine. For the new therapeutic strategies, it is indispensable to provide cells with a local environment that enhances and regulates their proliferation and differentiation for cell-based tissue regeneration [45].

Recently, nanostructured biomaterials having physical nano-features such as nanocrystals, nanofibres, nanosurfaces, nanocomposites, etc., have gained much interest in regenerative medicine [46]. Nanotechnology, and in particular the use of nanomaterials, can mimic surface properties of natural tissues while nanomaterials have been highlighted as promising candidates for improving traditional tissue engineering materials [47]. Different kinds of nanostructured materials (carbon nanostructures, metal, ceramic and polymeric nanoparticles, cellulose micro- and nano-crystals, etc.) are currently used to modulate the properties of biodegradable polymers in order to develop new dense films or scaffolds with multifunctional characteristics. Nanoscale materials can have chemical, physical and biological properties that differ from their larger counterparts and these properties can be exploited to influence cell attachment, proliferation and differentiation [48]. D'Angelo et al. recently demonstrated that nanocomposite fibrous mats of poly(L-lactic acid) (PLLA) loaded with different content of calcium-deficient nanohydroxyapatite (d-HAp) have an active role in inducing human multipotent (hBM-MSCs) and murine pluripotent (iPSCs and ESCs) stem cell differentiation toward the osteogenic lineage in the absence of exogenous soluble differentiating agents, proving the possibility to modulate the polymer matrix structural and functional properties due to the introduction of specific nanostructures.

Although research on cellulose, and in particular cellulose nanocrystals, has experienced a tremendous increase in a variety of fields over the last two decades, it has only recently attracted the interest of researchers in the field of tissue engineering and regenerative medicine, both as scaffolds or as drug-delivery systems. In fact, most of the studies addressing this topic have been published during the last five years, indicating that within the TE field, CNC have a large unexploited potential despite their many well-known attractive properties and possible applications in this context, as will be discussed later in this review.

Recently, Dugan et al. [49] reviewed the literature on bacterial cellulose scaffolds and discussed in detail the studies on the biocompatibility of CNC, addressing its potential in this field. Moreover, Fang et al. [50] demonstrated that bacterial cellulose/hydroxyapatite (HA) nanocomposite scaffolds were biocompatible and could promote cell proliferation and differentiation in vitro. More recently, Saska et al. [51] also described the preparation of membranes composed by BC reinforced with HA and evaluated the biological behavior of these membranes for the regeneration of noncritical bone defects in rat tibiae, over a time period of

16 weeks. The work revealed that the membranes are effective for bone regeneration since they accelerated new bone formation. Furthermore, bionanocomposites porous architectures based on BC were prepared in a collagen solution demonstrating the biocompatibility of the proposed formulations [52,53].

These recent results suggest that the cellulose-based materials have important potential in bone TE applications. For these reasons, the literature on nanocomposites based on both natural and synthetic origin polymer matrices relevant for TE field is reviewed here. Particular emphasis is given to functional materials obtained through different fabrication strategies usually employed to obtain scaffolds with various formats, namely, dense films and membranes (mostly considered as a prequel for three-dimensional (3D) scaffolds development), porous constructs (micro/nanofibres mats, foams and sponges), and hydrogels (Figure 2).

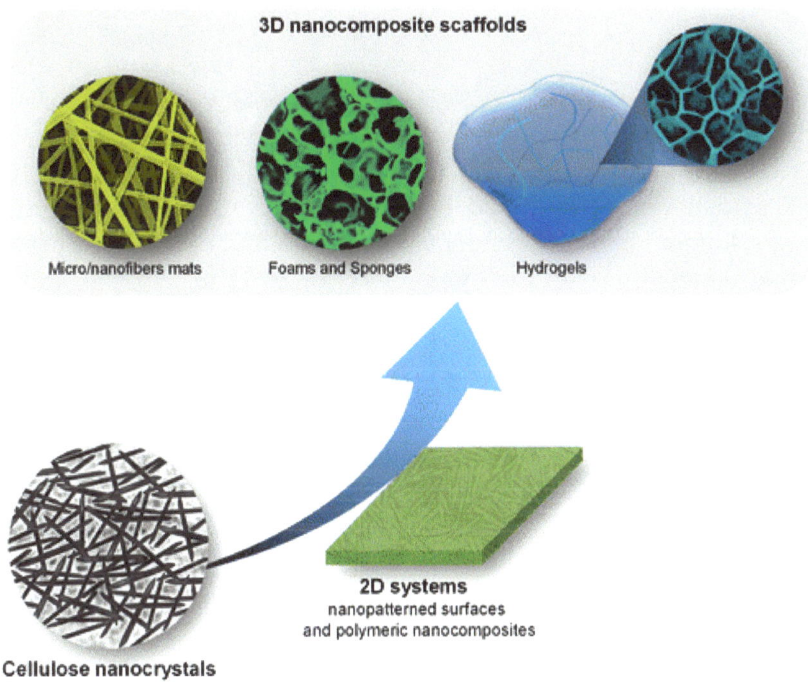

Figure 2. Summary of the current utilization of CNC on nanocomposite processing for tissue engineering applications. Reprinted with permission from Rui M. A. Domingues, Manuela E. Gomes, and Rui L. Reis [54].

4. Cellulose Nanocrystals Based Systems

4.1. Cellulose Nanocrystal Based Nanocomposites: Dense Films and Porous Structures

Considering their safety and efficacy, cellulose structures and in particular, cellulose nanocrystals (CNC), have attracted increasing attention in biomedical applications as reinforcements in composite materials. Dense bi-dimensional (2D) films and tri-dimensional (3D) porous scaffolds with different properties, architectures and pore sizes can be produced from synthetic or naturally occurring materials and cellulose nanofibres, cellulose nanocrystals and bacterial nanocellulose are widely used as bio-based nanoreinforcements in several polymeric matrices owing to their superior mechanical properties [11]. The high surface-area-to-volume ratio of the nanofibers favor cell adhesion, proliferation, migration and differentiation, which are useful properties for tissue engineering applications [55]. The development of nanofibres can be used to design nanocomposites able to mimic the architecture of natural human tissue. Moreover, toxicity tests conducted on cellulose nanostructures indicate that they are non-toxic to cells [56], and does not give serious environmental concerns [57,58]. Cellulose fiber reinforced polymer composites have received much attention because of their low density, nonabrasive, combustible, nontoxic, low cost and biodegradable properties. A lot of research studies have been performed on the use of cellulose fibers as a reinforcing materials although certain drawbacks, such as incompatibility with the hydrophobic polymer matrix, the tendency to aggregate and water-swellable nature of cellulose, reduce the potential of natural fibers as reinforcements in polymers. Moreover, considering, in particular, CNC as reinforcements, their interfacial adhesion with the selected polymer matrix is one of the most important factors affecting nanocomposite final properties. In order to enhance the dispersion of CNC and increase the interfacial strength between the two phases, various methods have been used, such as magnetic field alignment within the polymer [59], use of plasticizers [60] and surfactants [61], or specific surface modifications [62].

A quite extensive range of polymer matrices, both synthetic and natural, has been used to produce nanocomposite dense films and porous membranes containing CNC. Recently, Pooyan et al. have developed a cellulose-based nanohybrid material in which CNC were dispersed within a cellulose acetate propionate matrix to form a 3D rigid percolating network, aimed at fabricating a potential scaffold candidate in TE of small diameter vascular grafts [63]. CNC extracted from microcrystalline cellulose were suspended in acetone prior to the composite fabrication in order to ensure a uniform distribution. Thin films were produced by solvent casting and, according to the authors, the obtained bio-nanocomposites exhibited excellent mechanical performance at body temperature, claiming that CNC impart significant strength and directional rigidity even at very low concentrations.

In a multifunctional nanocomposite approach, Rescignano et al. [64] described the production of a new generation of hybrid bio-nanocomposites dense films based on a poly(vinyl alcohol) (PVA) matrix reinforced with a combination of cellulose nanocrystals and poly (D,L lactide-co-glycolide) (PLGA) nanoparticles (NPs) loaded with bovine serum albumin fluoresceinisothiocynate conjugate (FITC-BSA). The authors demonstrated that the studied bio-nanocomposite films are suitable to vehiculate biopolymeric nanoparticles to adult bone marrow mesenchymal stem cells successfully, representing a new tool for drug delivery strategies. In a different ternary nanocomposite system based on PVA matrix, George et al. combined bacterial CNC and silver (Ag) nanoparticles (1 wt %) as nanofillers [65]. In addition to the antimicrobial properties imparted by the Ag nanoparticles, a synergistic effect on the improvement of the mechanical properties of the PVA films was obtained by combining these two nanomaterials; while CNC increased the modulus and tensile strength of the films, the Ag nanoparticles decreased its brittleness. These effects could be useful in the development of multifunctional hybrid materials for TE applications. Taking into account the use of natural polymer, Li et al. have developed collagen-based nanocomposite films reinforced with CNC for potential applications in skin TE. The films were prepared by a solvent-casting method. The authors proved that the addition of CNC to collagen-based nanocomposite films improved their bulk ordering and stability and led to an increase in the swelling capacity, which is highly desirable for the proposed application, as it enables the absorption of a large amount of exudates in the early stages of wound healing. The mechanical properties of the composite films were also significantly improved when compared with those of the pure collagen film while the composite formulations showed no obvious cytotoxicity facilitating cell adhesion and displaying an excellent biocompatibility [66].

The fabrication of ternary fibrous mats based on poly(lactic) acid (PLA), cellulose nanocrystals (CNC), both pristine (p-CNC) and modified with a commercial surfactant (s-CNC)) and silver nanoparticles by electrospinning is reported by Cacciotti et al. [67]. The addition of Ag and/or pristine p-CNC fillers did not remarkably affect the fiber morphology and the average size dimension. On the other hand, more interestingly, the authors proved that the presence of surfactant-modified s-CNC promoted the self-assembling of submicrometric fibers in bundles randomly organized in polygonal structures, such as pentagons and hexagons, resulting in a honeycomb-like network (Figure 3). The study demonstrated that it is possible to modulate the thermal, mechanical properties and especially morphological properties of PLA fibers, properly selecting the cellulose-based reinforcements able to induce specific topography useful for the final applications.

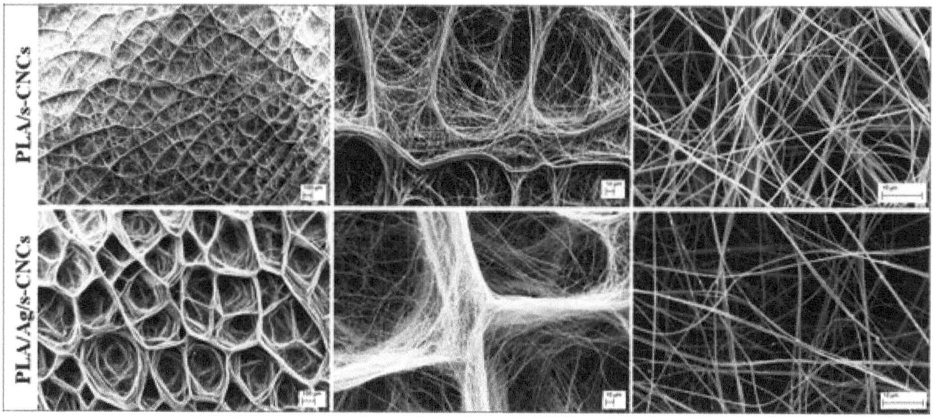

Figure 3. SEM micrographs of electrospun PLA/s-CNC and PLA/Ag/s-CNC mats. Reprinted with permission from Ilaria Cacciotti, Elena Fortunati, Debora Puglia, Josè Maria Kenny and Francesca Nanni [67].

Natural silk fibroin/CNC-based porous nanocomposites have also been prepared by electrospinning. Huang et al. [68] produced electrospun silk fibroin nanofiber mats reinforced with CNC extracted from Morus alba L. branch bark (20–40 nm in diameter and 400–500 nm in length). The authors reported that CNC were homogeneously dispersed and aligned along the fiber axis in the silk fibroin matrix while an increase in mechanical response was detected. Similar enhanced mechanical properties were achieved with bacterial cellulose nanocrystal as nanofillers in silk fibroin nanofibers [69].

The increase of knowledge on the cellulose-based components acted as a leverage regarding its practical use in the biomedical, pharmaceutical and/or cosmetic markets. Wound dressing systems based on pure oxidized regenerated cellulose (ORC) have been developed such as Tabotamp® and as mixtures from collagen and ORC named Promogran®. Tabotamp® is a thin gauze layer used in acute wounds like trauma, surgical injuries and burns while Promogran® is a commercial spongy collagen matrix and cellulose introduced in the US and European markets for the treatment of exuding diabetic and ulcer wounds [11]. Moreover, bacterial cellulose-based porous membranes have been applied for guided bone regeneration in bone defects, periodontal lesions and as a resorbable barrier membranes for preventing the invasion of fibroblast cells and fibrous connective tissue into bone defects [11].

The summarized recent research studies concerning cellulose-based materials and their uses as products in biomedical or pharmaceutical fields will contribute to sustain these market developments.

4.2. Cellulose Nanocrystals Based Hydrogels

Hydrogels are polymeric materials, chemically or physically crosslinked, characterized by a three-dimensional (3D) and elastic network capable to swell or de-swell when immersed in aqueous solutions. In particular, chemically crosslinked hydrogels are prepared either through water-soluble polymer crosslinking or by converting hydrophobic into hydrophilic polymers, which in turn are then crosslinked to form a network [70]. Lately, a particular class of polymer hydrogels has gained more interest in the scientific community; the so-called smart or stimuli-responsive hydrogels, which are able to change their size and shape in response to an external stimulus, such as temperature, pH, ionic force, pressure, electric and magnetic field. The change in solubility or the degree of swelling are due to a fine balance among competing interactions such as electrostatic forces and hydrophobic dehydration [71]. Moreover, some systems have been developed to combine two or more stimuli-responsive mechanisms into one polymer system. Recently, dual stimuli-responsive or ternary stimuli-responsive polymer hydrogel microspheres were prepared and applied in various fields, especially in controlled release drug delivery systems [72]. Hydrogels were the first biomaterials to be rationally designed for use in humans. They are biocompatible, their aqueous environment can protect cells and fragile drugs (such as peptides, proteins, etc.), they present good transport of nutrients to cells and metabolic products from the cells, they may be easily modified with cell adhesion ligands, or they can be injected in vivo [73]. All these properties make it an interesting candidate for biomedical and pharmaceutical applications [74]. The main areas in which hydrogels are used as biomaterials are in contact lenses, synthetic wound coverings, drug-delivery systems, permselective membranes, and in organ and tissue replacements, such as skin, tendon, cartilage, heart valve stents and bone [75]. However, because of the random nature of the crosslinking reactions produced by a large number of organic crosslinker polymer, hydrogels exhibit poor mechanical properties, which strongly limit their use in structural applications. For such a reason, different nanofillers, such as silicates [76], ceramics [77], metals [78], magnetic particles [79] and graphene [80] have been introduced into the hydrogel matrices, thus obtaining the corresponding nanocomposites. The incorporation of nanoparticles into the hydrogel 3D matrix, producing a class of materials known as nanocomposite hydrogels (NHPs), has been a widely investigated strategy to improve some existing physical properties or to provide them with new physical or chemical features. Considering the excellent dispersion of cellulose nanocrystals in water, they are obvious candidates to prepare NHPs, having many advantages compared to other polymeric or metal nanoparticles [54]. Cellulose-based hydrogels represent an important material class in biomedical fields due to their biocompatibility and biodegradability. Several water-soluble cellulose derivatives can be used as mono-components of multi-component systems and they can be used also as reinforcement phases in some other biodegradable polymers to form hydrogel networks with specific properties in terms of swelling capability and sensitivity to external stimuli [81]. However, the current trend in the design of cellulose-based hydrogels is associated with the use of non-toxic crosslinking agents or crosslinking chemical treatments, to further improve the safety of both the final product and the manufacturing process [11], and few recent studies have investigated the incorporation of CNC with particular surface modifications in the preparation of NPHs, thus acting as both filler and cross-linker to reinforce hydrogel systems. Zhou et al. [82] demonstrated that the CNC acted as reinforcing agents, and also as multifunctional cross-linkers, accelerating the formation of hydrogels and increasing their effective cross-linking density. Moreover, chemically cross-linked gelatin/CNC hydrogels were prepared using oxidized CNC as cross-linkers [83]. The authors proved that the dialdehyde groups of oxidized CNC reacted with the free amine groups of gelatin to cross-link the hydrogel framework and the final properties of these materials were dependent on the amounts of CNC aldehyde groups.

Recently, Sanna et al. [71] reported on the synthesis and characterization of thermoresponsive poly(N-vinylcaprolactam), PNVCL, nanocomposite hydrogels containing nanocrystalline cellulose produced by a frontal polymerization technique, which is a convenient, easy and low-energy-consuming method of macromolecular synthesis. The authors proved that the presence of CNC resulted in a significant increase of the mechanical properties even at very low CNC concentrations, as confirmed by rheological tests indicating that the nanocellulose has a great potential to reinforce PNVCL polymer hydrogels. Finally, Yang et al. arguably reported the first injectable hydrogels reinforced both physically and covalently with CNC, based on a carboxymethyl cellulose (CMC)/dextran system [84]. Their approach was based on coextruding aldehyde functionalized CNC with dihydrazide-modified CMC and aldehyde-modified dextran solutions through a double-barrel syringe. The produced NPHs and their components revealed no evident cytotoxicity to NIH 3T3 fibroblast cells. The authors hypothesize that these CNC reinforced injectable polysaccharide hydrogels are of potential interest for TE applications where longer term dimensional stability and enhanced mechanical strength are desirable [84].

5. Concluding Remarks

Over the past few years, a growing interest has emerged in applying cellulose structures, and in particular cellulose nanocrystals, as biomaterials for the development of advanced functional bionanocomposites which could find a wide range of potential applications in TE. This Chapter has attempted to provide a general overview of the potential of CNC in the design of these functional nanomaterials, through various examples involving different approaches and processes. Recent studies demonstrated that different types of nanomaterials presenting desired properties and functions could be produced from CNC. Research on nanocomposite films, membranes and hydrogels with potential application in TE strategies was summarized.

The use of lignocellulosic materials in drug-delivery devises has been studied for several years while the use of cellulose derivatives as excipients in drug formulations is already a standard in the pharmaceutical industry. In the future, the appearance of commercial formulations that are able to control the release rate and timeframe of different drugs, is expected. However, it is necessary to attract the industry to this growing market opportunity capable of creating sustainable technological improvements in bionanocomposites.

The positive results obtained by the research activity concerning increased performance of the bionanocomposites, will contribute to sustain these market developments although it is important to validate the most sustainable supplies of lignocellulosic materials. In addition, it is also relevant to create mechanisms to contribute to the certification of these nanostructured biomaterials according to common methods and standards.

Conflicts of Interest: The authors declare no conflict of interest.

References

1. Klemm, D.; Kramer, F.; Moritz, S.; Lindström, T.; Ankerfors, M.; Gray, D.; Dorris, A. Nanocelluloses: A new family of nature-based materials. *Angew. Chem. Int. Ed.* **2011**, *50*, 5438–5466. [CrossRef] [PubMed]
2. Siro, I.; Plackett, D. Microfibrillated cellulose and new nanocomposite materials: A review. *Cellulose* **2010**, *17*, 459–494. [CrossRef]
3. Krassig, H.A. *Cellulose: Structure, Accessibility and Reactivity*; Polymer Monographs 11; Huglin, M.B., Ed.; Gordon and Breach Science Publishers: Amsterdam, The Netherlands, 1993; p. 376.
4. Tingaut, P.; Zimmermann, T.; Sèbe, G. Functionalization of Microfibrillated Cellulose (MFC) and cellulose nanowhiskers fort the elaboration of functional materials. *J. Mater. Chem.* **2012**, *22*, 20105–20111. [CrossRef]
5. Habibi, Y.; Chanzy, H.; Vignon, M.R. TEMPO-mediated surface oxidation of cellulose whiskers. *Cellulose* **2006**, *13*, 679–687. [CrossRef]
6. Brinchi, L.; Cotana, F.; Fortunati, E.; Kenny, J.M. Production of nanocrystalline cellulose from lignocellulosic biomass: Technology and applications. *Carboydr. Polym.* **2013**, *94*, 154–169. [CrossRef] [PubMed]
7. Lin, S.P.; Calvar, I.L.; Catchmark, J.M.; Liu, J.R.; Demirci, A.; Cheng, K.C. Biosynthesis, production and applications of bacterial cellulose. *Cellulose* **2013**, *20*, 2191–2219. [CrossRef]

8. Armentano, I.; Dottori, M.; Fortunati, E.; Mattioli, S.; Kenny, J.M. Biodegradable polymer matrix nanocomposites for tissue engineering: A review. *Polym. Degrad. Stab.* **2010**, *95*, 2126–2146. [CrossRef]
9. Kamel, S. Nanotechnology and its applications in lignocellulosic composites, a mini review. *Express Polym. Lett.* **2007**, *1*, 546–575. [CrossRef]
10. Darder, M.; Aranda, P.; Ruiz-Hitzky, E. Bionanocomposites: A new concept of ecological, bioinspired, and functional hybrid materials. *Adv. Mater.* **2007**, *19*, 1309–1319. [CrossRef]
11. Fernandes, E.M.; Pires, R.A.; Mano, J.F.; Reis, R.L. Bionanocomposites from lignocellulosic resources: Properties, applications and future trends for their use in the biomedical field. *Prog. Polym. Sci.* **2013**, *38*, 1415–1441. [CrossRef]
12. Armentano, I.; Bitinis, N.; Fortunati, E.; Mattioli, S.; Rescignano, N.; Verdejo, R.; Lopez-Manchado, M.A.; Kenny, J.M. Multifunctional nanostructured PLA materials for packaging and tissue engineering. *Prog. Polym. Sci.* **2013**, *38*, 1720–1747. [CrossRef]
13. Fortunati, E.; Armentano, I.; Zhou, Q.; Iannoni, A.; Saino, E.; Visai, L.; Berglund, L.A.; Kenny, J.M. Multifunctional bionanocomposite films of poly(lactic acid), cellulose nanocrystals and silver nanoparticles. *Carboydr. Polym.* **2012**, *87*, 1596–1605. [CrossRef]
14. Fortunati, E.; Peltzer, M.; Armentano, I.; Jiménez, A.; Kenny, J.M. Combined effects of cellulose nanocrystals and silver nanoparticles on the barrier and migration properties of PLA nano-biocomposites. *J. Food Eng.* **2013**, *18*, 117–124. [CrossRef]
15. Eichhorn, S.J.; Dufresne, A.; Aranguren, M.; Capadona, J.R.; Rowan, S.J.; Weder, C.; Thielemans, W.; Roman, M.; Renneckar, S.; Gindl, W.; et al. Review: Current international research into cellulose nanofibres and nanocomposite. *J. Mater. Sci.* **2010**, *45*, 1–33. [CrossRef]
16. Dahman, Y. Nanostructured biomaterials and biocomposites from bacterial cellulose nanofibres. *J. Nanosci. Nanotechnol.* **2009**, *9*, 5105–5122. [CrossRef] [PubMed]
17. Williams, D.F. On the mechanisms of biocompatibility. *Biomaterials* **2008**, *29*, 2941–2953. [CrossRef] [PubMed]
18. Nadia, H.; Mohd, C.I.M.A.; Ishak, A.; Zulkifli, M.H.; Noriah, J. Swelling of bacterial cellulose–acrylic acid hydrogels: Sensitivity towards external stimuli. *Sains Malays.* **2009**, *38*, 785–791.
19. Oshima, T.; Taguchi, S.; Ohe, K.; Baba, Y. Phosphorylated bacterial cellulose for adsorption of proteins. *Carbohydr. Polym.* **2011**, *83*, 953–958. [CrossRef]
20. Wang, J.H.; Gao, C.; Zhang, Y.S.; Wan, Y.Z. Preparation and in vitro characterization of BC/PVA hydrogel composite for its potential use as artificial cornea biomaterial. *Mater. Sci. Eng. C* **2010**, *3*, 214–218. [CrossRef]
21. Habibi, Y.; Lucia, L.A.; Rojas, O.J. Cellulose nanocrystals: Chemistry, self-assembling, and applications. *Chem. Rev.* **2012**, *110*, 3479–3500. [CrossRef] [PubMed]
22. John, M.J.; Sabu, T. Biofibres and biocomposites. *Carbohydr. Polym.* **2008**, *71*, 343–364. [CrossRef]
23. Fengel, D.; Wegener, G. *Wood Chemistry, Ultra-Structure, Reactions*; Walter de Gruyter: Berlin, Germany; New York, NY, USA, 1984.

24. Moon, R.J.; Martini, A.; Nairn, J.; Simonsen, J.; Younblood, J. Cellulose nanomaterials review: Structure, properties and nanocomposites. *Chem. Soc. Rev.* **2011**, *40*, 3941–3994. [CrossRef] [PubMed]
25. Kalia, S.; Dufresne, A.; Cherian, B.M.; Kaith, B.S.; Avérous, L.; Njuguna, J.; Nassiopoulos, E. Cellulose-based bio- and nanocomposites: A review. *Int. J. Polym. Sci.* **2011**, *2011*, 1–35. [CrossRef]
26. Kalia, S.; Kaith, B.S.; Kaur, I. Pretreatments of natural fibres and their application as reinforcing material in polymer composites: A review. *Polym. Eng. Sci.* **2009**, *49*, 1253–1272. [CrossRef]
27. Herrick, F.W.; Casebier, R.L.; Hamilton, J.K.; Sandberg, K.R. Microfibrillated cellulose: Morphology and accessibility. *J. Appl. Polym. Sci. Appl. Polym. Symp.* **1983**, *37*, 797–813.
28. Turbak, A.F.; Snyder, F.W.; Sandberg, K.R. Microfibrillated cellulose, a new cellulose product: Properties, uses, and commercial potential. *J. Appl. Polym. Sci. Appl. Polym. Symp.* **1983**, *37*, 815–827.
29. Zimmermann, T.; Bordeanu, N.; Strub, E. Properties of nanofibrillated cellulose from different raw materials and its reinforcement potential. *Carbohydr. Polym.* **2010**, *79*, 1086–1093. [CrossRef]
30. Boldizar, A.; Klason, C.; Kubat, J.; Näslund, P.; Saha, P. Prehydrolyzed cellulose as reinforcing filler for thermoplastics. *Int. J. Polym. Mater.* **1987**, *11*, 229–262. [CrossRef]
31. Henriksson, M.; Henriksson, G.; Berglund, L.A.; Lindstrom, T. An environmentally friendly method for enzyme-assisted preparation of microfibrillated cellulose (MFC) nanofibers. *Eur. Polym. J.* **2007**, *43*, 3434–3441. [CrossRef]
32. Pääkkö, M.; Ankefors, M.; Kosonen, H.; Nykänen, A.; Ahola, S.; Österberg, M.; Ruokolainen, J.; Laine, J.; Larsson, P.T.; Ikkala, O.; et al. Enzymatic hydrolysis combined with mechanical shearing and high-pressure homogenization for nanoscale cellulose fibrils and strong gels. *Biomacromolecules* **2007**, *8*, 1934–1941. [CrossRef] [PubMed]
33. Berglund, L. Cellulose-based nanocomposites. In *Natural Fibers, Biopolymers, and Biocomposites*; Mohanty, A., Misra, M., Drzal, L., Eds.; CRC Press: Boca Raton, FL, USA, 2005; pp. 807–832.
34. Fortunati, E.; Puglia, D.; Monti, M.; Peponi, L.; Santulli, C.; Kenny, J.M.; Torre, L. Extraction of cellulose nanocrystals from Phormiumtenax fibres. *J. Polym. Environ.* **2012**, *21*, 319–328. [CrossRef]
35. Favier, V.; Canova, G.R.; Cavaille, J.Y.; Chanzy, H.; Dufresne, A.; Gauthier, C. Nanocomposite materials from latex and cellulose whiskers. *Polym. Adv. Technol.* **1995**, *6*, 351–355. [CrossRef]
36. Siqueira, G.; Tapin-Lingua, S.; Bras, J.; da Silva Perez, D.; Dufresne, A. Morphological investigation of nanoparticles obtained from combined mechanical shearing, and enzymatic and acid hydrolysis of sisal fibers. *Cellulose* **2010**, *17*, 1147–1158. [CrossRef]
37. Hsieh, Y.C.; Yano, H.; Nogi, M.; Eichhorn, S.J. An estimation of the Young's modulus of bacterial cellulose filaments. *Cellulose* **2008**, *15*, 507–513. [CrossRef]
38. Chanliaud, E.; Burrows, K.M.; Jeronimidis, G.; Gidley, M.J. Mechanical properties of primary plant cell wall analogue. *Planta* **2002**, *215*, 989–996. [CrossRef] [PubMed]

39. Kim, D.Y.; Nishiyama, Y.; Kuga, S. Surface acetylation of bacterial cellulose. *Cellulose* **2002**, *9*, 361–367. [CrossRef]
40. Petersen, N.; Gatenholm, P. Bacterial cellulose-based materials and medical devices: Current state and perspectives. *Appl. Microbiol. Biotechnol.* **2011**, *91*, 1277–1286. [CrossRef] [PubMed]
41. Backdahl, H.; Risberg, B.; Gatenholm, P. Observations on bacterial cellulose tube formation for application as vascular graft. *Mater. Sci. Eng. C* **2011**, *31*, 14–21. [CrossRef]
42. Bodin, A.; Concaro, S.; Brittberg, M.; Gatenholm, P. Bacterial cellulose as a potential meniscus implant. *J. Tissue Eng. Regen. Med.* **2007**, *1*, 406–408. [CrossRef] [PubMed]
43. Klemm, D.; Schumann, D.; Kramer, F.; Heßler, N.; Hornung, M.; Schmauder, H.P.; Marsch, S. Nanocelluloses as innovative polymers in research and application polysaccharides II. *Adv. Polym. Sci.* **2006**, *2015*, 49–96.
44. Jagur-Grodzinski, J. Polymers for tissue engineering, medical devices, and regenerative medicine. Concise general review of recentstudies. *Polym. Adv. Technol.* **2006**, *17*, 395–418. [CrossRef]
45. Armentano, I.; Fortunati, E.; Mattioli, S.; Rescignano, N.; Kenny, J.M. Biodegradable Composite scaffolds: A strategy to modulate Stem cell behaviour. *Recent Pat. Drug Deliv. Formul.* **2013**, *7*, 9–17. [CrossRef] [PubMed]
46. Dvir, T.; Timko, B.P.; Kohane, D.S. Nanotechnological strategies for engineering complex tissues. *Nat. Nanotechnol.* **2011**, *13*, 13–22. [CrossRef] [PubMed]
47. Zhang, L.; Webster, T.J. Nanotechnology and nanomaterials: Promises for improved tissue regeneration. *Nano Today* **2009**, *4*, 66–80. [CrossRef]
48. D'Angelo, F.; Armentano, I.; Cicciotti, I.; Tiribuzi, R.; Tiribuzi, R.; Quattrocelli, M.; Del Gaudio, C.; Fortunati, E.; Saino, E.; Caraffa, A.; Cerulli, G.G.; et al. Tuning multi-/pluri-potent stem cell fate by electrospun poly(L-lactic acid)-calcium-deficient hydroxyapatite nanocomposite mats. *Biomacromolecules* **2012**, *13*, 1350–1360.
49. Dugan, J.M.; Gough, J.E.; Eichhorn, S.J. Bacterial cellulose scaffolds and cellulose nanowhiskers for tissue engineering. *Nanomedicine* **2013**, *8*, 287–298. [CrossRef] [PubMed]
50. Fang, B.; Wan, Y.Z.; Tang, T.T.; Tang, T.T.; Gao, C.; Dai, K.R. Proliferation and osteoblastic differentiation of human bone marrow stromal cells on hydroxyapatite/bacterial cellulose nanocomposite scaffolds. *Tissue Eng. Part A* **2009**, *15*, 1091–1098. [CrossRef] [PubMed]
51. Saska, S.; Barud, H.S.; Gaspar, A.M.M.; Marchetto, R.; Ribeiro, S.J.L.; Messaddeq, Y. Bacterial cellulose-hydroxyapatite nanocomposites for bone regeneration. *Int. J. Biomater.* **2011**, *2011*, 1–8. [CrossRef] [PubMed]
52. Zhijiang, C.; Guang, Y. Bacterial cellulose/collagen composite: Characterization and first evaluation of cytocompatibility. *J. Appl. Polym. Sci.* **2011**, *120*, 2938–2944. [CrossRef]
53. Mathew, A.P.; Oksman, K.; Sain, M. Mechanical properties of biodegradable composites from poly lactic acid (PLA) and microcrystalline cellulose (MCC). *J. Appl. Polym. Sci.* **2005**, *97*, 2014–2025. [CrossRef]
54. Domingues, R.M.A.; Gomes, M.E.; Reis, R.L. The potential of cellulose nanocrystals in tissue engineering strategies. *Biomacromolecules* **2014**, *15*, 2327–2346. [CrossRef] [PubMed]

55. Vasita, R.; Katti, D.S. Nanofibres and their applications in tissue engineering. *Int. J. Nanomed.* **2006**, *1*, 15–30. [CrossRef]
56. Roman, M.; Dong, S.P.; Anjali, H.; Lee, Y.W. Cellulose Nanocrystals for Drug Delivery. In *Polysaccharides Materials: Performance by Design*; ACS Symposium Series; Edgar, K.J., Heinze, T., Buchanan, C., Eds.; American Chemical Society: Washington, DC, USA, 2010; Volume 1017, pp. 81–91.
57. Kovacs, T.; Naish, V.; O'Conner, B.; Blaise, C.; Gagne, F.; Hall, L.; Trudeau, V.; Martel, P. An eco-toxicologial characterization of nanocrystalline cellulose (NCC). *Nanotoxicology* **2010**, *4*, 255–270. [CrossRef] [PubMed]
58. Kümmerer, K.; Menz, J.; Schubert, J.; Thielmens, W. Biodegradability of organic nanoparticles in the aqueous environment. *Chemosphere* **2011**, *82*, 1387–1392. [CrossRef] [PubMed]
59. Pooyan, P.; Kim, I.T.; Jacob, K.I.; Tannenbaum, R.; Garmestani, H. Design of a cellulose-based nanocomposite as a potential polymeric scaffold in tissue engineering. *Polymer* **2013**, *54*, 2105–2114. [CrossRef]
60. Li, R.; Zhang, Y.; Zhu, L.; Yao, J. Fabrication and characterization of silk fibroin/poly(ethylene glycol)/cellulose nanowhisker composite films. *J. Appl. Polym. Sci.* **2012**, *124*, 2080–2086. [CrossRef]
61. Fortunati, E.; Peltzer, M.; Armentano, I.; Torre, L.; Jiménez, A.; Kenny, J.M. Effects of modified cellulose nanocrystals on the barrier and migration properties of PLA nano-biocomposites. *Carbohydr. Polym.* **2012**, *90*, 948–956. [CrossRef] [PubMed]
62. Rueda, L.; Saralegi, A.; Fernandez-d'Arlas, B.; Zhou, Q.; Alonso-Varona, A.; Berglund, L.A.; Mondragon, I.; Corcuera, M.A.; Eceiza, A. In situ polymerization and characterization of elastomeric polyurethane-cellulose nanocrystal nanocomposites. Cell response evaluation A. *Cellulose* **2013**, *20*, 1819–1828. [CrossRef]
63. Pooyan, P.; Tannenbaum, R.; Garmestani, H.J. Mechanical behavior of a cellulose-reinforced scaffold in vascular tissue engineering. *Mech. Behav. Biomed. Mater.* **2012**, *7*, 50–59. [CrossRef] [PubMed]
64. Rescignano, N.; Fortunati, E.; Montesano, S.; Emiliani, C.; Kenny, J.M.; Martino, S.; Armentano, I. PVA bio-nanocomposites: A new take-off using cellulose nanocrystals and PLGA nanoparticles. *Carbohydr. Polym.* **2014**, *99*, 47–58. [CrossRef] [PubMed]
65. George, J.; Sajeevkumar, V.A.; Ramana, K.V.; Sabapathy, S.N. Augmented properties of PVA hybrid nanocomposites containing cellulose nanocrystals and silver nanoparticles. *J. Mater. Chem.* **2012**, *22*, 22433–22439. [CrossRef]
66. Li, W.; Guo, R.; Lan, Y.; Zhang, Y.; Xue, W.; Zhang, Y. Preparation and properties of cellulose nanocrystals reinforced collagen composite films. Biomed. *Mater. Res. Part A* **2014**, *102*, 1131–1139. [CrossRef] [PubMed]
67. Cacciotti, I.; Fortunati, E.; Puglia, D.; Kenny, J.M.; Nanni, F. Effect of silver nanoparticles and cellulose nanocrystals on electrospun poly(lactic) acid mats: Morphology, thermal properties and mechanical behavior. *Carbohydr. Polym.* **2014**, *103*, 22–31. [CrossRef] [PubMed]
68. Huang, J.; Liu, L.; Yao, J. Electrospinning of *Bombyxmori* silk fibroin nanofiber mats reinforced by cellulose nanowhiskers. *J. Fibers Polym.* **2011**, *12*, 1002–1006. [CrossRef]

69. Park, D.J.; Choi, Y.; Heo, S.; Cho, S.Y.; Jin, H.J. Bacterial cellulose nanocrystals-embedded silk nanofibers. *J. Nanosci. Nanotechnol.* **2012**, *12*, 6139–6144. [CrossRef] [PubMed]
70. Slaughter, B.V.; Khurshid, S.S.; Fisher, O.Z.; Khademhosseini, A.; Peppas, N.A. Hydrogels in regenerative medicine. *Adv. Mater.* **2009**, *21*, 3307–3329. [CrossRef] [PubMed]
71. Sanna, R.; Fortunati, E.; Alzari, V.; Nuvoli, D.; Terenzi, A.; Casula, M.F.; Kenny, J.M.; Mariani, A. Poly(*N*-vinylcaprolactam) nanocomposites containing nanocrystalline cellulose: A green approach to thermoresponsive hydrogels. *Cellulose* **2013**, *20*, 2393–2402. [CrossRef]
72. Garbern, J.C.; Hoffman, A.S.; Stayton, P.S. Injectable pH- and temperature-responsive poly(*N*-isopropylacrylamide-co-propylacrylic acid) copolymers for delivery of angiogenic growth factors. *Biomacromolecules* **2010**, *11*, 1833–1839. [CrossRef] [PubMed]
73. Allan, S.H. Hydrogels for biomedical applications. *Adv. Drug Deliv. Rev.* **2002**, *54*, 3–12.
74. Liang, X.; Kozlovskaya, V.; Chen, Y.; Zavgorodnya, O.; Kharlampieva, E. Thermosensitive multilayer hydrogelsof poly(*N*-vinylcaprolactam) as nanothin films and shapedcapsules. *Chem. Mater.* **2012**, *24*, 3707–3719. [CrossRef] [PubMed]
75. Millon, L.E.; Wan, W.K. The polyvinyl alcohol-bacterial cellulose systemas a new nanocomposite for biomedical applications. *J. Biomed. Mater. Res. Part B* **2006**, *79*, 245–253. [CrossRef] [PubMed]
76. Loizou, E.; Butler, P.; Porcar, L.; Kesselman, E.; Talmon, Y.; Dundigalla, A.; Schmidt, G. Large scale structures in nanocomposite hydrogels. *Macromolecules* **2005**, *38*, 2047–2049. [CrossRef]
77. Liang, L.; Liu, J.; Gong, X. Thermosensitive poly(*N*-isopropylacrylamide)-clay nanocomposites with enhancedtemperature response. *Langmuir* **2000**, *16*, 9895–9899. [CrossRef]
78. Stuart, C.M.A. Supramolecular perspectives in colloidscience. *Colloid Polym. Sci.* **2008**, *286*, 855–864. [CrossRef]
79. Liu, T.Y.; Hu, S.H.; Liu, K.H.; Shaiu, R.S.; Liu, D.M.; Chen, S.Y. Instantaneousdrug delivery of magnetic/thermally sensitive nanospheresby a high-frequency magnetic field. *J. Controlled Release* **2008**, *126*, 228–236. [CrossRef] [PubMed]
80. Sanna, R.; Alzari, V.; Nuvoli, D.; Scognamillo, S.; Marceddu, S.; Mariani, A. Polymer hydrogels of 2-hydroxyethylacrylate and acrylic acid obtained by frontal polymerization. *J. Polym. Sci. Part A Polym. Chem.* **2012**, *50*, 1515–1520. [CrossRef]
81. Sannino, A.; Demitri, C.; Madaghiele, M. Biodegradable cellulose-basedhydrogels: Design and applications. *Materials* **2009**, *2*, 353–373. [CrossRef]
82. Zhou, C.; Wu, Q.; Yue, Y.; Zhang, Q.J. Application of rod-shaped cellulose nanocrystals in polyacrylamide hydrogels. *Colloid Interface Sci.* **2011**, *353*, 116–123. [CrossRef] [PubMed]
83. Wang, Y.; Chen, L. Impacts of nanowhisker on formation kinetics and properties of all-cellulose composite gels. *Carbohydr. Polym.* **2011**, *83*, 1937–1946. [CrossRef]
84. Yang, X.; Bakaic, E.; Hoare, T.; Cranston, E.D. Injectable polysaccharide hydrogels reinforced with cellulose nanocrystals: Morphology, rheology, degradation, and cytotoxicity. *Biomacromolecules* **2013**, *14*, 4447–4455. [CrossRef] [PubMed]

Chitin Nanofibrils-Chitosan Composite Films: Characterization and Properties

Galina Tishchenko, Pierfrancesco Morganti, Marco Stoller, Ivan Kelnar, Jana Mikešová, Jana Kovářová, Jindřich Hašek, Radomír Kužel, Miloš Netopilík, Ludmila Kaprálková, Milena Špirková, Ewa Pavlová, Eliška Chanová, Libuše Brožová, Michal Pekárek, Libor Kobera, Zdenka Sedláková and Dana Kubies

Abstract: In this research, an attempt to prepare a prototype of fully biodegradable nanocomposite films for food packaging was made by using commercial chitosan (CS) and chitin nanofibrils (CN). If food products could be packaged in these innovative films instead of paper (cellulosic) ones, the speed of deforesting in Europe will be diminished, thus improving the ecological situation and climate. In the paper, properties of the CS/CN films are described. The prototype films have been prepared from aqueous viscous CS/CN slurries by casting technique. The effect of many operating variables on morphology and properties of the nanocomposite CS/CN films such as sorption of water vapors, oxygen permeability, contact angle, mechanical stability and thermostability is analyzed and discussed. The examined variables are CN content, type of plasticizers (polyglycerols) and their content, type of metal ions such as Mg^{+2}, Ca^{+2}, Ba^{+2} and type of various additives (gelatin, lignin and monolignols) or surface modifiers (polylactides, polyglycerol polyricinoleate). Special attention was paid to the study of the effect of the storage time of the slurries on their aging and mechanical characteristics of CS/CN films.

---

1. Introduction

This research solves several particular problems of the global European project Bioeconomy, designed to become the basis of a new lifestyle for future generations. The goal of Bioeconomy is to create a more sustainable future, where all natural resources will be used in the most rational and efficient way for existence of a "zero waste" society. As a concrete example of Bioeconomy in action, the utilization of waste from the fishing industry can be considered. The production of chitin nanofibrils (CN) and chitosan (CS) from exoskeletons of crustaceans, which are accumulated in huge quantities in fisheries daily, improves not only the ecological situation but also creates additional sources of renewable feedstock. New products based on CS and its derivatives and based on CN have been already applied in medicine, dermatology, cosmetics, and material science.

Nature is the most skillful designer of myriad nanocomposites. Crustaceans' exoskeletons (the polysaccharide-protein-mineral nanocomposites) are typical examples [1]. The integuments of crustaceans contain the linear water-insoluble homopolymer of β-(1→4)-linked *N*-acetyl-D-glucosamine (GlcNAc) in the form of nanofibrils composed mainly from α-chitin with antiparallel orientation of GlcNAc chains. Each chitin nanofibril, about 3 nm in diameter and 300 nm long [1], is formed by a bundle of 20–30 α-chitin chains interconnected with multiple hydrogen bonds and hydrophobic interactions. CN, as is well known, are nanocrystals [2–4], which reinforce the non-crystalline proteins in carapaces just as the steel reinforcement strengthens concrete panels. The surface of CN serves for proteins enveloping them perhaps as a template. Micro- and nano-crystals of calcium carbonate and phosphate are distributed chiefly between protein molecules and probably affect their mobility and spatial arrangement [1], finally enhancing their rigidity.

In material science, mainly chitosan (deacetylated chitin) has been of interest since its discovery in 1811 by Henri Braconnot. Owing to solubility in acidified water at pH 4–4.5, a lot of remarkable properties of chitosan, including its excellent film-forming ability, have been discovered and recently exploited in many applications [5]. In contrast to rigid CN, CS chains bearing protonated amine groups in glucosamine (GlcN) rings are repelled and the chains acquire flexibility. However, the hydrophobic parts of CS chains are less flexible due to intra-chain hydrogen bonds between the atom HO3 of one sugar unit and the O5 atom of the next monosaccharide in the same chain that reduces the conformational variability. The chains with block-distribution of GlcNAc rings form the micelle-like aggregates which are interconnected by almost fully deacetylated CS chains stretched by electrostatic repulsion. The molecular dynamic simulations have shown that in such aggregates, the density of chain packaging is higher than that of aggregated chains with uniformly distributed acetylated rings. These phenomena are enhanced by increasing the acetylation degree (DA) affecting the viscosity of CS solutions [6]. After evaporation of plasticizing water molecules, the dissociation of acid molecules decreases, the protonation of CS chains is suppressed, and their repulsion becomes so negligible that they come together forming homogeneous and transparent CS films. The dried flexible CS films are stable so long as they are not in contact with water, which can promote their swelling and even complete dissolution. The non-protonated CS films are stable in water at neutral pH. They are obtained when counterions and free acid molecules, which remained inside after drying the slurries, are removed from the dried films. This is usually achieved by dipping the dried CS films in the alkaline solution followed by rinsing with water and drying. For preparing chitosan films stable in aqueous solutions in a wide pH range, the CS chains must be crosslinked with covalent bonds using bifunctional agents (e.g., glutaric dialdehyde, genipin and, etc.). Both procedures make the practical realization for production of CS films at large scale difficult.

An alternative simple method based on "physical crosslinking" of CS chains with chitin nanofibrils is described in this paper. It has been developed for preparing biodegradable water-insoluble-at-neutral-pH nanocomposite CS/CN films for use in food packaging.

2. Experimental Methods

2.1. *Solid State ^{13}C CP/MAS NMR*

1D solid-state NMR spectra were measured using a Bruker Avance 500 NMR spectrometer. Magic angle spinning (MAS) frequency of the sample was 10 kHz. In all cases the dried samples were placed into the ZrO_2 rotors and stored under silica-gel to prevent rehydration. Amplitude-modulated cross-polarization (CP) with duration 1 ms was used to obtain ^{13}C CP/MAS NMR spectra with 5 s recycle delay. The ^{13}C scale was calibrated with glycine as external standard (176.03 ppm—low-field carbonyl signal).

2.2. *Fourier Transform Infrared Spectroscopy (FT-IR)*

FT-IR spectra were recorded using a spectrophotometer Perkin-Elmer Paragon 1000PC and Attenuated total reflection technique Specac MKII Golden Gate Single Reflection ATR system with a diamond crystal; the incidence angle was 45°. The 16 scan-spectra with 4 cm^{-1}-resolution were recorded in the range of wavenumbers: 4400–450 cm^{-1} and evaluated using Spectrum 2.00 software. The samples were directly applied without modification to the diamond crystal and measured.

2.3. *Size Exclusion Chromatography (SEC)*

The commercial chitosan (Giusto Faravelli SpA, Italy) (DA = 22%) was analyzed by SEC with dual light scattering-concentration detection: a light-scattering photometer (DAWN DSP-F, Wyatt Technology Corp.) measuring at 18 angles of observation and a differential refractometer (Shodex RI 71) were the detectors. The mobile phase was the acetate buffer, 0.15 M ammonium acetate/0.2 M acetic acid buffer (pH 4.5) and chitosan solutions (1–5 $mg{\cdot}mL^{-1}$) were filtered through 0.2-μm pore size disposable syringe filters (Watrex) before injection. Two columns PL Aquagel OH-MIXED B was the separation system.

For studied chitosan samples, molecular weight was also determined by measuring the intrinsic viscosity $[\eta]$. The viscosity-average molecular weight M_η was calculated using the Mark-Houwink equation with the constants for chitosan in the acetate buffer experimentally obtained by Yomota et al. [7]. In the range of molecular weights $1.15 \times 10^3 < M < 1.59 \times 10^6$, the constants K and a equal to 0.199/mL·g^{-1} and 0.59, respectively, were used in calculations. For measuring the intrinsic viscosity, the classic Ubbelohde viscometer modified for foaming solutions with the capillary diameter of 0.42 mm was used [8].

2.4. Rheological Measurements

The rheological experiments were performed at room temperature, using a rheometer Physica MCR 501 (Anton Paar GmbH, Austria), equipped with an anti-slipping parallel-plate geometry. Viscoelastic properties in oscillatory shear flow were measured in the linear viscoelasticity region. The start-up tests at small shear frequencies were performed to investigate the dependence of solutions viscoelasticity on time. The measurements were started after 3 s pre-shearing at the shear rate 0.03s^{-1}, to ensure the same shear history for the measured samples. The homogeneous slurries with CS/CN proportion 70/30 wt. % containing 25 wt. % of a plasticizer from the total content of CS and CN in the slurry were tested.

2.5. Scanning Electron Microscopy (SEM) and Transmission Electron Microscopy (TEM)

SEM was used to examine the surface structures of CS/CN films and their fractures. The observation was performed on field-emission gun scanning electron microscope Quanta 200 FEG (SEM; produced by FEI, Czech Republic) in the secondary electron mode at accelerating voltages of 5 kV. The samples were sputtered with platinum (~5 nm; Vacuum sputter coater SCD 050, Balzers) to avoid charging and sample damage due to electron beam.

TEM observations of CN dispersions were performed on a Tecnai G2 Spirit Twin 120kV (FEI, Czech Republic). The solution of bromophenol blue (0.1 wt. %) was used as a color marker to increase the contrast of the chitin nanofibrils. A drop (2 μL) of the solutions was put onto a copper TEM grid (300 mesh) coated with a thin carbon film transparent for electrons. The excess of solution was sucked out by touching the bottom of the grid with filtering paper. This fast removal of the solution was performed after 1 min to minimize oversaturation during the drying process. The sample was left to dry completely at ambient temperature and then observed with a TEM microscope using bright field imaging.

2.6. Atomic Force Microscopy (AFM) of CS/CN Films in the Dried State

For characterizing the film surface at the micrometer and nanometer scale, atomic force microscopy (AFM) was applied for obtaining information about the surface topography and about the film properties, such as the surface heterogeneity. An Icon Dimension instrument, Bruker equipped with the SSS-NCL probe, Super Sharp Silicon™ - SPM-Sensor (NanoSensors™ Switzerland at the spring constant 35 Nm^{-1} with resonant frequency $\approx$ 170 kHz) was used. Measurements were performed under ambient conditions using the tapping mode AFM technique. The scans covered the sizes from 0.35 × 0.35 to 30 × 30 μm^2. The upper and bottom surfaces of the samples were also analyzed.

2.7. AFM of CS/CN Films in the Swollen State

AFM characterization of CS/CN films was performed on Atomic Force Microscope Dimension ICON (Bruker). All images were acquired as topographical scans in Peak Force Tapping mode in water using silicon nitride tip ScanAsyst-Fluid (Bruker) with typical spring constant k = 0.7 N/m and scan rates within the range of 0.7–0.9 Hz. Prior to testing the CS supports CN adsorbed on their surface, they were double rinsed with fresh Milli-Q water to remove an excess of no adsorbed CN.

2.8. Sorption of Water Vapors

Water vapor sorption isotherms were determined by a gravimetric method using the sorption balance IGA-003 (Hiden Isochema, UK). In the used isothermal static procedure, the source of water vapors was liquid water added to the IGA reservoir. The film sample was loaded into the microbalance, weighed, and evacuated until constant weight was reached. After determining the mass of the dry sample, it was equilibrated at the isothermal temperature (25 °C). During the measurements, the pressure of water vapors was increased step by step to achieve the constant values equal to 5, 10, 15, 20, 25 or 30 mbar. At each pressure value, the saturated state of water vapors was reached. Values of each relative pressure at 25 °C were calculated using the software program, which also restricted the maximum pressure below condensation of water vapors. The amount of adsorbed water vapors (S_W) was evaluated as a percentage from the mass of a dry film.

2.9. Contact Angle Measurements

The hydrophobicity of CS/CN films was evaluated by the values of contact angles measured using an optical device OCA20 system (DataPhysics, Germany). The measurements were performed using the sessile drop method in the static mode. The samples have been prepared as follows: film disks (1.8 cm in diameter) were placed on a glass microscope slide and their borders were fixed in four points using double-sided Scotch tape (1 mm in diameter). Before measuring, all film samples were cleaned with compressed air. Both surfaces of the studied films were analyzed at least twice if difference between the measured values was high. Three to five 30 μL-drops of water were applied to each film disk at the rate of 5 μL/s. Values of the contact angles were calculated based on the Young-Laplace fitting.

2.10. Thermogravimetric Analysis (TGA)

Thermogravimetric analysis (TGA) and differential scanning calorimetry (DSC) were used to characterize the thermal behavior of CN and CS/CN films. TGA measurements were performed on Perkin Elmer Thermogravimetric Analyzer Pyris 1 in air atmosphere to determine among others also the thermo-oxidative stability of the studied samples. The temperature interval of measurements was usually 30–800 °C, with gradual temperature rise of 10 °C/min and air flow 50 mL/min on samples of about 10 mg.

2.11. Measuring the Mechanical Characteristics of CS/CN Films

Tensile tests were carried out at ambient temperature using an Instron 5800 apparatus at a crosshead speed of 1 mm/min. At least eight specimens were tested for each sample. The stress-at-break, σ_b (variation coefficient < 2%), elongation at break, ε_b (variation coefficient < 5%), and Young's modulus, E (variation coefficient < 6%), were evaluated. Test specimens according to ISO with a length of working part 10 mm and width 2 mm were cut from cast films with thickness < 0.1 mm. The samples were conditioned in a hermetic box at 43% RH and 22 °C for 5 days before testing.

2.12. X-ray Diffraction Studies

Small Angle X-ray Scattering (SAXS) experiments were performed using a 3-pinhole camera (Molecular Metrology SAXS System now Rigaku). A X-ray beam produced by the microfocus X-ray tube (Bede Microsource) operating at 45 kV and 0.66 mA (30 W) was monochromized and focused by multilayer spherical optics (Osmic Confocal Max-Flux). SAXS patterns were measured using a multi-wire, gas-filled area detector with an active area diameter of 20 cm (Gabriel design). The probed q region was from 0.049 to 10.3 nm^{-1}, $q = (4\pi/\lambda)\sin(\vartheta)$, where $\lambda = 0.154$ nm is the wavelength and 2ϑ is the scattering angle. Data were always merged from two overlapping measurements in high- and low-resolution regions. Calibration was performed using Silver Behenate powder. Measurements took up to 2 h, depending on the quality and intensity of diffraction from sample.

3. Results and Discussion

3.1. Chitin Nanofibrils

The aqueous dispersions of CN (Figure 1a,b) resemble milky fluids. Their color may be different from slightly yellowish to brownish due to some admixture of carateno-protein components remaining bound to CN surface.

Figure 1. (**a**,**b**) Aqueous commercial dispersions of chitin nanofibrils (MAVI Sud Srl, Italy), (**c**) Clarification of commercial CN dispersions by dialysis against distilled water through 3.5-kDa SpectraPor dialysis cellulosic film. The precipitate and supernatant of CN were separated by centrifugation at 5000 rpm for 30 min.

The commercial CN dispersions (MAVI Sud Srl, Italy) contain sodium benzoate (0.5 wt. %) protecting them against microbial growth. As usual, the chitin nanofibrils do not precipitate at storage for at least a half a year but at their clarification by dialysis against distilled water (Figure 1c), a part of CN has been precipitated due to changing in the charge state of the CN surface after removal of the surface-adsorbed molecules of sodium benzoate. Initial pH of CN dispersions is shifted from acidic values (1.8–2) to more neutral (about 5.2–5.5), which are more favorable for aggregation of CN facilitating their precipitation.

The characterization of chitin nanofibrils in the swollen state was performed using AFM in the peak force tapping mode in water. In both precipitate (Figure 2a,c) and supernatant (Figure 2b,d), the CN had a spindle-like shape. They were slightly longer and thinner in the precipitate (Figure 2a,c) than in the supernatant (Figure 2b,d). The shape of CN in the swollen state was quite similar to that in the dried state (Figure 3a,b).

In AFM (Figure 2), SEM (Figure 3a) and TEM (Figure 3b) the average length of both swollen and dried CN did not exceed about 400–500 nm but their average diameter was about twice as small (about 35–40 nm) as that of the swollen CN (75–80 nm). The diameter of a single CN was considerably less (about 3 nm) [1]. It seems that the method of production of CN [4] has allowed the isolation of the chitin nanoparticles consisting of a higher quantity of bundles of nanofibrils.

We already know that the dimensions of CN have been affected by temperature and some salts such as $CuSO_4$ and $Cu(CH_3COO)_2$. The effect of copper salts on the structure of CN has been enhanced with time and depends on the type of added salt. The copper sulfate molecules were able to fix on the CN surface (Figure 3b) without their destruction. When the CN supernatant was heated at 60 °C for 30 min in the presence of copper acetate, the formation of bundles of single chitin chains was observed (Figure 3c).

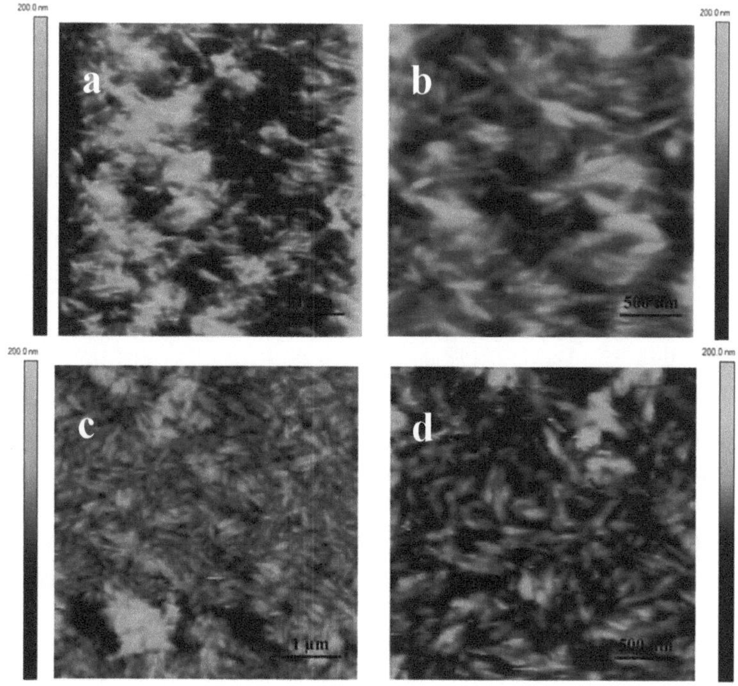

Figure 2. AFM images of CN in the swollen state in precipitate (**a**,**b**) and in supernatant (**c**,**d**). The scan size: 5 × 5 µm (**a**,**c**) and 2.5 × 2.5 µm (**b**,**d**). The CN were previously adsorbed on the surface of CS films during a day.

Figure 3. (**a**) SEM and (**b**,**c**) TEM images of dried CN: (**a**) initial (reprinted from [9]); (**b**) with surface-adsorbed $CuSO_4$, and (**c**) heated at 60 °C for 30 min in the presence of Cu $(CH_3COO)_2$.

This is the clear evidence that some protein molecules enveloping the chitin chains are tightly fixed on the surface of CN. In the absence of metal salts, there were no essential changes of CN dimensions even during heating of CN supernatant free of metal salts at 60 °C for 1.5 h. Although the CN became thinner and somewhat shorter (Figure 4c,d vs. Figure 4a,b), their structure was not destroyed, and the reinforcing ability did not change.

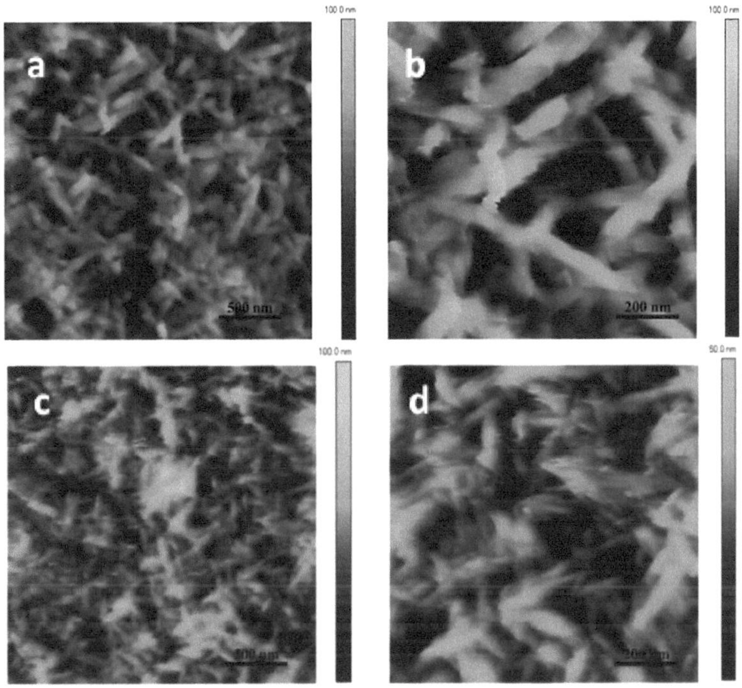

Figure 4. AFM topography images of CN in water: (**a**,**b**) initial supernatant, (**c**,**d**) after its heating at 60 °C for 1.5 h. The scan size: (**a**,**c**)—2.5 × 2.5 μm, (**b**,**d**)—1.0 × 1.0 μm. The CN were previously adsorbed on the surface of CS films during a day.

The crystallinity of CN [4] for all the samples has been tested by SAXS. The character of the X-ray diffraction pattern (Figure 5) is typical for a well-ordered crystalline phase.

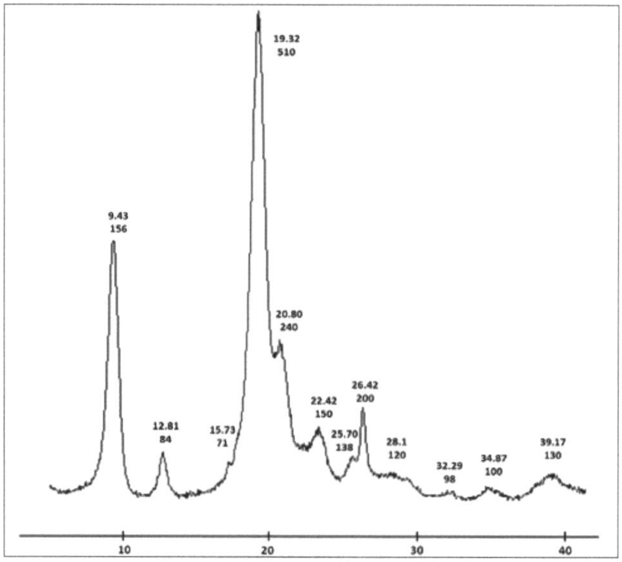

Diffraction angle (in 2θ degrees)

Figure 5. The X-ray diffraction pattern of α-chitin nanofibrils. The upper and bottom digital labels (e.g., 9.43 and 156) at diffraction peaks denote the respective diffraction angle (in 2θ degrees for $CuK\alpha_{12}$) and X-ray intensity (in arbitrary scale), respectively.

Broadening of peaks corresponds to crystal imperfections due to different lengths of polysaccharide chains forming CN and due to small crystalline domains of nanofibrils. Positions of peaks uniquely identify the crystalline phase of α-chitin. This agrees (Table 1) with other observations [10].

Table 1. Characteristic peaks of the α-chitin nanofibrils in the X-ray diffraction pattern.

Diffraction Angles (in 2θ Degrees) at the Main Peaks in X-ray Spectrum of α-Chitin							Reference
1	2	3	4	5	6	7	
9.4	12.8	19.3	20.8	22.4	26.4	32.3	Figure 5
9.2	12.6	19.2	20.6	23.2	26.2	32.2	[10]

The measured diffraction pattern of dried α-chitin (Figure 5) corresponds to the well-ordered crystalline polymer. It remains unchanged, i.e., the crystalline quality and average size of nanofibrils are identical no matter if the diffraction angle (in 2θ degrees) has been recorded on aqueous CN dispersion or dried CN flakes, on a CN monolith or CS/CN plasticized or non-plasticized films. However, heating the aqueous CN dispersion in the presence of nitrates of such metals as Cu^{+2}, Be^{+3} and Ag^{+} almost completely destroyed the crystallinity of CN.

3.2. Chitosan

As a rule, four production steps (demineralization, deproteinization, discoloration and deacetylation) are common for all manufacturers of chitosan from carapaces of crustaceans [9,11]. Nevertheless, the quality of various commercial chitosans can differ essentially because of the difference in the type of crustacean carapaces and in processing conditions (temperature, reaction time, concentration of alkali, pretreatment of the chitin and its concentration, particle size, concentration of dissolved oxygen and intensity of stirring). The variation of these parameters during the processing of chitin affects the properties of CS such as molecular weight and polydispersity of molecular distribution, the DA and solubility, content of mineral salts, and admixtures.

Usually, the targeted use of any packaging film determines the choice of raw materials, the quality of which must be sufficient for realization of this goal. One of the possible applications of completely biodegradable CS/CN films is, initially, one-off packaging for sandwiches, sliced cheese, or sausages, and, perhaps, packaging of food products with a short shelf life. Undoubtedly, the cost of such packaging film should be minimized because its price is included in the cost of packed products.

In our study, the cheap commercial CS of technical grade (Giusto Faravelli SpA, Italy) with molecular weight (M_W) 1425 $\pm$ 35 kDa determined by SEC and DA 21% was mainly used for preparation of CS/CN films. In some rheological experiments, the high-quality CS with M_W 425 kDa and DA 11% (HMC$^+$ GmbH, Germany) was also tested.

Preliminary investigations have shown that the yellowish CS powder of the technical grade with sizes of particles from 0.1 mm up to 6 mm was not completely soluble at ambient temperature in distilled water acidified with acetic acid to pH = 4. After filtering the CS solution, the undissolved particles were retained on the surface of a non-woven polyester Histar filter (Figure 6a).

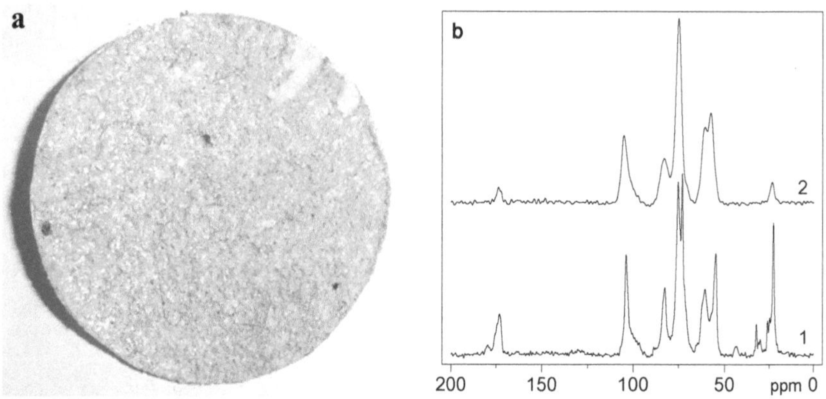

Figure 6. (**a**) Insoluble (at pH 4.0) CS particles retained on the non-woven Histar filter; (**b**) 13C CP/MAS NMR spectra of insoluble and soluble CS (curves 1 and 2, respectively) at pH 4.3.

As was determined by the solid state 13C CP/MAS NMR (Figure 6b), the insoluble particles with the DA value equal to 63% could be identified as chitin, since it is generally accepted that if the DA of the product is above 50 percent, it is characterized as chitin [12].

When prepared from raw chitosan solution, the composite CS/CN films contained additional microparticles (CM). In the SEM images (Figure 7a–c), these films looked unattractive from a commercial point of view because of both high roughness of their surfaces and morphological heterogeneity. Moreover, the films containing chitin microparticles together with CN were less mechanically stable (Table 2).

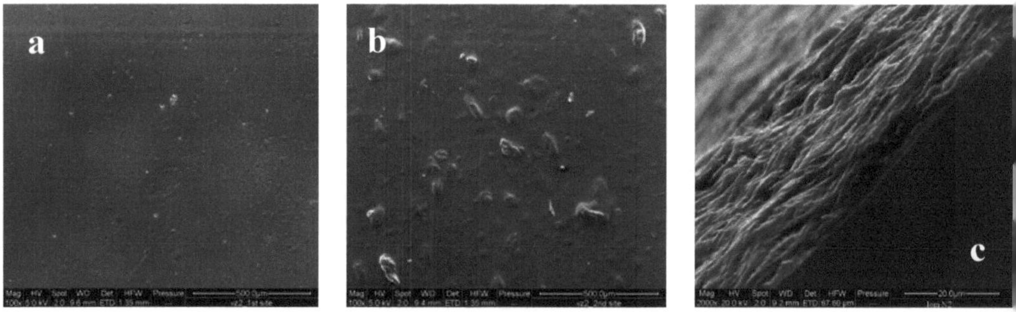

Figure 7. SEM images of (CS/CM)/glycerol film with the formulation: (40/60)/30 wt. %. (**a**,**b**) upper and bottom surfaces, respectively, (**c**) fracture of a film.

Table 2. Mechanical characteristics of the chitosan films containing CN or CM. Y—the Young's modulus; σ—the maximum tensile stress; ε—the strain at break of the film.

Film	Formulation	Y, MPa	σ, MPa	ε, %
(CS/CN)/glycerol	(40/60)/30 wt. %	3093 ± 780	61.6 ± 4.4	9.3 ± 2.0
(CS/CM)/glycerol		211 ± 35	21.6 ± 1.8	19.5 ± 2.5

Therefore, to standardize the preparation process of the CS/CN films, the commercial CS powder was sieved through a set of standard sieves and two fractions with particles' dimensions less than 80 μm and (80–140 μm) were used.

Both CS and composite CS/CN films prepared from the sieved CS powder looked rather homogeneous. Nevertheless, they still contained some microparticles undissolved at pH 4.3 at the ambient temperature. To dissolve them completely, the CS solution had to be heated at least for an hour at 60 °C.

3.3. Rheological Behavior of CS/CN Slurries

The CS/CN slurry consisting of two phases—liquid (aqueous chitosan solution) and solid (chitin nanofibrils)—behaved as a structural viscoelastic liquid, the rheological properties of which depend on multiple interactions between the components [13].

In both aqueous CS solution and CN dispersion, there are both repulsive and attractive forces. In CS solution at pH 4–4.5, repulsion of protonated GlcN rings in some parts of CS chains dominates over attraction of GlcNAc rings in other parts of chains that prohibit the fast aggregation of CS molecules. The CN dispersions tend not to settle also for very long time, due to steric hindrance and repulsion of protonated GlcN rings existing on the surface of nanofibrils. Both systems, if they stand undisturbed during some time, are stabilized through the networks of multiple hydrogen bonds and hydrophobic interactions combining all constituents in each system. In CS/CN slurry, new interactions between CS chains and CN become dominant because of the high surface potential at the liquid/solid interface [14]. All these reasons are valid for freshly prepared CS solutions and slurries if their testing was carried out for two days but no longer, since the longer-term storage has caused irreversible changes of their rheological properties.

The rheological behavior of pure CS solution, CS/CN slurries without and with added plasticizers (glycerol, polyglycerol-3 or PEG-600) or saturated solutions of calcium, magnesium or barium hydroxides was investigated in detail [15]. These investigations were the basis for optimization of the composition and concentration of CS/CN slurries that, in turn, has led to improvement of the mechanical properties of composite films.

It was found that in all tested systems, the self-assembly process has ended by formation of *reversible thixotropic gels*, the characteristic feature of which was the loss of its internal microstructure in shear and the subsequent recovery in rest. Moreover, the CN appeared to be a strong gelling agent accelerating the self-assembly of slurries. In contrast, the addition of plasticizers retarded the formation of gel. Thus, e.g., for pure CS solution, the gel point has not been achieved even after 3.5 h-rest, whereas 28 min and 80 min were needed for gel formation of non-plasticized and PEG-600-plasticized CS/CN slurries, respectively [15]. The gel point of the slurries containing ions of the alkali-earth metals depended on the type of introduced metal ion. Ions of Ca^{+2} had the highest effect on the rate of gelation: in two glycerol-plasticized CS/CN slurries with and without Ca^{+2} ions, the gel has been formed after 12 min and 22min in rest, respectively. Ions of Mg^{+2} had delayed the formation of the plasticized gel up to 49 min. The effect of added Ba^{+2} on gelling the glycerol-plasticized CS/CN slurry was considerably stronger. The same pre-shearing conditions were insufficient for breaking the internal microstructure of the gel when saturated solution of barium hydroxide instead of calcium or magnesium hydroxides were added into CS/CN slurries (Figure 8).

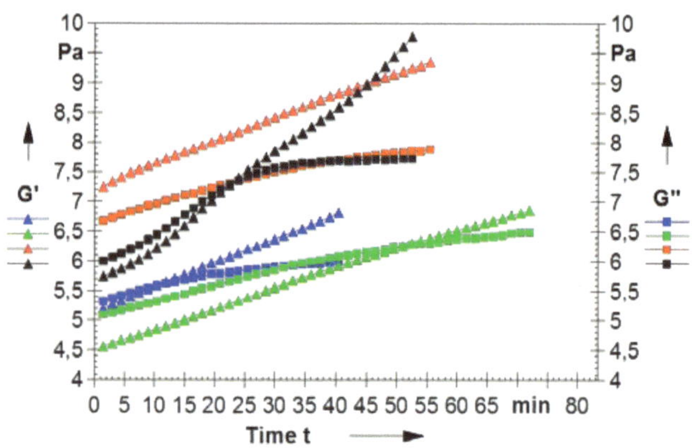

Figure 8. A self-organization process in the chitosan slurries—the storage G′ and loss G″ dynamic moduli are plotted as functions of time: the slurry CS/CN/glycerol without metal ions (▲, ■), with Ca^{+2} ions (▲, ■), with Mg^{+2} ions (▲, ■), with Ba^{+2} ions (▲, ■).

In contrast to the loss modulus G″ being a measure of the energy dissipated in the slurry under the applied shearing deformation, the storage modulus G′ characterizes the elastic deformation, at which the energy stored in a slurry under shear is then expended to return its parameters to the initial state when the shearing force is removed. For slurries containing transition metal ions, the values of the storage modulus G′ increase in the range: CS-Mg^{+2} < CS-Ca^{+2} < CS < CS-Ba^{+2} that can be interpreted as an increase in springiness of the slurries.

The found thixotropic behavior of the slurry is a key factor at casting the CS/CN slurry on a template [15]. It characterizes the ability of the slurry to be easily applied to a surface during casting owing to breaking down of the microstructure of gel and rebuilding its viscosity in rest so that the coating does not drip and does not spread over the surface [13].

Among the most important rheological aspects describing the quality of coating is considered the *leveling*, which refers to the ability of slurry to flow laterally and diminish differences in thickness of adjacent areas of the coating, thus improving the smoothness, uniformity, and mechanical properties of CS/CN films.

The time lag for leveling of the cast slurry, after which its viscosity is restored, depends on viscosity of the slurry: the higher the viscosity, the slower leveling of coating. The time lag leveling was about 20 min for the slurries with the CS/CN proportion (70/30) wt. % plasticized with glycerol (30 wt. %), which have been used in preparation of CS/CN films.

An equally important parameter describing the rheological behavior of the slurry is the *yield stress* indicating the maximum value of the shear stress, at which the disruption of internal structure of the thixotropic CS/CN slurry will happen.

It was shown that some efforts (pre-shearing) must be applied to CS/CN slurries for disruption of their internal microstructures to force them to flow. The value of critical stress (*yield stress*) was about twice as high for the non-plasticized CS/CN slurry (6.8 Pa) than for plasticized ones with glycerol (3.9 Pa) or PEG-600 (3.1 Pa) if commercial CS (HMC^+ GmbH, Germany) with low-molecular-weight (Mw 374 kDa) and low DA (11%) was used (Figure 9).

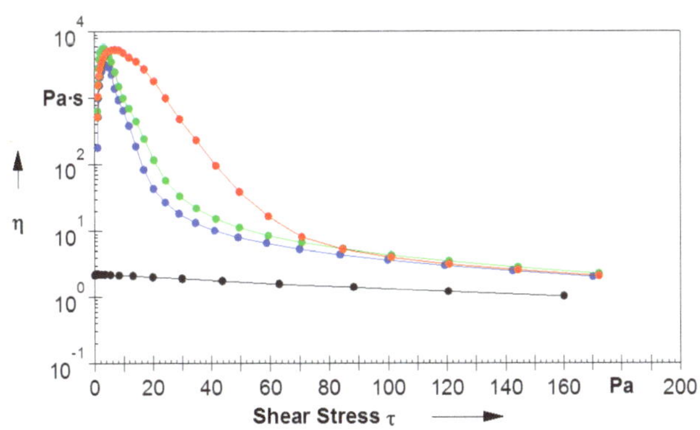

Figure 9. Steady shear viscosity vs. shear stress of the pure HMC^+ chitosan solution (●) and the slurries: CS/CN (●), CS/CN/PEG (●), CS/CN/glycerol (●). The (CS/CN)/plasticizer proportion was (65/35)/30 wt. %.

Values of the yield stress of non-plasticized CS/CN slurry have increased about three times (up to 18.2 Pa) if the CS (Giusto Faravelli SpA, Italy) with higher molecular weight (1425 kDa) and DA (20%) was used. For CS/CN slurries plasticized with glycerol and polyglycerol-3, the yield stress was less and equal to 14.5 Pa independently on the plasticizer type (Figure 10).

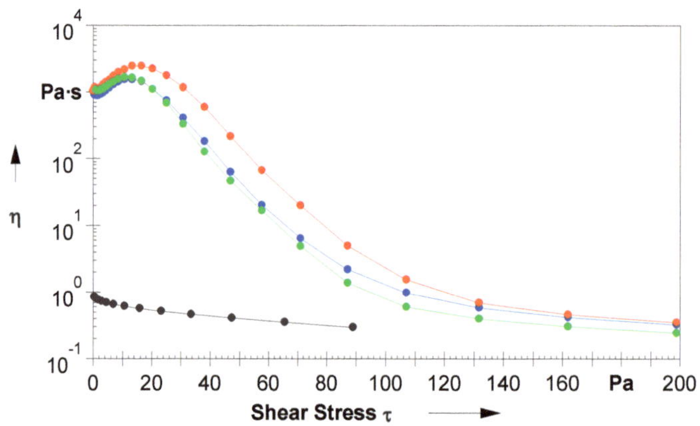

Figure 10. Steady shear viscosity as a function of the shear stress of the CS (Giusto Faravelli SpA, Italy) solution (●) and the slurries: CS/CN (●), CS/CN/glycerol (●), CS/CN/polyglycerol-3 (●). The (CS/CN)/plasticizer proportion was (70/30)/25 wt. %.

In both slurries, the CN dispersion (MAVI Sud Srl, Italy) with DA = 95% and pH = 5.12 has been mixed with CS solution. The slurries behaved as "fully" elastic fluids below the yield stress that had been manifested in their ability to absorb the shear energy as solids, which remain stationary, until the yield stress was not achieved. Above the yield stress, the internal microstructure has broken, and the slurries started to flow.

From a practical point of view, the gentle joggle or pulsed sonication of the CS/CN slurries could be effective for preventing the formation of their internal microstructure. It is reasonable to consider this recommendation for obtaining homogeneous coatings in the large-scale production of films using casting on the templates.

In this study, the homogeneous composite CS/CN films have been prepared from the slurry containing 70 and 30 wt. % of CS and CN, respectively, and 30 wt. % glycerol of total amount of CS and CN. The rheological parameters of this slurry (Figure 10, green curve) were determined as follows: the yield stress equal to about 15 Pa corresponded to the maximum steady shear viscosity about 1×10^3 Pa s.

Rheological testing provided useful information about rheological changes in time of the CS/CN slurries that allowed determination of the limit of their storage. The viscoelastic characteristics of the slurries had steadily getting worse. After their storage at 6 °C for 19 weeks, the slurries started to behave similarly to Newtonian fluids, the characteristic feature of which is the independence of their viscosity on the shear conditions (Figure 11).

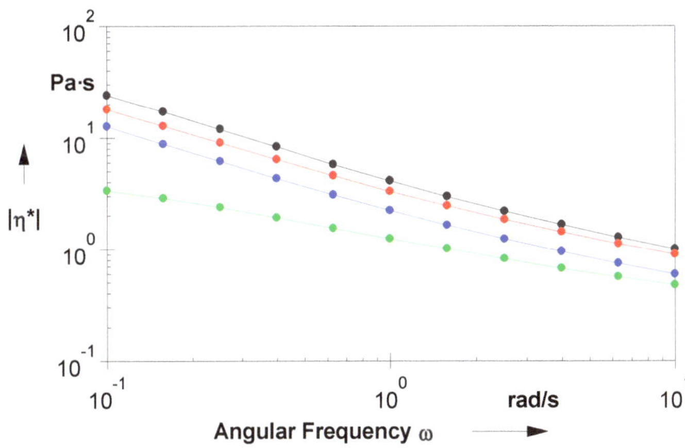

Figure 11. The absolute value of complex viscosity as a function of angular frequency of the slurry CS/CN/glycerol (70/30)/25 wt. % measured after various storage times: 1 day (●), 2 weeks (●), 6 weeks (●) and 19 weeks (●).

The absence of the rheological features typical for the physical network inside the slurries has indicated the clear evidence that CS chains had degraded into short fragments unable to form the extended microstructure of a thixotropic gel.

3.4. Composite CS/CN Films

The composite CS/CN films have been prepared by casting technique [16]. The used preparation procedure consisted of four steps (Figure 12).

Bearing in mind the effect of aging of CS/CN slurries on their rheology, its effect on the mechanical stability of the films has been also checked. It was found (Table 3) that all mechanical characteristics of the films worsened in strength the more significantly the longer the slurries were stored. Therefore, the slurries were always used as quickly as they were prepared.

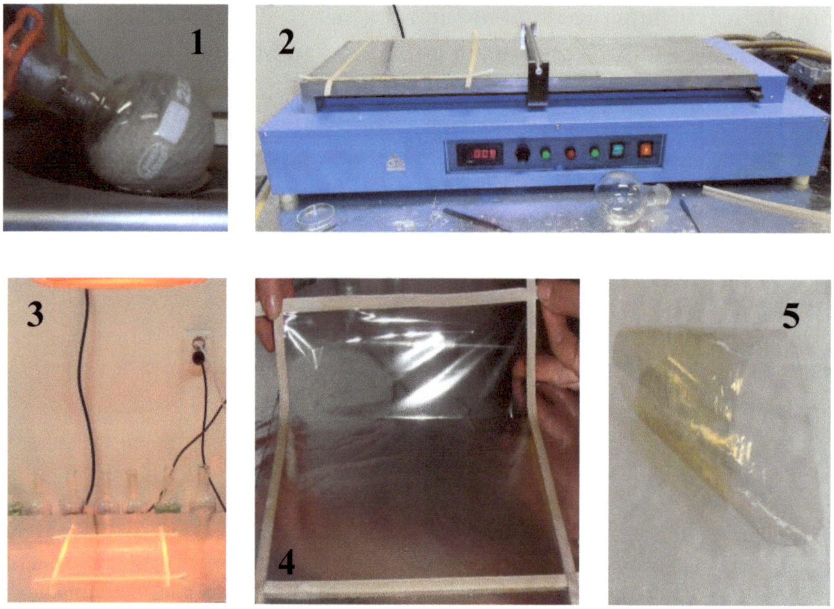

Figure 12. Visualization of the preparation steps of composite CS/CN films by casting technique. (**1**) deairing and concentrating the slurry using a rotary vacuum evaporator; (**2**) the slurry cast on a Dura Lar (Grafix Co., USA) support placed on a coater MSK AFA L800 (MTI Corp., USA) using a Doctor Blade; (**3**) drying the cast slurry by its heating at 50–55 °C using a source of the infrared light; (**4**) taking off the dried film from a support; (**5**) a died composite CS/CN film.

Table 3. Effect of the storage of a slurry on mechanical properties of (CS/CN)/glycerol films with the formulation: (70/30)/30 wt. %. I—used as prepared, pH = 4.25; II, pH = 4.43 and III, pH = 4.35—storage in fridge for 2 weeks and 1 month, respectively; The slurry II and III contained Ca^{+2} ions. The slurries II and III were additionally concentrated by vacuum evaporation prior to their casting on the supports.

Parameter	I	II	III
Y, MPa	2784 ± 582	460 ± 224	201 ± 44
σ, MPa	57.6 ± 5.9	37.8 ± 5.7	31.2 ± 4.9
ε, %	10.6 ± 2.4	25.2 ± 4.6	28.2 ± 3.2

It should be noted that vacuum evaporation used for mixing the aqueous solutions of the components, their deairing and concentrating allowed the preparation of very homogeneous slurries with desirable viscosity, which was close to the gel point but did not achieve it. This procedure took about a half of hour.

Evaporation of an excess of water from the slurry (about 25% from its weight) was carried out in a water bath with temperature at 60 °C.

The right choice of a support, on which the slurry is cast, becomes very important, if not the main factor in preparing the films of large dimensions. To avoid any defect in the dried film during its removal from the support, it is extremely important to use such a support, which will ensure easy removal of the dried film from its surface.

Some researchers were successful in preparation of experimental CS-based films having small dimensions when poly(vinyl chloride) sheets [17], polystyrene [18] or polypropylene plates [19], metallic [20] or non-stick trays [21], acrylic [22] or even Teflon [23] plates were used. Sometimes, the CS solutions were cast onto plastic [24] or glass [25] Petri dishes.

In our study, the CS/CN slurries have exhibited good adhesion to poly(vinyl terephthalate) films (Dura Lar, Grafix. Co., USA) and to stainless steel plates (Figure 13) and, at the same time, have ensured rather easy removal of the dried films from their surface by applying minimal efforts.

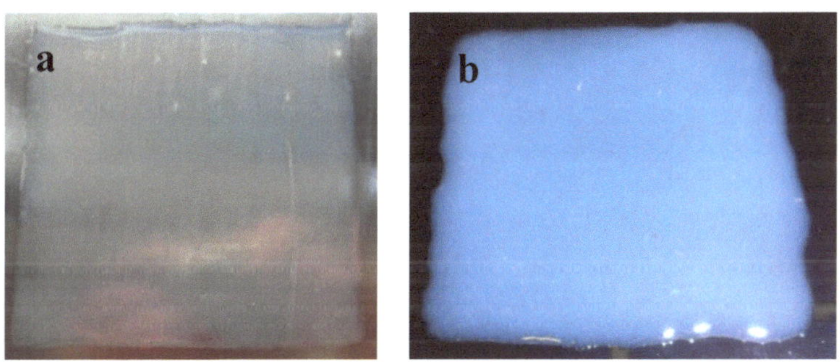

Figure 13. An optical view of the (CS/CN)/glycerol slurry at the proportion of components: (70/30)/30 wt. % cast on: (**a**) Stainless steel plate, (**b**) Dura Lar film.

The cast slurries on these supports formed uniform layers with thickness of about 0.75 mm. The thickness of dried CS/CN films plasticized with glycerol was equal to 46 ± 16 μm. For determination of the optimal formulation of the CS/CN slurry, the content of CN was increased from 3 wt. % to 40 wt. % (from the CS content). Chitin nanofibers introduced in CS solution promoted creating new spatial arrangement of chitosan chains in the dry CS/CN films, which differed considerably from the original one existing in the chitosan itself after its drying (Figure 14a–d). In aqueous acidified solution, the protonated CS chains preferred to interact with functional groups on the surface of rigid chitin nanofibrils than with each other. During the fixing on the surface of chitin nanofibers, the CS chains are packed more densely than in solution, acquiring a more energetically favorable spatial arrangement. The reinforcing effect of CN on CS phase was ensured by a high surface potential of chitin nanofibrils with a highly developed surface (180 m^2 per 1 g of dried CN) [4]. The compatibility of CS with CN was excellent owing to similarity of their chemical structures differing only in the quantitative proportion of GlcN and GlcNAc rings and their distribution in polysaccharide chains. From the structural point of view, the CS/CN films can be considered to be composites consisting of one polymer. That is why there was no phase separation after increasing the content of CN in CS solution up to 80 wt. %. This newly formed interconnected spatial structure differed from the previous one in CS by the apparent new heterogeneity at the nano and micro level.

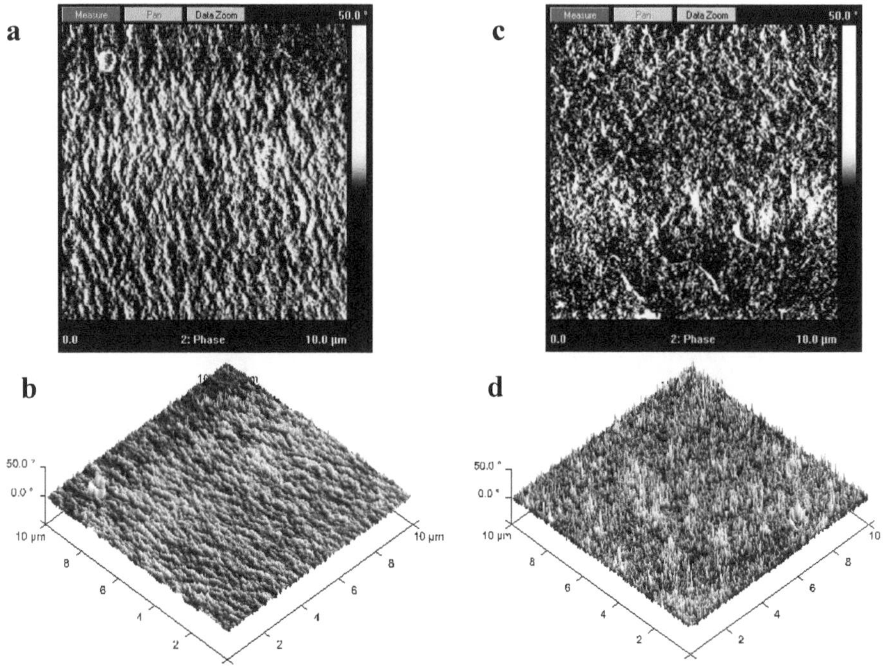

Figure 14. AFM phase images of the dry non-plasticized films: (**a**,**b**) CS; (**c**,**d**) (CS/CN) film at the proportion of components equal to (60/40) wt. %.

The increase in CN content in CS phase increased the stiffness of CS/CN films since the elastic deformation of the films has required applying stronger tensile stress at increase in CN content to a definite value (Figure 15). The ultimate tensile stress increased with CN content and achieved its maximum value for the CS/CN film with 10 wt. % of CN in CS solution. After further increase in CN content up to 40 wt. %, values of the maximum tensile stress decreased gradually. The effect of CN content on elongation of CS/CN films was the most perceptible. The value of the strain at break of a film with 40 wt. % of CN was about half of that value of a chitosan film free of CN.

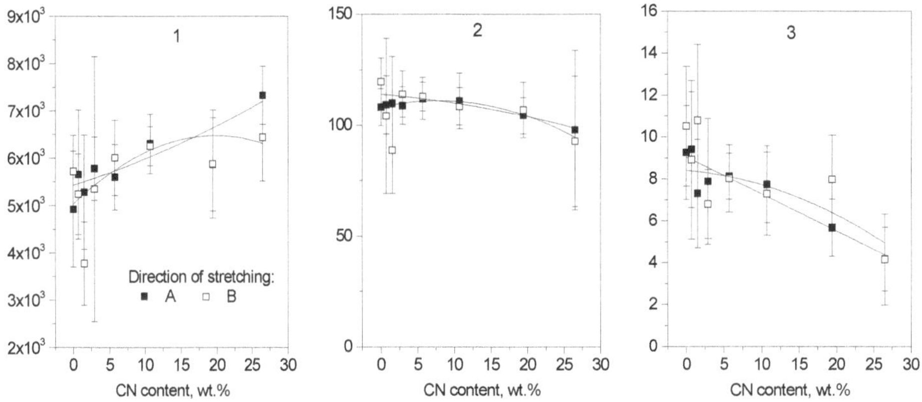

Figure 15. Effect of CN content in CS phase on the mechanical properties of non-plasticized films measured in two mutually perpendicular directions A and B of the applied tension. (1) Young's modulus; MPa, (2) Maximum stress, MPa; (3) Strain at break, %.

When the CN content exceeded 10 wt. %, the saturation limit of CS phase with CN was reached. An excess of CN was pushed out from CS phase and formed a surface-adjacent layer enriched with chitin nanofibrils (Figure 16).

Figure 16. Images of the upper surface of a (CS/CN)/glycerol film at the proportion of components (70/30)/25 wt. % in the dry and swollen state in water. (**a**) SEM; (**b**) AFM images.

The bonds between chitosan chains and the surface groupings of chitin nanofibrils were so stable that the CS/CN films did not dissolve in water despite some excess of free molecules of acetic acid, which have remained inside the films after their drying. After dipping in water, the dimensions of CS/CN films reinforced with 30 wt. % of CN have increased only a little despite the presence of plasticizing glycerol molecules weakening interactions between the components in the films (Figure 17). However, an increase in the stiffness of the CS/CN films, especially those with high CN content, resulted in a decrease in their elasticity. Such film was more brittle.

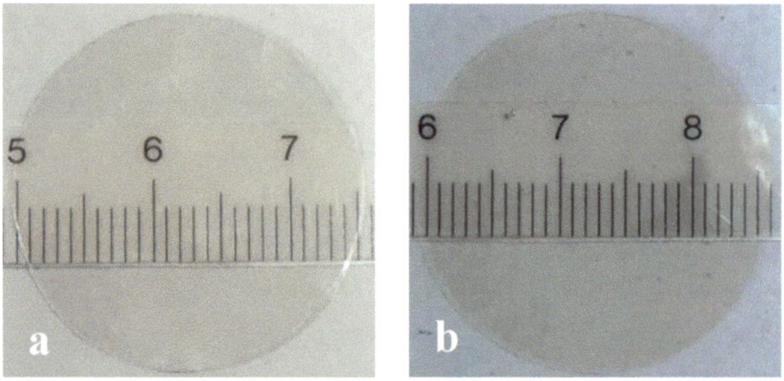

Figure 17. Dimensions of a (CS/CN)/glycerol film at the proportion of components (70/30)/25 wt. %. (**a**) dry film; (**b**) swollen film.

This problem is usually overcame by using plasticizers [26]. Small molecules of plasticizers penetrated between chitosan chains and weakened the bonds between them (Table 4). The comparison of the effect of the tested plasticizers on mechanical characteristics has shown that the elasticity of the CS/CN films increased considerably at the presence of a plasticizer and depended on both its quantity and its type.

For a film with the CS/CN proportion (85/15) wt. %, values of the elastic modulus decreased with increase in the plasticizer content in a great extent. Simultaneously, values of the ultimate tensile stress decreased for this film by about half but the values of its strain at break increased by about three-fold. It seems that it is not reasonable to increase the content of plasticizers higher than 20–30 wt. % because of dramatic worsening of all mechanical characteristics of CS/CN films.

The temperature of drying the CS/CN films was the next important parameter affecting their morphology (Figure 18) and mechanical characteristics (Table 5). The internal heterogeneity of CS/CN films became more pronounced when a cast slurry was dried using the infrared irradiation (Figure 18c,d) for half an hour instead of its drying at the ambient temperature (Figure 18a,b) for 12 h. The differences in morphology of the dried films were observed on the micro level only. No noticeable differences in the appearance of the films were observed.

Table 4. Effect of type and content of a plasticizer on mechanical characteristics of CS/CN films with the formulation (85/15 wt. %). Support: Dura Lar film.

Plasticizer, wt. %		Y, MPa	σ, MPa	ε, %
no	0	5720 ± 350	107.5 ± 7	6.7 ± 2.4
Glycerol	20	1390 ± 600	43.0 ± 5	20.4 ± 4
_"_	30	300 ± 50	35.4 ± 4	27.6 ± 4
_"_	40	50 ± 12	17.0 ± 6	36.0 ± 4.5
pG-2	20	1835 ± 177	48.6 ± 8	25.4 ± 4.7
_"_	30	700 ± 65	36.7 ± 5	32.5 ± 5.7
_"_	40	98 ± 33	25.6 ± 5.8	44.3 ± 7.4
pG-3	20	1390 ± 100	33.5 ± 6	26.7 ± 8
_"_	30	379 ± 47	25.0 ± 5	35.0 ± 6
_"_	40	65 ± 8	24.4 ± 4	50.0 ± 4
pG-4	20	2320 ± 110	41.4 ± 4.9	13.7 ± 5
_"_	30	1380 ± 50	30.0 ± 4.5	17.0 ± 5
_"_	40	370 ± 57	28.6 ± 6	31.6 ± 7

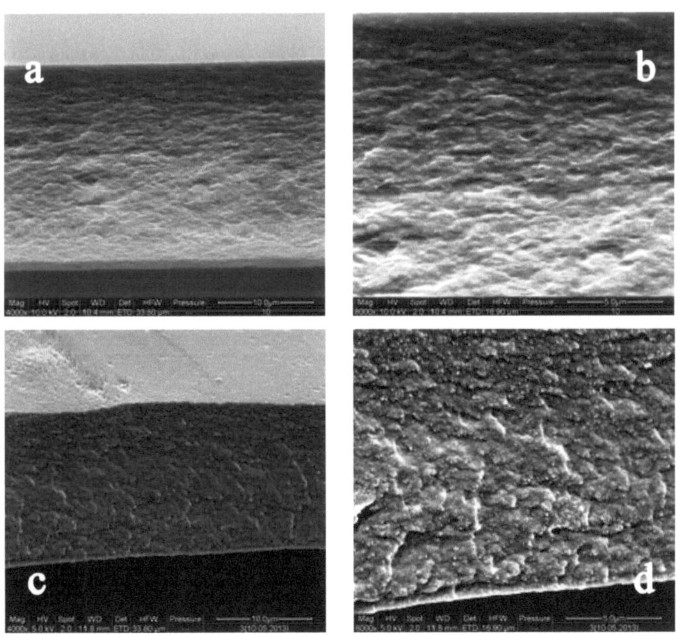

Figure 18. SEM images of fractures of (CS/CN)/glycerol films with the formulation (85/15)/30 wt. % dried at: (**a**,**b**) ambient temperature and (**c**,**d**) IR irradiation at 55 °C.

Table 5. Effect of the temperature of drying on the mechanical properties of (CS/CN)/glycerol films with the formulation (85/15)/30 wt. %.

Parameter	22–25 °C	50–55 °C
Y, MPa	1462 ± 306	2495 ± 380
σ, MPa	25.0 ± 3.6	45.0 ± 4.5
ε, %	12.6 ± 3.0	12.0 ± 3.0

Under exposition of the tested polysaccharides' films to UV light or microwave irradiation for 8 days and 30 min, respectively, the oxidative reactions were triggered inside the films. The fact of their existence was detected by analysis of UV spectra of the irradiated films (Figure 19).

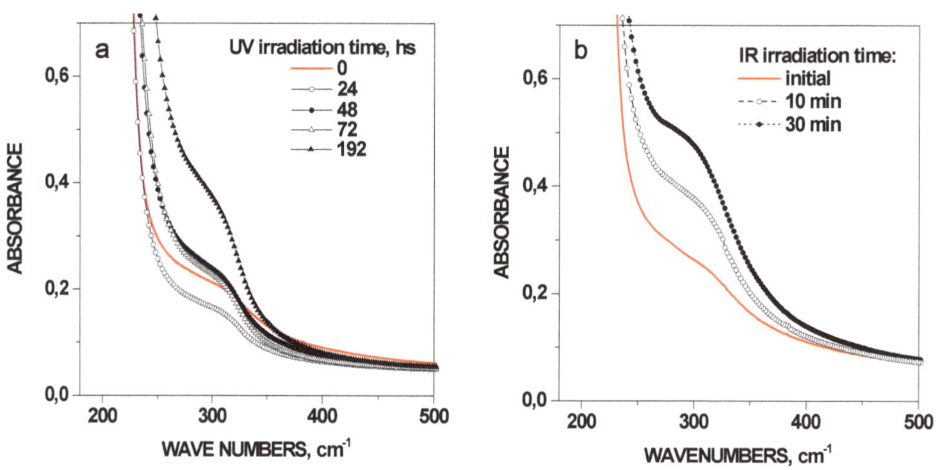

Figure 19. UV spectra of (CS/CN)/glycerol films after: (**a**) UV irradiation (0.5 N/m^2) for 192 hs and (**b**) microwave irradiation up to 30 min with the 1000 W-magneton power. The formulation of the slurries: (**a**) (40/60)/30 wt. %, (**b**) (20/80)/30 wt. %.

It is reasonable to suggest that drying the films using IR irradiation was also accompanied by accelerating the oxidative reactions in CS/CN films resulting in changes in packaging of CS chains that is manifested in morphological changes of the films (Figure 20c,d).

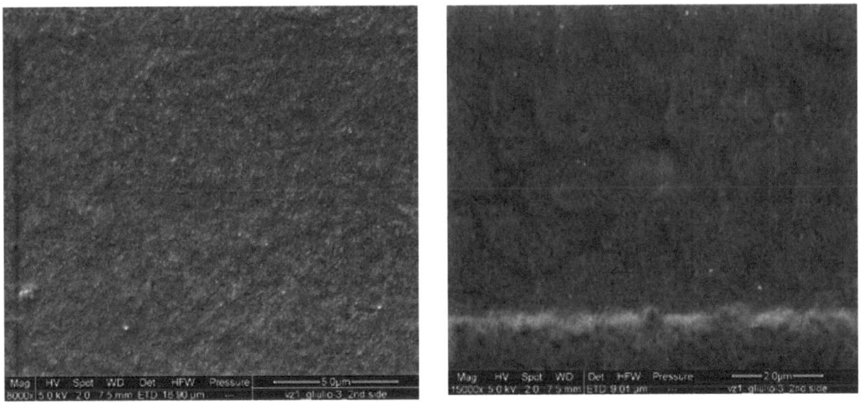

Figure 20. SEM images of the bottom surface of (CS/CN)/glycerol film with the formulation (70/30)/25 wt. % obtained by casting the slurry on a stainless steel plate.

Investigations of the degradation of CS/CN films at the elevated temperatures by thermogravimetric analysis (Table 6) has revealed that CN consisting of the densely packed chitin chains required higher temperatures than raw CS powder or reprecipitated CS. It should be noted that a sample of reprecipitated CS has lost 40% of its mass within a very narrow temperature interval with its maximum at 306 °C. This value was about 12 °C higher than the temperature maximum on a TGA curve of the raw CS. It means that to improve the thermostability of raw CS, the reprecipitation could be used to remove low-molecular-weight oligosaccharides and increase the fraction of high-molecular-weight CS. The decomposition of chitin nanofibrils occurred at temperature approximately 50 °C higher than that of CS.

It is interesting that two maximums on TGA curves of CS/CN films have been observed. The decomposition temperature of the films rose to higher values (for the first and second decomposition steps from about 304 to 311 °C and from 361 to 371 °C, respectively) with the increase in CN content from 15 wt. % to 35 wt. %. The loss of the mass of a sample decreased from 30% to 23% for the first degradation step. Simultaneously, at the second degradation step, the loss of the mass of a sample increased from about 20% to 26%. For CS/CN films having the same formulation (85/15)/30 wt. % but containing polyglycerols of different type, the higher temperature was required for decomposition of the film plasticized with polyglycerol-4 having longer molecules. For CS/CN films plasticized with polyglycerol-3, the decomposition temperature rose with an increase in the content of plasticizer at both steps of their degradation. The detailed description of TGA and DSC analysis of the composite films based on CS and CN will be published elsewhere.

Table 6. TGA analysis of CS, CN, and composite CS/CN films. T1, T2, T3 – temperature, °C. ΔW1, ΔW2, ΔW3 – loss of a sample's weight, wt. %.

Sample	Composition	T1	ΔW1	T2	ΔW2	T3	ΔW3
CS raw		294.4	47	–	–	–	–
CS *		306.0	40	–	–	–	–
CN		358.2	66	–	–	–	–
CS/CN	85/15	303.7	30	360.9	19.5	–	–
–"–"–"–	75/25	306.5	28	371.0	22.0	–	–
–"–"–"–	65/35	310.9	23	370.9	25.5	–	–
(CS/CN)/pG-2	(85/15)/30	261.5	22.5	315.4	23.5	384.5	18.0
(CS/CN)/pG-3	–"–"–"–	303.4	40.0	374.6	19.5	–	–
(CS/CN)/pG-4	–"–"–"–	304.4	41.0	385.2	21.0	–	–
(CS/CN)/pG-3	(85/15)/20	314.1	37	384.8	18.0	–	–
–"–"–"–	(85/15)/30	312.0	41.0	384.6	17.5	–	–
–"–"–"–	(85/15)/40	304.9	55.5	401.1	17.5	–	–
–"–"–"–	(85/15)/50	305.7	63.0	400.0	12.0	–	–

Analyzing the SEM images in Figure 20, everyone can clearly see that the (CS/CN)/glycerol films have denser upper and bottom layers and moreover, the film surface facing a support seems to be considerably denser than that contacting with air.

It was observed (Table 7) that not only the type of the surface (upper or bottom) of CS/CN film but also the chemical nature of a support on which it was formed remarkably affected the values of contact angles of both film surfaces. As a rule, the values of contact angles of the bottom surfaces were higher than those of the upper ones exposed to air. In contrast to the transparent glycerol-plasticized CS/CN films formed on the Dura Lar supports from poly(vinyl terephthalate), those formed on the stainless steel plates looked matte because of higher surface roughness of the latter support (Figure 20).

Table 7. Contact angles of the surfaces of glycerol- or pG-3-plasticized CS/CN films.

Surface \ Support	DL	SS	DL	SS	DL	SS
	Non-plasticized		30% Glycerol		30% pG-3	
Upper	102.2 ± 4.4	105.6 ± 2.7	105.2 ± 5.1	106.5 ± 3.5	103.2 ± 4.1	104.7 ± 1.8
Bottom	107.1 ± 3.8	125.2 ± 4.8	107.0 ± 3.8	115.4 ± 2.6	106.5 ± 3.9	109.7 ± 0.6

The chemical nature of a support, on which the slurry was cast, influenced also the composition of chemical groupings on the formed film surface. The X-ray photoelectron spectroscopy analysis of both surfaces of glycerol-plasticized CS/CN films, which differed in formulations (Table 8), has revealed that the hydrophilic groupings (–C–O–, –C–O–C– and –N–C=O–) were predominantly located on the upper surface of the films. In contrast, the bottom surfaces formed on the glass plates silanized with ethyltrimetoxysilane had more hydrophobic groupings (–C–C– and –C–H–).

Table 8. Binding energy of chemical groupings in the spectra of C 1 s electrons.

Binding Energy, eV	(CS/CN)/Glycerol Proportion, wt. %				Chemical Groupings
	(80/20)/20		(60/40)40		
	Upper	Bottom	Upper	Bottom	
285.0	29	37	28	33	–C–C–; –C–H–
285.6	9	10	10	10	$-C-NH_2$
286.6	52	46	50	47	–C–O–
288.2	10	7	12	10	–C–O–C–; –N–C=O–

The surface wettability of CS/CN films with water (Figure 21) depended on both chemical groupings on the surface of films and on the density of packaging of their structural elements.

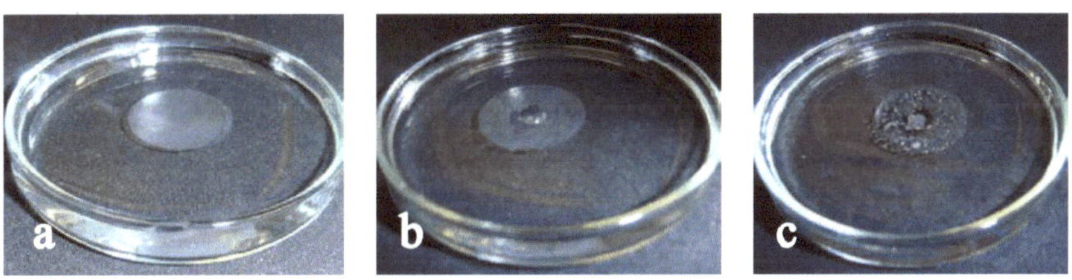

Figure 21. Wettability of a (CS/CN)/glycerol film with the formulation (75/25)/30 wt. %: (**a**) just after placing on the water surface; (**b**,**c**) after 5-min- or 2-h contact with water, respectively.

The suction of water inside such films occurred when the content of an admixture of low-molecular-weight chitooligosaccharides together with some excess of the acetic acid molecules were high and these components have transferred into water creating the channels inside the films dipped into water. This phenomenon was often observed when CS solution was heated at 60 °C for one hour to dissolve small CS microparticles for obtaining homogeneous slurry or when the slurry has been stored in the fridge for two days.

It should be noted that despite good wettability, these films did not dissolve in water and did not change their size during at least 10 days of contact with water.

Both high permeability and adsorption of water (Table 9) have been also detected for the CS/CN films containing the ions of alkali earth metals such as Ca^{+2}, Mg^{+2} or Ba^{+2}. Freshly prepared saturated solutions of hydroxides of these metals free of carbon dioxide were added to CS/CN slurries with the aim of neutralizing an excess of the acetic acid used for dissolving of CS at preparation of the slurries.

Table 9. Adsorption of water vapors with CS/CN films.

Film	Components	Support	Composition wt. %	S_W, wt. % Air-Dried [1]	Swollen [2]	$\frac{S_{W2}}{S_{W1}}$
1	(CS/CN)/PEG-600	DL	(70/30)/25	7.6	25.2	3.3
2	(CS/CN)/glycerol	–"–"–	(75/25)/30	5.3	24.8	4.7
3	(CS/CN)/pG-3	–"–"–	(85/15)/30	7.4	20.8	2.8
4	–"–"–"–	–"–"–	(75/25)/30	6.8	33.5	4.8
5	–"–"–"–	–"–"–	(65/35)/30	5.5	32.5	6.1
6	(CS/CN)/glycerol, Ca^{+2}	DL	(70/30)/25	45.2	38.1	0.8
7	–"–"–"–	SS	–"–"–	43.9	35.7	0.8
8	(CS/CN)/glycerol	DL	(70/30)/10	8.3	22.3	2.7
9	–"–"–"–	SS	–"–"–	6.8	24.0	3.5
10	(CS/CN)/glycerol, PGPR	DL	(70/30)/30	7.2	20.0	2.8
11	–"–"–"–	SS	–"–"–	6.0	25.1	4.2
12	(CS/CN)/glycerol, PLA	DL	(70/30)/30	10.3	22.2	2.2
13	–"–"–"–	SS	–"–"–	8.4	21.8	2.6
KFC	cellulose			5.4	13.0	2.4

Comparison of the air-dried CS/CN films has shown that the water content was always higher in the films prepared using Dura Lar supports than in those obtained using the stainless-steel plates. The content of water in CS/CN films plasticized with polyglycerol-3 decreased with the increase in CN content from 15 to 35 wt. %. The CS/CN films with the surface-adsorbed polyglycerol polyricinoleate or polylactide had lower water content if the stainless-steel plates instead of Dura Lar ones have been used as the support for casting the slurry. The glycerol-plasticized CS/CN films containing calcium ions retained the highest amount of water in the air-dried state. Most of this water was bound with metal ions in their hydration shells. At equilibrium with water vapors at 25 mbar, the saturation state was not achieved for these films and they contained only 80% of water from its content in the air-dried films. In the sorption experiments, the sorption of water proceeded slowly by the films and took several hours until equilibrium with water vapors was achieved at each partial pressure. Being equilibrated with water vapors at 25 mbar, the swollen CS/CN films contained from 3 to 5 times more water than the air-dried ones. In the moistened state, the CS/CN films prepared by casting the slurry on the stainless-steel plates contained more water than those obtained by using Dura Lar supports. This difference can be explained by the structural differences of the films.

These are the structural features of the films that determine the dependence of the change in the moisture vapor transmission rate through a film on its equilibrium state at each change of the partial pressure of water vapors (Table 10).

Table 10. Moisture vapor transmission rate for the (CS/CN)/glycerol film with the formulation (70/30)/30 wt. %. The film contained 0.234 mmol of Ca^{+2} ions.

Δp, mbar	**0–0.45**	**0.45–5.2**	**5.2–10.4**	**10.4–15.4**	**15.4–20.3**	**20.3–25.3**
MVTR, g/(m^2 × 24 h)	4.1	31.6	65.1	111.5	186.1	180.1
ΔMVTR/Δp	9.1	6.1	6.3	7.2	9.2	7.1

At low water vapor pressures, the rate of MVTP change (ΔMVTR/Δp value) was high since water molecules have adsorbed onto the outer surfaces and on the surface of the available pores of a dry film free from any adsorbed water at zero pressure of water vapors in the apparatus chamber at the beginning of the sorption process.

When the partial water vapor pressures achieved 15.4 mbar the second increase in the rate of MVTR changing occurred due to the changes of packaging the CS chains and the increase in the distance between them in the moistened film.

The difference in distance between CS chains in the dry and swollen films influenced their permeability for gases (Table 11). The permeability of hydrogen, oxygen and nitrogen through a swollen glycerol-plasticized CS/CN film was about triple higher than that of the dry film. The molecules of carbon dioxide penetrated through the swollen film about five times faster compared with the dry film. Comparison of the kinetic diameters and molecular mass of gases with the rate of their transfer through a CS/CN film allows us to conclude that the decisive factors defining the film permeability (P) for gases are their diffusivity (D) and solubility (S) controlled by numerous interactions of the diffusing molecules inside the CS/CN films.

Table 11. Permeability of a (CS/CN)/glycerol film with the formulation: (70/30)/25 wt. % for gases.

Film	**P**	**H2**	**O2**	**N2**	**CO2**	**CH4**
Dry	barrer	0.04	0.012	0.005	0.142	0.01
Swollen		0.137	0.038	0.013	0.761	–
Kinetic diameter	Å	2.89	3.46	3.64	3.30	3.8
Molecular mass	D	2	32	28	44	18

Analysis of the transfer of oxygen through the composite films with the same content of CS and CN but containing various plasticizers has elucidated the correlation between the structure of the composite films and their permeability for gases.

The most permeable for oxygen were the films plasticized with PEG-600 (Table 12). The permeability and the diffusion coefficient of oxygen in these films were about three and five orders, respectively, higher than those of the films plasticized with glycerol. In contrast, the solubility of oxygen in the former films was about three orders lower than that in the latter ones. High diffusivity together with low solubility of oxygen in the PEG-plasticized films ensured their high permeability. The observed differences in the transport performance of the films were determined by the difference in their spatial structures, on which the different plasticizers were affected in different ways.

Table 12. Oxygen transport through dry plasticized CS/CN films with the formulation: (70/30)/30 wt.% * 1 barrer = 1× 10^{-10} cm^3 (STP) × cm /(cm^2 × s × cm Hg) [27,28].

Film	P		D	S
	10^{18} mol/(m × Pa × S)	barrer*	10^{15} m^2/s	10^6 mol/(m^3 × Pa)
(CS/CN)/PEG-600	172	0.512	1.21 × 10^6	0.14
(CS/CN)/glycerol	0.18	0.00054	2.25	80.7
(CS/CN)/pG-2	1.28	0.0038	14.4	88.7
(CS/CN)/pG-3	1.21	0.0036	25.9	46.8
(CS/CN)/pG-4	0.13	0.00037	2.43	51.7

For decreasing wettability and permeability of CS/CN films for water vapors, the hydrophobization of their surfaces was carried out.

This procedure was performed by dipping the films into dioxane solution of polylactide (PLA) with molecular mass of 7.7 kDa or polyglycerol polyricinoleate (PGPR) for half an hour. Both surfaces of the plasticized CS/CN film treated in such a way have acquired their water-repellent properties (Figure 22) and simultaneously their mechanical characteristics were considerably improved.

The effect of surface modifiers on the mechanical characteristics of (CS/CN) films containing various low- and high-molecular-weight substances are summarized in Tables 13–16. From the results (Table 13), we can get an idea of the effect of glycerol, gelatin, and CN on the magnitudes of the elastic modulus (Y), the ultimate stress (σ) and the strain at break (ε) of the CS-based films. In contrast to glycerol being the disintegrating agent for the CS phase, gelatin molecules have combined the CS chains promoting their tighter packaging but the introduction of glycerol in the CS/gelatin slurry-2 caused another disintegration of the formed intermolecular bonds.

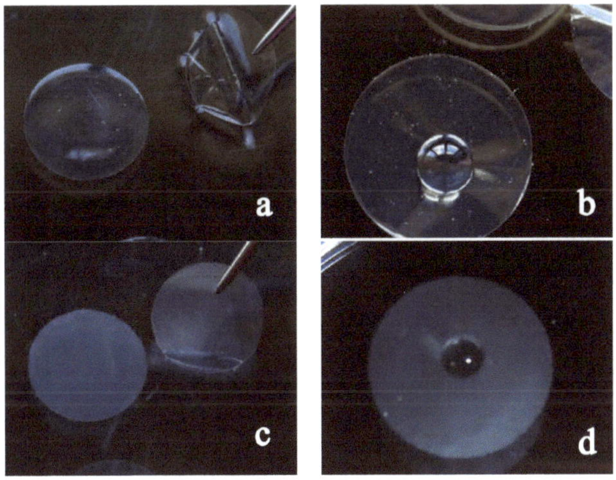

Figure 22. Photos of PLA-modified (CS/CN)/glycerol film with the formulation: (70/30)/30 wt. %. The slurry was cast on: (**a**,**b**) Dura Lar support; (**c**,**d**) stainless steel plate.

Table 13. Mechanical characteristics of CS-based films with different formulations.

Slurry	Components	Formulation	Y, MPa	σ, MPa	ε, %
1	CS/glycerol	70/30	1442 ± 55	34.0 ± 3.5	18.8 ± 2.9
2	CS/gelatin	90/10	3702 ± 129	69.9 ± 3.3	4.9 ± 2.2
3	(CS/gelatin)/glycerol	(90/10)/30	1714 ± 79	35.5 ± 4.2	16.7 ± 4.4
4	[(CS, gelatin)/CN]/glycerol	[(90/10)/25]/15	3000 ± 164	55.3 ± 5.0	14.3 ± 2.5
5	–"–"–"–"–"–	[(90/10)/25]/25	1263 ± 544	36.9 ± 6.1	16.2 ± 4.4
6	–"–"–"–"–"–	[(90/10)/30]/30	325 ± 48	33.3 ± 7.1	18.7 ± 3.9

Table 14. Mechanical characteristics of (CS/CN/glycerol films with the formulation: (70/30)/30 wt. % with 0.234 mmol $Ca(OH)_2$ and 0.1 mmol $Ca_3(PO_4)_2$ or with 10 wt. % gelatin from CS amount.

Additive / Parameter	Gelatin			$Ca(OH)_2$, $Ca_3(PO_4)_2$		
	None	PLA	PGPR	None	PLA	PGPR
Y, MPa	634 ± 95	3993 ± 447	4032 ± 264	176 ± 42	670 ± 246	1363 ± 868
σ, MPa	36.1 ± 4.3	61.1 ± 8.6	58.8 ± 1.9	20.3 ± 5.8	53.8 ± 11.2	42.9 ± 6.7
ε, %	20.3 ± 2.2	11.8 ± 2.9	12.2 ± 1.4	19.7 ± 5.8	20.6 ± 5.2	15.4 ± 2.7
θ, °	93.2 ± 1.3	95.1 ± 3.1	99.3 ± 3.2	94.1 ± 0.1	96.8 ± 2.4	100.9 ± 2.4
S_W, %	23.8	ND	ND	ND	37.2	ND

Chitin nanofibers, which are powerful reinforcing agents, have leveled the disintegrating effect of the plasticizer at its small amount in slurry 4 (Table 13). However, all mechanical characteristics of the composite films have dramatically worsened with increasing content of glycerol in slurries 5 and 6.

The mechanical characteristics of the composite films obtained from slurry 6 were practically restored after their modification with polylactide (PLA) or polyglycerol polyricinoleate (PGPR) (Table 14).

Table 15. Mechanical characteristics of (CS/CN)/glycerol films with the formulation: (70/30)/30 wt. % with 5 wt. % monolignols or 1 wt. % nanolignin from CS amount. Surface modifier: PLA—polylactide, 7.7 kDa or PGPR—polyglycerol polyricinoleate.

Additive / Parameter	Monolignols			Nanolignin		
	None	**PLA**	**PGPR**	**None**	**PLA**	**PGPR**
Y, MPa	293 ± 66	3503 ± 457	3236 ± 472	455 ± 56	3883 ± 405	4027 ± 400
σ, MPa	26.3 ± 3.6	57.1 ± 9.8	45.9 ± 9.8	24.1 ± 3.7	51.4 ± 4.9	50.9 ± 4.9
ε, %	20.8 ± 3.2	13.1 ± 2.7	14.7 ± 1.6	14.6 ± 2.4	10.9 ± 2.2	12.9 ± 2.0
θ, °	96.0 ± 1.7	96.5 ± 1.2	104.9 ± 3.2			
S_W, %	Not determined		21.5			

Table 16. Mechanical characteristics of (CS/CN)/glycerol film with the formulation: (70/30)/30 wt. % with 0.178 mmol Mg^{+2}, 0.234 mmol Ca^{+2} or 0.321 mmol Ba^{+2} ions.

Modifier / Parameter	PGPR			PLA		
	Mg^{+2}	Ca^{+2}	Ba^{+2}	Mg^{+2}	Ca^{+2}	Ba^{+2}
Y, MPa	3987 ± 134	2807 ± 186	2535 ± 506	3940 ± 200	3597 ± 271	3034 ± 262
σ, MPa	67.2 ± 7.2	55.1 ± 5.6	49.6 ± 9.8	65.1 ± 4.7	62.9 ± 5.2	55.9 ± 5.6
ε, %	9.9 ± 2.4	13.0 ± 2.5	15.6 ± 2.4	9.5 ± 2.4	11.8 ± 1.5	13.7 ± 2.5

The mechanical properties of CS/CN films prepared from slurry 6 (Table 13) have been improved considerably after modification of the films with PGPR or PLA. The values of the elastic modulus increased six-fold compared with unmodified film. The values of the ultimate tensile stress increased about twice while simultaneously decreasing the strain at break of both modified films.

Strengthening of the films containing mineral substances has also occurred but the values of their elastic modulus and strain at break changed considerably less after modification with PLA and PGPR. In both cases, the increase in the values of contact angles was observed, especially for films modified with PGPR. It should be noted that the adsorption of PLA and PGPR within the plasticized CS/CN films was probably accompanied by the extraction of the molecules of both acetic acid and glycerol from the films into dioxane.

Practically the same changes in mechanical characteristics of CS/CN films containing monolignols and nanolignin occurred after their modification with PLA or PGPR (Table 15).

It is interesting that the order of strengthening of CS/CN films containing ions of the alkali earth metals coincided with the order of increasing their ionic radii equal to 66 pm, 99 pm and 134 pm for Mg^{+2}, Ca^{+2} and Ba^{+2}, respectively. The highest and lowest values of the elastic modulus and the ultimate tensile stress have been observed for PGPR- and PLA-modified CS/CN films containing Mg^{+2} and Ba^{+2} ions. The most strengthened films with Mg^{+2} ions have the lowest values of strain at break (Table 16).

Commercial wrapping paper from cellulose used for food packaging and PLA- or PGPR-modified plasticized CS/CN films named CHITOPACK have the comparable values of elastic modulus and ultimate tensile stress (Table 17). However, elasticity of CHITOPACK films are about 3–4 times higher than that of commercial paper such as KFC paper widely used in fast food restaurants for packaging of sandwiches.

Table 17. Comparison of the mechanical characteristics of the commercial films for food packaging with PLA- or PGPR-modified (CS/CN)/pG-3 films with the formulation: (70/30)/25 wt. %.

Film	Composition	Surface Modifier	Y, MPa	σ, MPa	ε, %
1	CHITOPACK	PGPR	4792 ± 496	81.2 ± 7.5	9.7 ± 2.8
2	–"–"–"–	PLA	4891 ± 594	74.7 ± 5.1	7.7 ± 2.2
3	Packaging paper	unknown	4489 ± 136	44.8 ± 5.7	2.1 ± 0.7
4	KFC paper	unknown	4891 ± 211	71.8 ± 6.8	2.3 ± 0.3
5	DOMOPACK	none	196 ± 23	15.6 ± 1.8	623.0 ± 111

4. Conclusions

Summarizing the results of the research, it should be concluded that the native linear polysaccharide polymer extracted from crustaceans' carapaces in the form of chitin nanofibrils (CN) has proved to be a potent "physical crosslinker" forming numerous stable noncovalent bonds with chitosan chains by such a way that the CS/CN films do not dissolve in contact with water for a long time without any additional crosslinking.

It was found that there was a limit of saturation of CS phase with CN at about 10 wt. %-content from CS amount in slurry. At higher CN content, the unbound CN were concentrated in the surface-adjacent layer with higher density of packaging components because some chitosan chains present between chitin nanofibrils.

The CS/CN slurries have good compatibility with various low- and high-molecular-weight substances such as ions of the alkali earth metals (Mg, Ca, Ba) and monolignols, nanolignin and gelatin. These substances exhibited some reinforcing effects on the CS chain community. In contrast, the plasticizers disintegrated the interactions between CS and CN in the films. Therefore, their content should be limited by 20–30 wt. %.

The surface modification of CS/CN films by sorption immobilization of PLA or PGPR exhibited great effect on improvement of both mechanical stability and hydrophobicity of CS/CN films. This finding opens the wide avenue for optimizing characteristics of the films in the manner similar to that used by nature in the creation of a polysaccharide-mineral nanocomposite having a hydrophobic layer consisting of hydrocarbon" molecules on the surface of crustaceans' carapaces.

Chitosan is a very sensitive and malleable polymer, the properties of which are strongly influenced by such factors as temperature, conditions and duration of storage, exposure to light, infrared, ultrasonic or microwave irradiation, intensity of stirring and acidity of its solutions, the microheterogeneity and chemical nature of a support used for casting the CS/CN slurries, the conditions and temperature of their drying, and subsequent hydrophobization of the obtained films. It should be emphasized, that in addition to factors mentioned above, it is extremely important to control the quality of raw materials in order to produce of the CHITOPACK packaging films with reproducible characteristics.

The formed CS/CN composite films seem to be promising candidates for production of completely biodegradable films comparable in their mechanical stability with commercial wrapping paper that is used nowadays for one-off packaging of some food products. Indisputably, the innovative CHITOPACK films will gain an extremely widespread in the future, especially, in the Floating Cities [29] owing to easy biodegradability of the films' components and their bio-compatibility with the Environment.

Author Contributions: Preparation of chitin nanofibrils/chitosan composite films, analysis of the results and coordination of experimental work (Galina Tishchenko); coordination of the research and preparation of chitin nanofibrils (Pierfrancesco Morganti and Marco Stoller); measuring and analysis of mechanical properties of the composite films (Ivan Kelnar and Ludmila Kaprálková); investigation of rheological properties of chitin nanofibrils/chitosan slurries (Jana Mikešová); thermogravimetric analysis and differential scanning calorimetry of the composite films (Jana Kovářová); investigation of the effect of temperature and copper ions on crystallinity of chitin nanofibrils (Jindřich Hašek and Radomír Kužel); determination of molecular weight of chitosans by size-exclusion chromatography (Miloš Netopilík); SEM and TEM of the surfaces and fractures of composite films (Ewa Pavlová); AFM analysis of the morphology of chitin nanofibrils and surfaces of the composite films in the swollen state in water (Eliška Chanová) and in the dry state (Milena Špirková); measuring the sorption of water vapors, gas permeability and contact angles of composite films (Libuše Brožová); FTIR spectroscopy of chitosan nanofibrils, chitosans and composite films (Michal Pekárek); the solid state 13C CP/MAS NMR of chitosan nanofibrils, chitosans and composite films (Libor Kobera); synthesis of polylactides (Dana Kubies); surface modification of the composite films by PLA and PGPR (Zdenka Sedláková).

Acknowledgments: We thank the EU for the financial support given to the European Research Projects: n-Chitopack, grant No. 315233 and Biomimetic, grant No. 282945.

Conflicts of Interest: We declare that Galina Tishchenko has no conflict of interest. Pierfrancesco Morganti works in the R&D Centre of Nanoscience, MAVI Sud Srl, Italy.

References

1. Raabe, D.; Al-Sawalmih, A.; Romano, P.; Sachs, C.; Brokmeier, H.G.; Yi, S.-B.; Servos, G.; Hartwig, H.G. Structure and crystallographic texture of arthropod bio-composites. *Mater. Sci. Forum* **2005**, *495–497*, 1665–1674. [CrossRef]
2. Swift, N.B. Biomineral Structure and Strength of Barnacle Exoskeleons. *Colgate Acad. Rev.* **2012**, *8*, 10.
3. Martin, J.W. *Concise Encyclopedia of the Structure of Materials*; Elsevier: Oxford, UK; Amsterdam, The Netherlands, 2007.
4. Morganti, P.; Muzzarelli, C. Production of Nanofibrillar Chitin. U.S. Patent 2008/0118563 A1, 2009.
5. Muzzarelli, R.A.A.; Morganti, P.; Morganti, G.; Palombo, P.; Palombo, M.; Biagini, G.; Belmonte, M.M.; Giantomassi, F.; Orlandi, F.; Muzzarelli, C. Chitin nanofibrils/chitosan glycolate composites as wound medicaments. *Carbohydr. Polym.* **2007**, *70*, 274–284. [CrossRef]
6. Franca, E.F.; Freitas, L.C.G.; Lins, R.D. Chitosan Molecular Structure as a Function of N-Acetylation. *Biopolymers* **2011**, *95*, 448–460. [CrossRef] [PubMed]
7. Yomota, C.; Miyazaki, T.; Okada, S. Determination of the viscometric constants for chitosan and the application of universal calibration procedure in its gel permeation chromatography. *Colloid Polym. Sci.* **1993**, *271*, 76–82. [CrossRef]
8. Netopilík, M.; Bohdanecký, M. Ubbelohde viscometer modified for foaming solutions of water soluble polymers. *Eur. Polym. J.* **1995**, *31*, 289–290. [CrossRef]
9. Morganti, P.; Tishchenko, G.; Palombo, M.; Kelnar, I.; Brozova, L.; Spirkova, M.; Pavlova, E.; Kobera, L.; Carezzi, F. *Marine Biomaterials: Characterization, Isolation and Applications*; Kim, S.-K., Ed.; CRC Press, Taylor & Francis Group: Boca Raton, FL, USA, 2013; pp. 681–716.
10. Ifuku, S.; Shervani, Z.; Saimoto, H. Chitin Nanofibers, Preparations and Applications. In *Nanotechnology and Nanomaterials, Advances in Nanofibers*; Maguire, R., Ed.; InTechOpen: Rijeka, Croatia, 2013; pp. 85–101.
11. Youn, D.K.; No, H.K.; Kim, D.S.; Prinyawiwatkul, W. Decoloration of chitosan by UV irradiation. *Carbohydr. Polym.* **2008**, *73*, 384–389. [CrossRef]
12. Younes, I.; Rinaudo, M. Chitin and Chitosan Preparation from Marine Sources. Structure, Properties and Applications. *Mar. Drugs* **2015**, *13*, 1133–1174. [CrossRef] [PubMed]
13. TA Instruments. Understanding rheology of structured fluids. 2015. Available online: http://www.tainstruments.com/pdf/literature/AAN016_V1_U_StructFluids.pdf (accessed on 1 November 2017).
14. Adamson, A.W.; Gast, A.P. *Physical Chemistry of Surfaces*, 6th ed.; John Wiley & Sons, Inc.: Stanford, CA, USA, 1997.
15. Mikesová, J.; Hasek, J.; Tishchenko, G.; Morganti, P. Rheological study of chitosan acetate solutions containing chitin nanofibrils. *Carbohydr. Polym.* **2014**, *112*, 753–757. [CrossRef] [PubMed]
16. Mistler, R.E.; Twiname, E.R. *Tape Casting. Theory and Practice*; The American Ceramic Society: Westerville, OH, USA, 2000.

17. Ozcelik, B.; Brown, K.D.; Blencowe, A.; Daniell, M.; Stevens, G.W.; Qiao, G.G. Ultrathin chitosan–poly(ethylene glycol) hydrogel films for corneal tissue engineering. *Acta Biomater.* **2009**, *9*, 6594–6605. [CrossRef] [PubMed]
18. Martínez-Camacho, A.P.; Cortez-Rocha, M.O.; Ezquerra-Brauer, J.M.; Graciano-Verdugo, A.Z.; Rodriguez-Félix, F.; Castillo-Ortega, M.M.; Yépiz-Gómez, M.S.; Plascencia-Jatomea, M. Chitosan composite films: Thermal, structural, mechanical and antifungal properties. *Carbohydr. Polym.* **2010**, *82*, 305–315. [CrossRef]
19. Fimbeau, S.; Grelier, S.; Copinet, A.; Coma, V. Novel biodegradable films made from chitosan and poly(lactic acid) with antifungal properties against mycotoxinogen strains. *Carbohydr. Polym.* **2006**, *65*, 185–193.
20. Arvanitoyannis, I.S. Nakayama, A. Aiba Sei-ichi, Chitosan and gelatin based edible films: State diagrams, mechanical and permeation properties. *Carbohydr. Polym.* **1998**, *37*, 371–382. [CrossRef]
21. Bourtoom, T.; Chinnan, M.S. Preparation and properties of rice starch–chitosan blend biodegradable film. *LWT Food Sci. Technol.* **2008**, *41*, 1633–1641. [CrossRef]
22. Fernandes, S.C.M.; Freire, C.S.R.; Silvestre, A.J.D.; Neto, C.P.; Gandini, A.; Berglund, L.A.; Salmén, L. Transparent chitosan films reinforced with a high content of nanofibrillated cellulose. *Carbohydr. Polym.* **2010**, *81*, 394–401. [CrossRef]
23. Bhatt, A.S.; Bhat, D.K.; Santosh, M.S. Electrochemical properties of chitosan–Co3O4 nanocomposite films. *J. Electroanal. Chem.* **2011**, *657*, 135–143. [CrossRef]
24. Leceta, I.; Guerrero, P.; de la Caba, K. Functional properties of chitosan-based films. *Carbohydr. Polym.* **2013**, *93*, 339–346. [CrossRef] [PubMed]
25. Huang, D.; Mu, B.; Wang, A. Preparation and properties of chitosan/poly(vinyl alcohol) nanocomposite films reinforced with rod-like sepiolite. *Mater. Lett.* **2012**, *86*, 69–72. [CrossRef]
26. Kelnar, I.; Kovářová, J.; Tishchenko, G.; Kaprálková, L.; Pavlová, E.; Carezzi, F.; Morganti, P. Chitosan/Chitin nanowhiskers composites: Effect of plasticisers on the mechanical behaviour. *J. Polym. Res.* **2015**, *22*, 1–6. [CrossRef]
27. Stern, S.A. The "Barrer" Permeability Unit. *J. Polym. Sci.* **1968**, *6*, 1933–1934. [CrossRef]
28. Pauly, S. Permeability and Diffusion Data. In *Polymer Handbook*; Brandrup, J., Immergut, E.H., Grulke, E.A., Eds.; John Wiley and Sons: New York, NY, USA, 1999; pp. 435–446.
29. World's first floating city to be built off the coast of French Polynesia by 2020. Available online: www.independent.co.uk/news/science/floating-city-french-polynesia-2020-coast-islands-south-pacific-ocean-peter-thiel-seasteading-a8053836.html (accessed on 6 November 2018).

Baby Diapers Past and Present: A Critical Review

Pietro Febo and Alessandro Gagliardini

Abstract: Disposable baby diapers are a popular consumer product, the use of which contributes nearly 77 million tons of solid waste to landfills, with a degradation period of at least 500 years! Moreover, their use continues to increase worldwide. More than 200,000 trees are lost each year to their manufacture in the US market alone, with a consumption of 3.4 billion gallons of fuel oil, which contributes to the production of greenhouse gas emissions, furthering the environmental strain due to the Earth's climate changes. It is therefore necessary to change the way these products are consumed and to find more eco-compatible solutions, by using biodegradable polymeric plastics and/or re-usable cloth diapers. This chapter reports the historical evolution of baby diapers, its global market, the connected environmental problems, and the efforts that are being made in order to create bioactive and biodegradable baby diapers.

1. Introduction

The continuous increase of the world's population and the high level of human development has created a negative ecological footprint (i.d. 1.2 kilograms per person per day), with an annual cost in natural capital degradation estimated at US$4.7 trillion/year [1,2]. Thus, the current global waste levels are approximately 1.3 billion tons/year and are expected to increase to around 2.2 billion/year by 2025, i.e., from 1.2 to 1.42 kg per person per day in the next 15 years.

Managing waste properly is, therefore, essential not only for building sustainable and livable cities and developing countries, but also for reducing greenhouse emissions and natural environmental disasters, the latter of which have seen their frequency increase year by year. For this purpose, it is interesting to underline that the majority of waste is produced by food production, distribution and consumption, estimated to be around 0.5/1 billion tons/year and by disposable baby diapers that reached a worldwide figure of 3.5 million tons/year. Thus, it is necessary to create a sustainable chain of food and to produce baby diapers made of biodegradable and natural polymers in order to achieve zero waste, driving industrial changes towards a new techno-economic system, the so-called bio-economy.

Technological advances, in fact, are set out to replace finite resources and conventional industrial processes with procedures and components that are biologically derived. These innovative bio-based processes, expected to be more sustainable because of their use of renewable resources and their decreased levels of CO_2 emissions, should ensure economic growth as well as contributing to the achievement of environmental and climate goals.

In conclusion, promoting a more inclusive and sustainable industrialization by using natural biopolymer obtained from waste materials, will foster innovation and provide access to more affordable economic growth through the development of new markets and employment potential. New ways of producing biodegradable and reusable baby diapers could be part of these innovative processes.

2. The Diaper's Story

From time immemorial, there has been a need to protect babies due to their physiological needs; infants were supposed to be wrapped in swaddling bands in many societies and since antiquity. The swaddling bands were made of strips of linen or wool that were wrapped tightly around each limb and then crosswise around the body (Figure 1).

Figure 1. Different types of baby diapers used in different periods [3].

After their use, the diapers were seldom washed. They were usually just hung by the fireplace or outside to dry and then used again.

Between the end of the 19th and the beginning of the 20th century, infants in Europe and North America started wearing what could be considered the prototype of the modern diaper. A square or rectangle of linen, cotton flannel, or stockinet was folded into a rectangular shape and held in place with safety pins. Such new diapers were originally made with white cotton or linen fabric (or similar), which are absorbent natural materials (Figure 2).

Figure 2. Linen fabric and similar natural materials [3].

It seems that the first mass-produced cloth diapers were introduced by Maria Allen in 1887 in the United States [4–7].

At the beginning of the 20th century, many mothers had concerns related to the negative effects caused by diapers such as rashes or redness of the skin (Figure 3).

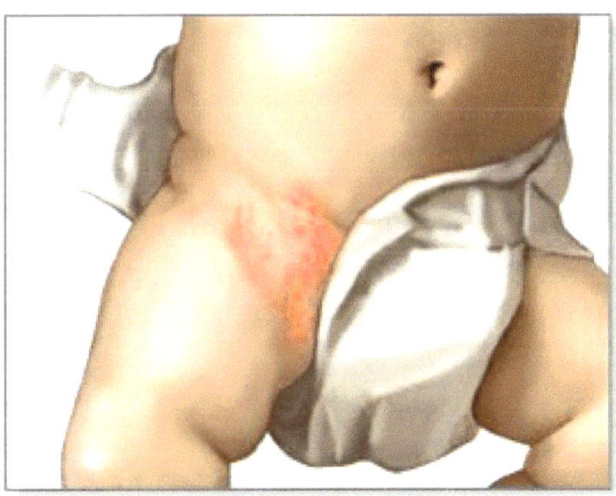

Diaper rash

Figure 3. Diaper rash (diaper dermatitis) [8,9].

Diapers were also associated with the presence of bacteria, viruses and fungi. Concerned mothers understood the need to eradicate, or at least be able to control them. The mothers began using boiled water in order to reduce the common rash problem. The identified process consisted in putting used diapers in a big pot of boiled water; however, this required great amounts of energy and time. Skin rash was a serious problem in those days.

It is presently still not yet clear who can be credited as the real inventor of the disposable diaper. Current knowledge establishes that the first disposable diaper concept was most probably made by using unbleached craped cellulose tissue just after the second world war in Europe and specifically in Paulistróm, Sweden.

A few years later in the United States, a Westport housewife named Marion Donovan invented the Boater, a waterproof covering for cloth diapers (Figure 4) [10].

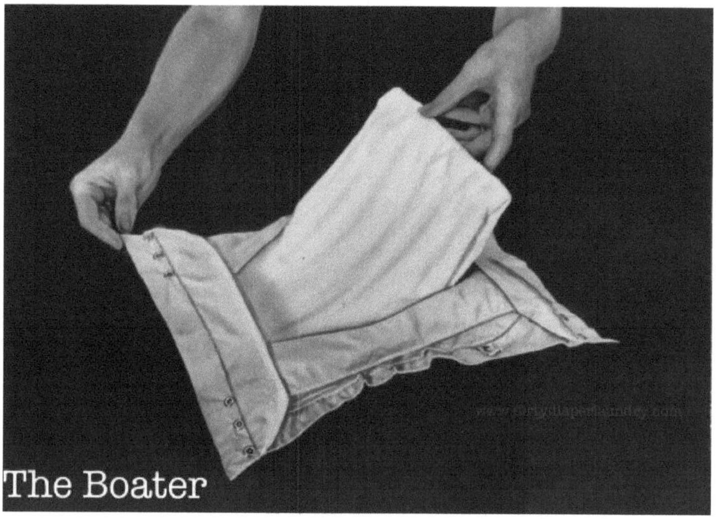

Figure 4. The Donovan Boater [11].

Her first model of the disposable diaper was made of shower curtain plastic into which a conventional cloth diaper was inserted. She obtained four granted patents for the designs, including the use of plastic snaps that replaced the traditional and dangerous safety pins.

Following this new innovative design, the first disposable diaper made with non-woven fabric was created in 1949 in the USA. In the same year, a British mother developed a two-piece disposable diaper. Following this, disposable diapers with a rectangular one piece diaper were invented, followed by the launch of the roll diaper (Figure 5).

Figure 5. The Roll Diaper [12]. For a timeline of major diaper development, see reference [13].

Following these years, the disposable diaper was still considered a luxury item and used only for special occasions such as vacation trips and the like.

The first truly disposable diapers were made using a very simple rectangular design. The absorbent core structure was made of several layers of tissue paper, and using a plastic film with no tape on the outside, which came with the product.

In 1957, Mölnlycke entered the market with a product made of paper pulp encapsulated in tissue and surrounded by a knitted net. The sanitary napkin, on the other hand, experienced a rapid growth in the European and North American markets. It was not until the end of this decade when Vic Mills, who worked for the Procter and Gamble company, invented "Pampers", as he was looking for better products to use for his baby grandson. The diaper was, however, not launched on the market until 1961 [14].

From the 1960s onwards, the disposable diaper evolved quickly as the industry was able to gauge mothers' needs. Tissue was replaced with pulp a decade after the first disposable sanitary napkins arrived on the market. In 1966, Pampers launched a new C-fold design and by 1969 they initiated a third sizing option. A typical commercial diaper machine ran at speeds of 150 diapers per minute.

The 1970s proved to be the literal baby boom for the disposable diaper industry in developed countries and even in some other, less developed areas of the world. Competition between Procter and Gamble and Kimberly Clark to gain control of the world diaper market resulted in quick diaper design improvements and lower prices for the consumer. In 1976, Kimberly Clark introduced its shaped Huggies diapers. Lateral elastomeric was used at the end of the decade by most producers in an attempt to improve the fit.

Then, in 1982, Unicharm introduced its concept of SAP (super-absorbent) [15] in Japan, following its use in sanitary napkins.

As of 1985 and to this day, there are huge and continuous developments in the diaper's structure and composition. Today's market allows for several companies producing different shapes and product designs. As reported in Table 1, the ratio of cellulose to pulp/SAP has significantly changed over the years in combination with the reduced thickness of the product.

Table 1. Development in the structure of baby diapers over the years.

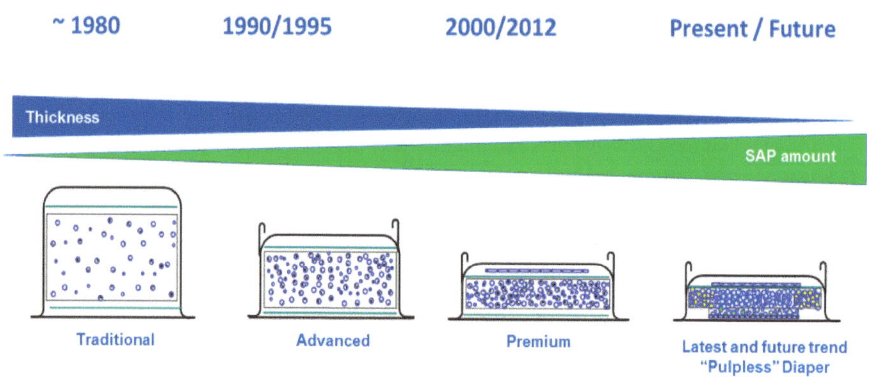

3. Technical Requirements and Performance

In today's global market, there are several types and structures of disposable diaper products designed to satisfy the needs of all babies in all countries (Figure 6).

Figure 6. Different types of baby diapers used currently and globally.

There are numerous different factors related to, among others, country, culture, religion and baby age that can affect the type of diaper. However, there are some needs that are universal:

- a meaningful ratio between product cost and quality; and
- no side-effects such as skin dryness or rashes.

These differences incite the different diaper manufacturers to constantly research and improve on their current designs in order to reflect customer needs.

In addition to the dryness performance of diapers, all companies are paying attention to developing and delivering a new generation of diaper products, paying particular attention to including minor details such as the softness, gentleness, sustainability and naturalness.

Figure 7. Safety and security of baby diapers according to publicity claims.

As reported in Figure 7, the safety and security of baby diapers are actually based on eliminating the use of some components considered to cause allergies and sensitivity problems. However, the real issue at hand is the necessity to make these products 100% bio-degradable, and to ensure that they do not use fossil materials for the sake of later generations. Thus, future baby diapers should be obtained from agricultural waste and the industrial by-products necessary to produce the natural bio-polymers that are indispensable to making disposable diaper components.

In this way, we could eliminate the many environmental problems linked to diapers and produce skin-friendly and environmentally-friendly products, thus safeguarding the ecosystem and biodiversity of our planet.

4. Environmental Problems Linked to Diapers

An infant requires up to 7000 diaper changes before leaving diapers behind. This typically requires 300–600 cloth diapers from birth to potty training, adding roughly 14 kg of cotton to landfills. In addition, the use of cloth diapers also entails other costs, such as greater water and energy use: about 76,000 liters of water are needed to launder diapers for one infant. The 450 billion disposable diapers used each year contribute to nearly 77 million tons of solid waste to landfills, and a disposable diaper takes at least 500 years to degrade [16], (Figure 8).

Figure 8. Disposable baby diapers in a landfill [17].

Efforts are being made to reduce these impacts. In 2000, the Mexican company Absormex created a disposable bioactive diaper that degraded 200 percent faster than ordinary disposables. The technology is based on a catalyst additive added to the plastic to enhance biodegradation [18,19]. However even if this diaper has been demonstrated to pass the ASTM D883-99 test, there are still many doubts about the times required for degradation under particular conditions, e.g., in anaerobic conditions.

Another approach currently under investigation is diaper recycling. The idea behind this is to separate a diaper's components (essentially plastic and organic compostable matter) in a process that neutralizes the potential biological hazard, e.g., by sterilizing the used diaper [20–23].

Recently, biodegradable and compostable bio-based resins such as Ecovio [20] and Mater-bi [21] have been commercialized. These resins could be a suitable substitute for the polyolefins and polyesters currently used in diapers.

5. What About the Global Baby Diaper Market?

Different market analyses [22,23] have projected that the global market for baby diapers should reach around US$60 billion by 2024 (Table 2), driven by the growing number of young mothers in the workforce. Further predictions indicate a decline in diaper prices as a result of the mass commoditization of the product, its increasing use to maintain hygiene and prevent rashes on babies' skin, the launch of smaller pack diapers, and a growing preference for diaper pants over open diapers.

Additionally, in developing countries, increasing birth rates, rapid urbanization and continuously improving economic conditions have fueled the growth of the baby diaper industry. Moreover, this market is also benefiting from growing awareness of the convenience offered by diapers in markets like India, South Africa, Mexico, Venezuela and Turkey, due to the ubiquity of parenting websites, baby blogs and social media. Thus, while economies such as the USA, the EU and Japan are characterized by maturing conditions due to declining birth rates, Asia Pacific represents the largest and fastest growing market worldwide. Furthermore, Europe was the largest regional producer market in 2013, followed by Asia Pacific, which is expected to have the fastest growth of 8.3% during the forecast period. Still, disposable diapers have exhibited the largest market share in North America owing to environmental regulations and the adoption of eco-friendly diapers in the American market.

As a result of the abovementioned factors, biodegradable diapers are expected to reach their highest growth rate during the forecast period [24].

Table 2. Baby diaper global market forecasts (2018–2024).

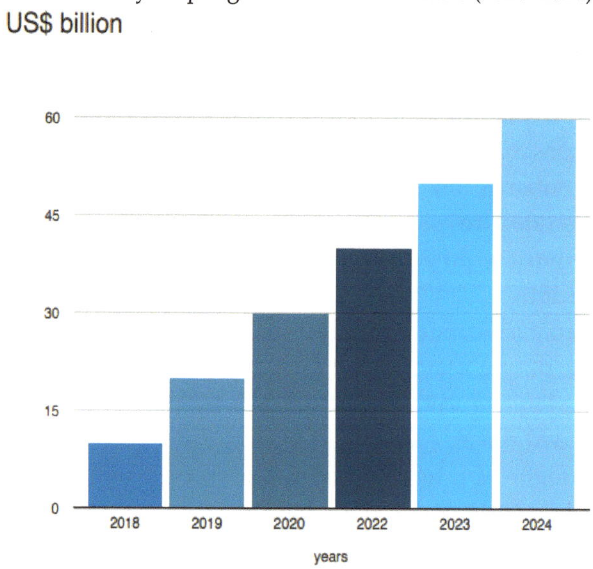

6. Conclusive Remarks

The demand for baby diapers depends highly on two factors [24]: fertility rates and the penetration of the diaper market across geographic areas. As an example, in Africa the fertility rate is high, but the market penetration rate is low, hence the sale of diapers in the region is also low.

In North America and Europe, the penetration of the diaper market is high, but the fertility rate has steadily decreased. However, the worldwide sensibility leaning toward more environmentally friendly practices is continually growing; therefore, we expect that the future market will see a higher growth of biodegradable and eco-friendly baby diapers.

Thus, it is necessary to increase research studies to recover and produce innovative bio-composites made by natural polymers produced through the use of waste materials and sustainable industrial processes.

This is our dream, together with all the scientists involved in the EU research project PolyBioSkin.

Acknowledgments: This chapter was realized thanks to the funding received from TEXOL and the Bio-Based Industries Joint Undertaken (BBI JU) by the European Union's HORIZON 2020 Research and Innovation programme, agreement No 745839.

Conflicts of Interest: The authors certify that they have no affiliations with or involvement in any organization or entity with any financial interest (such as honoraria; educational grants; participation in speakers' bureaus; membership, employment, consultancies, stock ownership, or other equity interest; and expert testimony or patent-licensing arrangements), or non-financial interest (such as personal or professional relationships, affiliations, knowledge or beliefs) in the subject matter or materials discussed in this manuscript.

References

1. Krafchik, B. History of Diapers and Diapering. *Int. J. Dermatol.* **2016**, *55*, 4–6. [CrossRef] [PubMed]
2. Kohno, G. History of Diapering in Japan. *Pediatrician* **1987**, *14*, 2–5. [PubMed]
3. The Disposable Diaper Industry Source. Available online: http://disposablediaper.net/?page_id=14 (accessed on 12 November 2017).
4. Hauch, R. Wrapped in diapers—A Brief History of the First Garment. *Kinderkrankenschwester* **2015**, *34*, 456–457. [PubMed]
5. Cole, W.G.; Lingeman, J.M.; Adolph, K.E. Go Naked: Diapers Affects Infant Walking. *Dev. Sci.* **2012**, *15*, 783–790. [CrossRef] [PubMed]
6. Ravanfar, P.; Wallace, J.S.; Pace, N.C. Diaper Dermatitis: a Review and Update. *Curr. Opin. Pediatr.* **2012**, *24*, 472–479. [CrossRef] [PubMed]
7. Robert, T. Marion Donovan, 81, Solver of the Damp-Diaper Problem. *New York City Times*, 18 November 1998.
8. Diaper rash. Available online: http://printer-friendly.adam.com/content.aspx?productId=117&pid=1&gid=000964&c_custid=758 (accessed on 12 November 2017).
9. The disposable diaper industry source—Diaper industry time line. Richer Investment. Available online: https://web.archive.org/web/20110723094929/ http://www.disposablediaper.net/content.asp?3 (accessed on 12 November 2017).
10. Pampers: The Birth of P&G's First 10-Billion-Dollar Brand. Available online: http://news.pg.com/blog/10-billion-dollar-brand/pampers-birth-pgs-first-10-billion-dollar-brand (accessed on 12 November 2017).

11. Dirty diaper laundry. Available online: http://dirtydiaperlaundry.com/marion-donovan-inventor-of-the-modern-cloth-diaper-the-boater-1946/ (accessed on 12 November 2017).
12. Pinterest. Available online: https://it.pinterest.com/kimrosas/historical-cloth-diapers/ (accessed on 12 November 2017).
13. Kabir, K. Synthesis of Fast-Swelling Superabsorbent Hydrogels: Effect of Crosslinker Type and Concentration on Porosity and Absorption Rate. *European Polymer Journal* **2003**, *39*, 1341–1348. [CrossRef]
14. World Watch Institute. Lifer-cycle studies—Diapers. *World Watch Magazine*. 2017, 20. No. 2. Available online: http://www.worldwatch.org/system/files/M-A%2007%20Life-cycle.pdf (accessed on 12 November 2017).
15. EPI. Available online: http://www.epi-global.com/en/epi-technology.php (accessed on 12 November 2017).
16. Sitnam, Elmo. Degradable Disposable Diaper. U.S. Patent 2001003797A1, 9 February 1993.
17. Diaper wrecker. Available online: http://diaperwrecker.com/tag/disposables/ (accessed on 12 November 2017).
18. Kim, K.S.; Cho, H.S. Pilot Trial on Separation Conditions for Diaper Recycling. *Waste Manag.* **2017**, *67*, 11–19. [CrossRef] [PubMed]
19. Somma, Marcello; Vaccaro, Giorgio; Michalek, Jan K.; Thomas, Theodore. Apparatus and Process for Sterilizing Absorbent Sanitary Products. EP Patent EP2596811B1, 29 May 2013.
20. Ecovio. Available online: https://www.plasticsportal.net/wa/plasticsEU/portal/show/content/products/biodegradable_plastics/ecovio (accessed on 12 November 2017).
21. Mater-Bi. Available online: http://www.novamont.com/eng/mater-bi (accessed on 12 November 2017).
22. Business Wire. Research and Markets. Global Baby Diapers Market to 2025 Strategy, Assessment and Forecasts. 2017.
23. Global Industry Analysts. Growing Emphasis on Baby Health, Well-being & Hygiene drives the Global Baby Disposable Diapers Market. 2018. Available online: https://www.strategyr.com/MarketResearch/ViewInfoGraphNew.asp?code=MCP-2531 (accessed on 12 November 2017).
24. Shende, S. *Baby Diapers Market by Product Type. Global Opportunity Analysis and Industry Forecast 2013–2020*; Report CO_14379; Allied Market Research: London, UK, 2014.

Production of Electrospun Nonwoven Materials as a Blending of Chitin Nanofibrils and Other Natural Polymers

Angelo Chianese and Paola Del Ciotto

Abstract: Electrospinning is an electrostatic fiber fabrication technique that has evinced much interest and attention in recent years due to its versatility and potential for applications in diverse fields. The sub-micron range spun fibers produced by this process offer various advantages, like high surface area to volume ratio, tunable porosity, and the ability to manipulate nanofiber composition in order to get a set of desired properties and function. This chapter deals with a preliminary investigation on the production of nonwoven materials obtained by the electrospinning of a blending of nanochitin fibrils and lignin, by using polyethylene oxide as the solvent. The adopted blend was carefully prepared as sol-gel material at a suitable temperature, mixing conditions, and time of ageing. The blends were characterized by the measurements of viscosity and electroconductivity. Many factors may influence the quality of the electrospun product, among them the main ones include the following: the applied voltage between the two electrodes, the distance between the tip and the collector, and the rotation of the collector. In this work, these operating parameters have been investigated using a preliminary factorial analysis experimental campaign. For this purpose, a pilot scale electrospinning machine (model Nanospider NS LAB 500 supplied by Elmarco) was used. The characterization of the produced electrospun nanofibers was performed by using the Field Emission Scanning Electron Microscope Auriga Zeiss. This instrument provides images of the nanofibers with an accuracy of less than 10 nm and allows the determination of the chemical composition by using the microanalysis device EDS 123 Mn-KeV supplied by Bruker.

---

1. Introduction

In recent years, a large number of research groups have carried out works focused on the development and production of new and improved wound dressings by synthesizing and modifying biocompatible materials [1,2].

In particular, efforts are devoted to the use of biologically derived materials such as chitin and its derivatives, which are capable of accelerating healing processes at the molecular, cellular, and systemic levels. Chitin is a readily available and inexpensive biological material obtained from invertebrate skeletons as well as the cell wall of fungi. It is a linear 1, 4-linked polymer composed of N-acetyl-D-glucosamine residues.

For its chain rigidity, chitin's dissolution in many solvents is a hard task, and for this reason, it is usually used as nanofibril, which are highly crystalline and rigid, thus allowing for the improvement of mechanical performance of composites. Based on infrared spectroscopy and x-ray diffraction data, chitin can be found in one of three crystalline forms: α-chitin, β-chitin, and γ-chitin. The molecules in orthorhombic α-chitin are arranged very tightly in an anti-parallel fashion. In this work, α-chitin, mainly present in shells of crabs, lobsters, and shrimps, was considered. High crystalline chitin is often called chitin nanocrystals, or chitin nanofibrils (CTN).

Chitin and its derivative, chitosan, are biocompatible, biodegradable, nontoxic, antimicrobial, and hydrating agents, and in general act as nanofillers in the reinforcement of both natural and synthetic composites [3]. Due to these properties, they show good biocompatibility and positive effects on wound healing. Previous studies have shown that chitin-based dressings can accelerate the repair of different tissues, thus facilitating the contraction of wounds and regulating the secretion of inflammatory mediators such as interleukin 8, prostaglandin E, interleukin 1 β, and others [4]. The effectiveness of three chitin nanofibril-based preparations, a spray (Chit-A), a gel (Chit-B), and a gauze (Chit-C), in healing cutaneous lesions was assessed macroscopically and by light microscopy immunohistochemistry [5]. These evaluations were compared to the results obtained using a laser co-treatment.

Ja.li Ji et al. [6] have underlined that chitin nanofibrils are an emerging novel filler. They have reinforcing effects on synthetic and natural fibers, and thus they can give rise to efficient scaffolds for tissue engineering. A number of techniques have been developed to fabricate nanofibrous tissue with unique properties. Among these techniques, electrospinning technology has become the most popular for the fabrication of tissue engineering scaffolds in recent years because it is a simple, rapid, efficient, and inexpensive method for producing nanofibers by applying a high voltage to electrically charged liquid [7,8]. Recently, Naseri et al. [9] successfully produced electrospun chitosan-based nanocomposite mats reinforced with chitin nanocrystals for wound dressing. For spinning solutions, chitosan and PEO were blended in a 1:1 ratio and used in a matrix, while CTN was used as reinforcement in order to produce final solutions of 3 wt % polymer in 50% aqueous acetic acid solvent.

The electrospinning process involves the application of a high voltage between a syringe filled with a polymer solution and a collector mounted at a fixed distance from the needle/syringe setup.

An electrical charge builds up on the surface of the solution that is attracted to the collector. The large potential difference overcomes the surface tension of the fluid droplet at the tip of the needle. Under specific conditions of voltage, flow rate, and distance, a jet of fluid is ejected from the needle and subjected to whipping and splaying instabilities due to stresses of electrostatic origin [10]. The solvent evaporates over the jet path, and polymer nanofibers are formed on the collector. Various factors affect the electrospinning process such as solution properties, process parameters (flow rate, voltage, electrode distance), and ambient conditions. Hence, different requirements should be met in order to have an efficient process [11,12].

In this work, chitin-based nanofibers are prepared by a sol-gel material produced by mixing an aqueous solution of chitin with polyethylene oxide (PEOX). The effect of the operating variables of the electrospinning machine was investigated in order to optimize the quality of the electrospun textile.

2. Experimental

2.1. Materials

The materials used in this study include: Chitin, purchased from Primex (Siglufjordur, Iceland), PEOX (polyethylene oxide), purchased from Amerchol (Dow Italia, Italy), Chitin nanofibrils (CN) and CN-Lignin complex purchased from MAVI sud S.r.l. (Aprilia, Italy).

The sol-gel mixture prepared for the electrospinning tests was obtained mixing the CN-Lignin complex (30.1% *w/w*) with deionized water (up to 100% *w/w*) at temperature of 15 °C for few minutes. Then PEOX (7% *w/w*) was added to the solution, under stirring until completely dissolved. This last step took 24 h to obtain a homogeneous gel without agglomerations. The properties of the sol-gel materials are as follows:

- pH: 10.52
- Viscosity: 8.4 P
- Conductivity: 7.8 mS

2.2. Electrospinning

The electrospinning process was performed by using the pilot scale machine Elmarco Nanospider NS LAB 500 based on the nozzle-less technology (S. Petrik, M. Maly, "Production Nozzle-Less Electrospinning Nanofiber Technology", Elmarco s.r.o.). The proof of concept of this technique is that a rotating drum is dipped into a bath of the liquid solution. The thin layer of solution is carried out on the drum surface and exposed to a high voltage electric field. If the voltage exceeds the critical value a number of electrospinning jets are generated. The jets are distributed over the electrode surface with periodicity. This is one of the main advantages of nozzle-less electrospinning, i.e., the number and location of the jets is set up naturally in their optimal positions.

The setting parameters of the machine were:

Voltage: 45–75 kV
Collecting electrode (CE): cylinder
Spinning electrode (SE): cylinder
Distance SE/CE: 10–16 cm
CE rotation: 2–8 rpm
Substrate material: Spunbond, 30 gsm, polypropylene 100% with antistatic treatment

The photo of the electrospinning set-up is reported in Figure 1.

Figure 1. The used electrospinning setup.

2.3. Rheological Measurements

Dynamic rheological properties of the sol-gel material were determined at 25 °C using the rotational rheometer supplied by Brookfield model HA DV-E 230. The applied operating conditions were: spindle code 05, speed of rotation 12 rpm, time 1 min

2.4. Fiber Diameter Characterization

The surface morphology of electrospun nanofibers was characterized by a field emission electron microscope–FESEM Auriga Zeiss, including microanalysis EDS 123 Mn-Kα eV(Bruker) and EBL –7 nm resolution (Raith). Samples cut from the electrospun material mounted on aluminum stubs were coated by an ultrathin layer of platinum for better conductivity during imaging. The samples were observed at magnifications between 100 and 40,000 times their original sizes to visually evaluate the electrospinnability and existence of beads.

Fiber diameters were also determined using Image-J image processing software. For each electrospun material, at least 100 fibers were considered from three different images to calculate the average diameter.

2.5. Factorial Experimental Plan

The main aim of the experimental work has been to determine the best set of the operating parameters of the electrospinning machine in order to optimize the quality of the produced fibers. The process parameters which could be chosen were: the distance between the electrodes, the voltage range applied along each run, and the rotational velocity of the cylindrical electrode. On the basis of a preliminary explorative work it was ascertained that the better operating range of these parameters are those reported in Table 1.

Table 1. The operating conditions of the electrospinning machine.

Electrodes Distance (cm)	Voltage (kV)	Rotational Velocity (rpm)
10	45–60	2
16	45 70/75 (MAX)	8

In order to minimize the experimental runs it was decided to carry out an experimental campaign based on the 2^3 factorial design. In fact, this technique can reduce the number of experiments to be performed by studying multiple factors simultaneously. Additionally, it can be used to find both main effects (from each independent factor) and interaction effects (when both factors must be used to explain the outcome).

Additionally, it can be used to find both main effects (from each independent factor) and interaction effects (when both factors must be used to explain the outcome). The operating conditions of the 8 runs of the experimental campaign are represented in Table 2.

Table 2. The operating conditions of the 8 experimental runs.

Run	Variables Level	Electrodes Distance (cm)	Voltage (kV)	Rotational Velocity (rpm)
(1)	- - -	10	45–60	2
a	+ - -	16	45–60	2
b	- + -	10	45-MAX	2
ab	+ + -	16	45-MAX	2
c	- - +	10	45–60	8
ac	+ - +	16	45–60	8
bc	- + +	10	45-MAX	8
abc	+ + +	16	45-MAX	8

All the operating variables but those ones indicated in Table 1 were maintained constant in all the runs. The investigation was focused on only one effect, i.e., the average diameter of the fibers. In Table 3 the obtained values of this variable, together with the relevant statistical parameters, are reported for all the experimental runs.

Table 3. Average diameters of the electrospun fibers obtained in the factorial campaign.

Run	(1)	a	b	ab	c	ac	bc	abc
Diameter, nm	193	146	184	148	163	144	167	138
Variance	5484	1898	2338	2822	2655	2509	3523	1171
E%	13	10	9	12	10	12	12	8

The average diameter was determined on the basis of more than 100 measurements. The obtained results show significant changes of the fibers diameter for the different operating parameters set, within the overall measured range 138–198 nm. The smaller diameter values are those obtained by adopting the upper value of the electrode distance. In order to evaluate in quantitative way the effect of each operating variable the results of the factorial experiment design were analyzed by means of the ANOVA method. This method is based on the comparison of variance of results corresponding to a single effect or a combination of effects with the variance of the experimental error. In this work, it was assumed that factor interactions are negligible. Under this assumption, the average estimate of high order interactions is considered an estimation of experimental error.

In Table 4 the variance in correspondence of each single effect, in background color, and the zero-variance corresponding to the interactions effect is reported.

Table 4. The variance of single effects and zero variance.

Effect	Variance	Zero Variance	Fisher Factor
A	2145		40.5
B	10		0.19
C	435		8.2
AB	0.71		
AC	151	53	
BC	4		
ABC	55		

Finally, the F-test was made by comparing the ratio between the variance of each single effect and the zero variance with the value of the F distribution at 95% of significance.

In the examined case this value of F distribution is equal to 7.71. By comparing the F-test with the F-distribution it is possible to state that the fiber diameter is strongly affected by the electrode distance, is not affected at all by the voltage and is only slightly affected by the rotational speed of the spinning electrode. The fibers obtained at the best conditions, that is that one of the run abc (upper value of all the three operating variables) is reported in Figure 2.

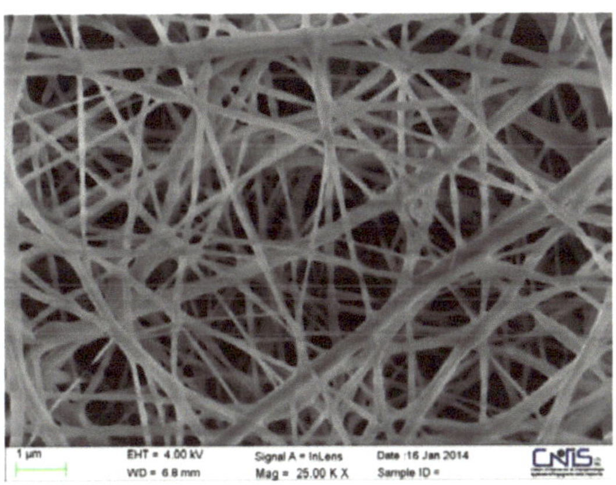

Figure 2. Image of the electrospun nonwoven tissue produced in run abc.

The fibers are elongated and quite regular without evidence of beads. It was not the case for some other runs performed at different operating conditions. For instance, for run b, carried out at lower values of the distance between pin and collector and of the voltage a quite bad tissue was obtained, as shown in Figure 3.

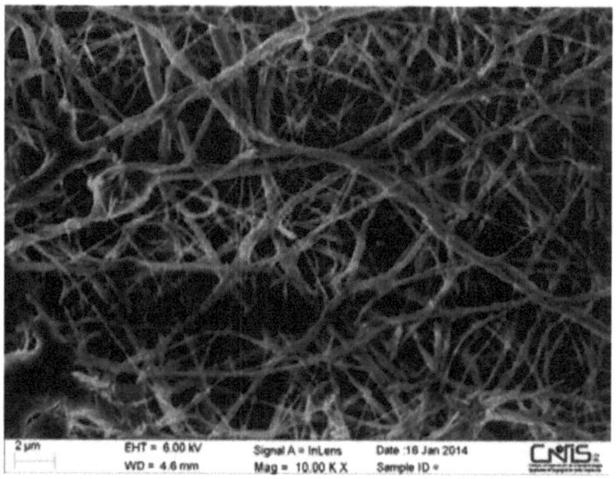

Figure 3. Image of the electron spun nonwoven tissue in run b.

The importance of the distance between tip and collector in order to minimize the beads and to obtain more regular fibers was noticed by other authors [13], in particular it was observed that beads are generated with too small and too large distance and a minimum distance is required to obtain uniform fibers. The applied voltage had no effect on the performances of the electrospinning. This is in agreement with the research work of Reneker and Chun [14], who has showed that there is not much effect of electric field on the fiber diameter with electrospinning of polyethylene oxide. As far as the effect of rotational speed is concerned it is well known that its increase resulted in more uniform and thinner fibers, potentially due to the higher stretching level imposed on them. In our particular case, the increase of the rotational speed from 2 to 8 rpm induces an improving of the electrospinnability of the tissue, even if it is not remarkable.

Obviously, the characteristics of the sol-gel material used for the electrospinning plays an important role. The used material was produced at a relatively low temperature, around 15 °C. If the sol-gel material is produced at a higher temperature, around 20 °C, its viscosity is lowered down to 7 P and the electrospun fibers, produced at the best operating conditions above reported, exhibit a larger size, i.e., of 191 nm. size. It has been found that with very low viscosity there is no continuous fiber formation and with very high viscosity there is difficulty in the ejection of jets from the polymer solution, thus there is a requirement of optimal viscosity for electrospinning. Fong et al. [15] have studied polyethylene oxide (PEOX) to study nanofiber formation at different viscosities and found that a range of viscosity between 1 and 20 poise is suitable for production of uniform nanofibers by electrospinning. The values of viscosity of the sol-gel material produced in this work in presence of PEO were well inside of the suitable range viscosity outlined by Fong et al. [15] and thus it's a confirmation of their results.

3. Conclusions

In this work, the influences of the operating conditions of a rotating electrospinning machine on the characteristics of a non-woven chitin-based tissue has been investigated. In order to minimize the experimental efforts, a factorial campaign of experiments have been designed and performed. The effects of the electrode distance, the voltage range and the rotational speed of the spinning electrode were considered. By means of preliminary experiments, the variables' operating ranges to be adopted were identified. The experimentation showed that the most important operating variable is, in our particular case, the distance between the electrodes, as its increase gave rise to a reduction of the fiber diameter down to less than 150 nm. A second significant effect was exhibited by the collected electrode rotational speed. By operating at the best set of operating variables, a very good electrospinnability of the tissue was obtained with parallel elongated fibers, but the size ratio of the produced fibers was quite high. Future work should be done to find out the best operating conditions and to improve the non-woven tissue, by an investigation within the domain of variables identified by this study.

Conflicts of Interest: The authors declare no conflict of interest.

References

1. Shibata, H.; Shioya, N.; Kuroyangi, Y. Development of new wound dressing composed of spongy collagen sheet containing dibutyryl cyclic AMP. *J. Biomater. Sci. Polym. Ed.* **1997**, *8*, 601–621. [CrossRef] [PubMed]
2. Ulubayram, K.; Nur Cakar, A.; Korkusuz, P.; Ertan, C.; Hasirci, N. EGF containing gelatin-based wound dressings. *Biomaterials* **2001**, *22*, 1345–1356. [CrossRef]
3. Zeng, J.-B.; He, Y.-S.; Li, S.-L.; Wang, Y.-Z. Chitin wiskers: An overview. *Biomacromolecules* **2012**, *13*, 1 11. [CrossRef] [PubMed]
4. Ferreira, S.H.; Cunha, F.Q.; Hyslop, S. Role of the inducible forms of ciclooxygenate and nitric oxide synthase in inflammatory pain. In *Inducible Enzymes in the Inflammatory Response*; Willoughby, D.A., Tomlinson, A., Eds.; Birkhauser Verlag: Basel, Switzerland, 1999; pp. 149–167.
5. Mattioli-Belmonte, M.; Zizzi, A.; Lucarini, G.; Giantomassi, F.; Biagini, G.; Tucci, G.; Orlando, F.; Provinciali, M.; Carezzi, F.; Morganti, P. Chitin nanofibrils linked to chitosan glycolate as spray, gel and gauze preparations for wound repair. *J. Bioact. Compat. Polym.* **2007**, *22*, 525–553. [CrossRef]
6. Ji, Y.L.; Wolfe, P.S.; Rodriguez, I.A.; Bowlin, G.L. Preparation of chitin nanofibril/polycaprolactone nanocomposite from a nonaqueous medium suspension. *Crabohydr. Polym.* **2012**, *87*, 2313–2319. [CrossRef]
7. Zhang, Y.Z.; Su, B.; Venugopal, J.; Ramakrishna, S.; Lim, C.T. Biomimetic and bioactive nanofi brous scaffolds from electrospun composite nanofibers. *Int. J. Nanomed.* **2007**, *2*, 623–638.

8. Teo, W.E.; Ramakrishna, S. Electrospun nanofibers as a platform for multifunctional, hierarchically organized nanocomposite. *Compos. Sci. Technol.* **2009**, *69*, 1804–1817. [CrossRef]
9. Naseri, N.; Algan, C.; Jacobs, V.; John, M.; Oksman, K.; Mathew, A.P. Electrospun chitosan-based nanocomposite mats reinforced with chitin nanocrystals for wound dressing. *Crabohydr. Polym.* **2014**, *109*, 7–15. [CrossRef] [PubMed]
10. Shin, Y.M.; Hohman, M.M.; Brenner, M.P.; Rutledge, G.C. Electrospinning: A whipping fluid jet generates submicron polymer fibers. *Appl. Phys. Lett.* **2001**, *78*, 1149–1151. [CrossRef]
11. Reneker, D.H.; Yarin, A.L. Electrospinning jets and polymer nanofibers. *Polymer* **2008**, *49*, 2387–2425. [CrossRef]
12. Shin, Y.M.; Hohman, M.M.; Brenner, M.P.; Rutledge, G.C. Experimental characterization of electrospinning: The electrically forced jet and instabilities. *Polymer* **2001**, *42*, 9955–9967. [CrossRef]
13. Bhardwaj, N.; Kundu, S.C. Electrospinning: A fascinating fiber fabrication technique. *Biotechnol. Adv.* **2010**, *28*, 325–347. [CrossRef] [PubMed]
14. Reneker, D.H.; Chun, L. Nanometre diameters of polymer, produced by electrospinning. *Nanotechnology* **1996**, *7*, 216–223. [CrossRef]
15. Fong, H.; Reneker, D.H. Elastomeric nanofibers of styrene–butadiene–styrene tri-block copolymers. *J. Polym. Sci. B Polym. Phys.* **1999**, *37*, 3488–3493. [CrossRef]

PART IV
Applications

Chitin-Hyaluronan Block Copolymeric Nanoparticles for Innovative Cosmeceuticals

Hong-Duo Chen, Li Yuan Hong, and Pierfrancesco Morganti

Abstract: Chitin (C8H13O5N)n is a low-cost available unbranched polysaccharide widely distributed in nature as supporting structure of the cell wall of fungi and exoskeleton of arthropod and insects. Chitin nanofibrils are easily metabolized by the body's endogenous enzymes and thus used in cosmetic dermatology and biotextiles. Hyaluronan is anionic, nonsulfated glycosaminoglycan which, as major component of the Extracellular Matrix (ECM) plays several biological roles, showing expensive age-related changes. A successful model of an innovative block-polymer nanoparticle (BPN) based on phosphatidylcholine, hyaluronan, and chitin nanofibrils entrapping amino acids, vitamins, and melatonin has been formulated. Both the in vitro and in vivo results obtained demonstrate the efficacy of the injected block-polymer nanoparticles in reducing skin wrinkling and ameliorating the signs of aging. Chitin nanofibrils, protecting both corneocytes and intracorneal lamellae, help to maintain cutaneous homeostasis, neutralize the activity of radicals and trap them in their structure, regularizing the correct cell turnover. The BPN seems to be useful for improving the activity of permanent fillers, rendering it useful as an anti-aging remedy for the plastic surgery armamentarium.

Raymond Reed and Albert Kligman brought about the concept of "cosmeceutical" more than 50 years ago [1,2]. According to these authors, *cosmeceuticals* are topical Cosmetic-Pharmaceutical hybrids intended to enhance the health and beauty of skin, being based on the use of biologically active ingredients with medicinal or drug-like benefits.

At this purpose, many substances, either chemically synthesized or extracted from plants or animals, are used as functional ingredients. Nowadays, many cosmetic products with biologically active ingredients have been developed and marketed, though there are discrepancies in relation to their regulations and approvals by the government [3–5]. However, a number of topical cosmeceutical treatments for conditions such as photoaging, hyperpigmentation, wrinkles, and hair damage have come into widespread use.

In the cosmeceutical arena, nanotechnology has played an important role. Using new techniques to manipulate matter at an atomic or molecular level, they have been at the root of numerous innovations, opening up new perspectives for the future of the cosmeceutical industry.

Nanotechnology-based cosmeceuticals, in fact, offer the advantage of diversity in products, increase the bioavailability of active ingredients, ameliorating the aesthetic appeal of cosmeceutical products with prolonged effects [6–8].

There is a great potential in the marine bioprocess industry to convert and utilize most marine resources and marine food by-products as valuable cosmeceutical ingredients. Current available potential cosmeceuticals from marine resources include: seaweed extract, phlorotannins, polysaccharides, carotenoid pigments, fucosterol, microalgae extract, collagen, bioactive peptides, chitooligosaccharide derivatives, enzymes, sea mud, sea water, and minerals. Potential health benefits of marine-derived cosmetic active ingredients on human skin include anti-aging, antioxidant, anti-wrinkling, anti-whitening, cytoprotective, anti-tyrosinase, anti-acne, anti-inflammatory, and UV photo protective effects [9].

Chitin (C8H13O5N)n is a long-chain glucose-derived of a N-acetylglucosamine widely distributed in nature. It is a modified polysaccharide which, containing nitrogen, is distributed in the cell wall of fungi and exoskeketon of arthropods and insects. This polymer available at low cost, has been shown to be bio- and eco-compatible, with a very low level of toxicity. At present, the world offshore disposal of this natural waste material is estimated to be around 250 billion tons per year.

Chitin is an underutilized resource and has the potential to supply a wide range of useful products if suitably recycled, thus contributing to sustainable growth and a greener economy [10,11]. In addition, it is a good inducer of defense mechanisms in plants, being a fertilizer that can improve overall crop yields. Chemically-modified chitin forms edible films as an additive to thicken and stabilize foods and pharmaceuticals. It also acts as a binder in dyes, fabrics, adhesives, a reproducible form of biodegradable plastic, and as a promising substrate for engineering human tissues. Chitin's flexibility and strength make it favorable as surgical thread. Its biodegradibility means it wears away with time as the wound heals and it might have some unusual properties that accelerate the healing of wounds in humans [12].

Recently, it has become possible to industrially produce pure chitin crystals, named "chitin nanofibrils" (CN) for their needle-like shape and nanostructured average size (240 × 5 × 7 nm) [11] (Figure 1).

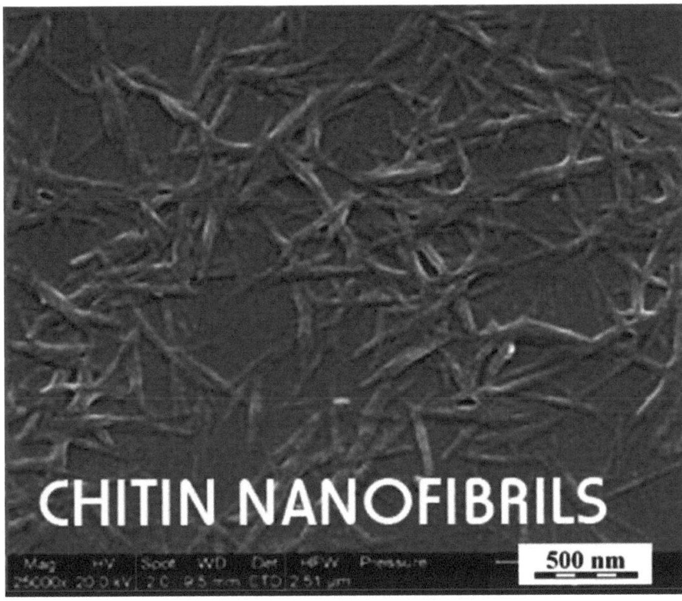

Figure 1. Chitin Nanofibrils at SEM.

Having a backbone like hyaluronic acid, chitin nanofibrils are easily metabolized by the body's endogenous enzymes and thus used in cosmetic dermatology and biotextiles. Moreover, because it occurs naturally and is considered as a safe raw material, it is safe to use. Chitin nanofibrils were recognized as a strong "gelling agent" [13].

As the nanofibril has an average size one-quarter that of a bacterium, 1 g of the product covers a surface area of 400 m^2. Many studies have shown that chitin nanofibrils can activate the proliferation of keratinocytes as well as fibroblasts, regulating not only collagen synthesis but also cytokine secretion and macrophage activity [11,13–15].

Chitosan is a natural biopolymer derived from the deacetylation of chitin and known to have various biological activities such as antifungal, antitumor, and antibacterial activity. It has been applied in various fields including wastewater treatment, agriculture, fabric and textiles, cosmetics, nutritional enhancement, and food processing.

Nanocomposite biomaterials based on chitosan and chitin are widely investigated for their antimicrobial activity, biocompatibility, and biodegradability [11,13]. P. Morganti's group has obtained interesting results showing how chitin nanofibrils can not only ameliorate the appearance of photoaged skin but also promote wound healing by reducing hypertrophic scar formation [6,15–20] (Figure 2).

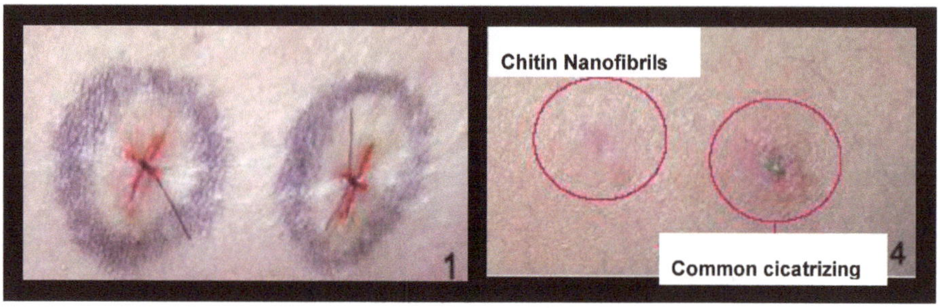

Figure 2. Cicatrizing activity of a gel based on chitin nanofibrils and chitosan (courtesy of P. Mezzana, MD [20]).

In vitro studies have shown how chitin nanofibrils can increase the reproduction of fibroblasts with a subsequent increase in collagen synthesis and in adenosine triphosphate production. In an in vivo double-blind study, skin hydration and superficial skin lipids were improved, with a simultaneous reduction in lipid peroxides and transepidermal water loss (TEWL) [21–23].

Hyaluronan is an anionic, non sulfated glycosaminoglycan distributed widely throughout connective, epithelial, and neural tissues. Hyaluronan is a polymer of disaccharides, composed of D-glucuronic acid and D-N-acetylglucosamine, linked via alternating β-1,4 and β-1,3 glycosidic bonds. Polymers of hyaluronan can range in size from 5000 to 20,000,000 Da in vivo [24,25].

Hyaluronan is found in many tissues of the body, such as skin, cartilage, and the vitreous humour. Therefore, it is well-suited to biomedical applications targeting these tissues. Native hyaluronan has a relatively short half-life, so various manufacturing techniques have been deployed to extend the length of the chain and stabilize the molecule for its use in medical applications [6,11,22,26].

The first hyaluronan biomedical product was developed in the 1970s and is approved for use in eye surgery. Hyaluronan has been used in attempts to treat osteoarthritis of the knee by injecting it into the joint [27–29].

Dry skin can be treated with a prescription skin lotion containing sodium hyaluronate as its active ingredient. In some cancers, hyaluronan levels correlate well with malignancy and poor prognosis. Hyaluronan is often used as a tumor marker for prostate and breast cancer and may also be used to monitor the progression of the disease [29]. Moreover, it may also be used postoperatively to induce tissue healing, notably after cataract surgery [30], as well as in the synthesis of biological scaffolds for wound-healing applications [31].

Hyaluronan is a common ingredient in skin-care products, so that hyaluronic acid fillers have been injected using a classic sharp hypodermic needle, cutting through nerves and vessels, with increasing risks of causing pain and bruising.

Nanotechnology is not only essential for marketing-oriented chemical companies, but is also a tool for developing science-based solutions for innovative therapeutics and cosmetics, enhancing well-being and addressing anti-aging issues. Nanomaterials and nanobiotechnology have the potential to radically change the way cosmetics and drugs deliver their benefits.

Specifically, nanoparticles are being developed to encapsulate a wide range of ingredients beneficial to the skin. To obtain nanoparticles, two principles approaches are used: (a) the bottom-up method in which nanoparticles are assembled from the molecular dimension; (b) the top-down approach that reduces larger particles through the use of physicochemical methods. In cosmetics, the top-down approach is more commonly used to produce different kinds of structures. Examples of such structures include nanosomes, cubosomes, niosomes, and liposomes [6,32].

A successful model has been formulated by P. Morganti's group: an innovative block-polymer nanoparticles (BPN) based on phosphatidylcholine, hyaluronan, and chitin nanofibrils entrapping amino acids, vitamins, and melatonin.

The specific formulation for each milliliter contained: hyaluronan salt 1 mg; phosphatidylcholine 3 mg; creatine 0.1 mg; caffeine 0.1 mg; ascorbyl tetraisopalmitate 0.5 mg; vitamin E 10 mg; chitin nanofibrils 1 mg; melatonin 0.1 mg; glucosamine 0.1 mg; glycine 0.1 mg; arginine 0.1 mg; sodium phosphate dibasic 2 mg; potassium di hydrogen phosphate 0.2 mg; sodium chloride 9 mg; sterile water for injection to 1 mL.

The injection was based on the mesotherapy technique, using 1 mL solution and a 30 g needle positioned at 45 to the skin surface. The injection rate was at all times less than 0.3 mL/min. A firm massage, with the index finger inside the mouth and the thumb outside, was then used to remove any unevenness. The 1mL quantity is sufficient to treat the entire face [5,6].

Both the in vitro and the in vivo results obtained demonstrate the efficacy of the injected block-polymer nanoparticles in reducing skin wrinkling and ameliorating the signs of aging. Subjects were satisfied with the general aspect of their skin, which appeared softer and more hydrated during the first month of treatment.

In line with their self-evaluation, the appearance of fine wrinkling was notably reduced and the consequent skin softness and firmness enhanced during the entire treatment period. The general amelioration remained during the regression period and 30 days after the interruption of the treatment (Figure 3).

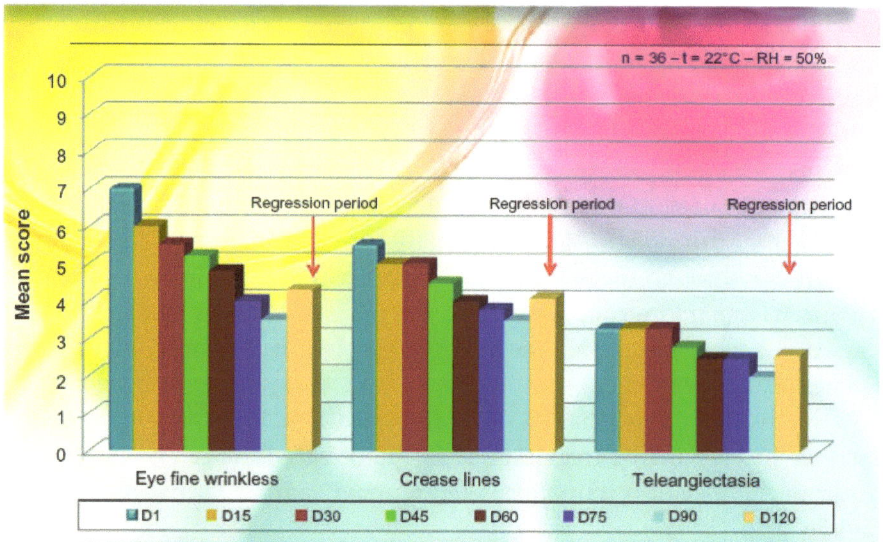

Figure 3. Dermatological mean evaluation on signs of photoaging after injective treatment with phosphatidylcholine-hyaluronic acid-chitin nanofibrils encapsulating active compounds (BPN). Note: All P values are highly significant as to baseline (p < 0.005). Abbreviations: BPN, block-polymer nanoparticles; RH, relative humidity.

In this formulation, the high content in linoleic acid of the phosphatidylcholine used allowed the active BPN to quickly reestablish the skin-barrier function. Thus, while the phosphatidylcholine fatty acids of the BPN composition contribute to balancing the disturbed composition and organization of lipids at the level of epidermal keratinocytes and consequently of corneocyte lamellae, the high level of linoleic acid should contribute to reintegrating the reduced level of ceramide 1, structural and stabilizing component of the stratum corneum [5,6,33].

Chitin nanofibrils, protecting both corneocytes and intracorneal lamellae, help to maintain cutaneous homeostasis, neutralize the activity of free radicals and trap them in their structure, and regularize the correct cell turnover [24,34–37]. All these activities are modulated and increased by the chitin nanofibrils/hyaluronan (CN-HA) encapsulation methodology made using the gelation method [38,39] (Figure 4).

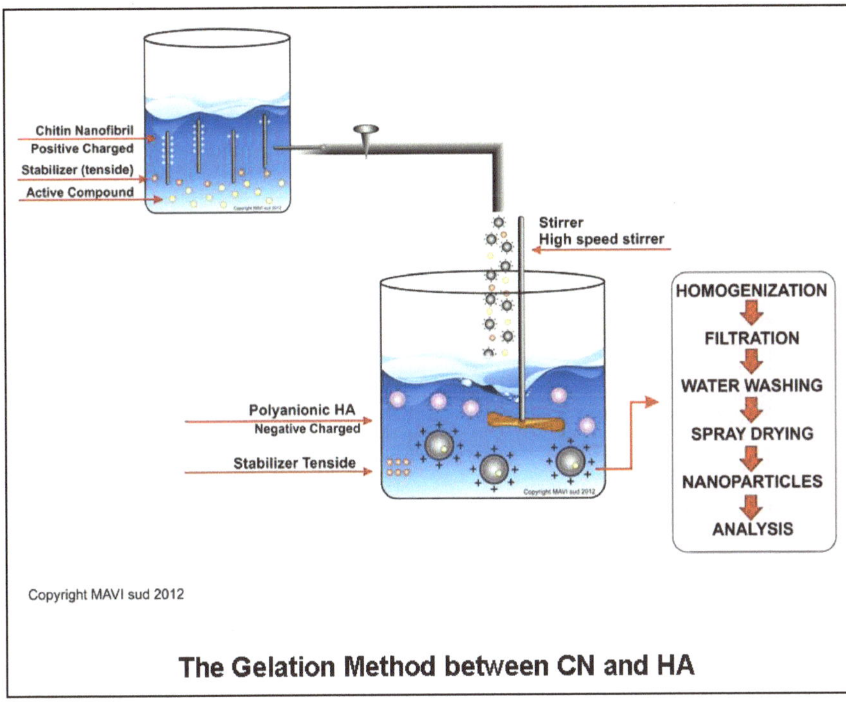

Figure 4. Gelation method for producing Chitin Nanofibril-Yaluronic acid (CN-HA) block polymeric nanoparticles.

In summary, the BPN encapsulating the active ingredient used seems to be useful in improving the activity of permanent fillers, rendering it useful as an antiaging remedy for the plastic surgery armamentarium.

In conclusion, this innovative biostimulating medical device should be used for wrinkle treatment and rejuvenating looks, as well as an adjuvant in soft-tissue augmentation and stretch-mark corrections [6,40].

Author Contributions: all the author contributed to write the paper in the same way.

Acknowledgments: We would like to thank The China Medical University for the help offered in setting up this paper.

Conflicts of Interest: We declare that Hong-Duo Chen and Li Yuan Hong have no conflict of interests. Pierfrancesco Morganti works as Head of the R&D Centre of Nanoscience, Mavi Sud, s.r.l, Italy.

References

1. Reed, R. E. The definition of "cosmeceuticals". *J Soc Cosm Chem* **1962**, *13*, 103–116.
2. Kligman, A.M. Why cosmeceuticals? *Cosm & Toil* **1993**, *108*, 37–40.
3. Mincea, M.; Negrulescu, A.; Ostafe, V. Preparation, modification, and applications of chitin nanowiskers: A Review. *Rev. Adv. Mater. Sci* **2012**, *30*, 225–242.

4. Gao, X.H.; Zhang, L.; Wie, H.; et al. Efficacy and safety of innovative cosmeceuticals. *Clin Dermatol.* **2008**, *26*(4), 367–374.
5. Morganti, P.; Guevara, L.; Palombo, P.; Palombo, M.; Cardillo, A.; Mezzana, P.; Fabrizi, G.; Svolacchia, F. A phosphatidylcholine hyaluronic acid chitin–nanofibrils complex for a fast skin remodeling and a rejuvenating look. *Clinical Cosmetic and Investigational Dermatology* **2012**, *5*, Pages 213–220.
6. Morganti, P. Use and potential of nanotechnology in cosmetic dermatology. *Clinical Cosmetic and Investigational Dermatology* **2010**, *3*, 5–13.
7. Dureja, H.; Kaushik, D.; Gupta, M.; et al. Cosmeceutical: an emerging concept. *Indian J Pharmacol* **2005**, *37*, 155–9.
8. Lohani, A.; Verma, A.; Joshi, H.; et al. Nanotechnology-Based Cosmeceuticals. *ISRN Dermatol.* **2014**, *2014*, 1–14.
9. Li, D.H.; Wu, Z.M.; Marini, N.; et al. Advanced carrier systems in cosmetics and cosmeceuticals: A review. *J Cosmet Sci.* **2011**, *62*, 549–563.
10. Kim, S.K. Marine cosmeceuticals. *J Cosmet Dermatol.* **2014**, *13*(1), 56
11. Morganti, P.; Carezzi F.; Del Ciotto, P.; Morganti, G.; Nunziata, M.L.; Xingua, G.; Hong-Duo, C.; Tishchenko, G; Yudin, V.E. Chitin NAnofibrils: A Natural Multifunctional Polymer: Physicochemical characteristics and safeness. In: *Nanobiotechnology*; Phenix, D.A., Waqar, A., Eds.; One Central Press: Manchester, UK, 2014, pp. 1–31.
12. Morganti, P.; Morganti, G.; Morganti, A. Transforming nanostructured chitin from crustacean waste into beneficial health products: a must for our society. *Nalnotechnol Sci Appl.* **2011**, *6*(4), 123–129.
13. Lee, J.-Y.; Choi, B.; Wu, B.; et al. Customized biomimetic scaffolds created by indirect three-dimensional printing for tissue engineering. *Biofabrication* **2013**, *5*(4), 10.
14. Mikesova, J.; Hasek, J.; Tishchenko, G.; et al. Rheological study of chitosan acetate solution containing chitin nanofibrils. *Carbohydr Polym.* **2014**, *112*, 753–757.
15. Nguyen, T.H.; Kwak, H.S.; Kim, S.M. Physicochemical and biofunctional properties of crab chitosan nanoparticles. *J Nanosci Nanotechnol.* **2013**, *13*(8), 5296–5304.
16. Morganti, P. Use and potential of nanotechnology in cosmetic dermatology. *Clin Cosmet Investig Dermatol.* **2010**, *24*, 5–13.
17. Mikesova, J.; Hasek, J.; Tishchenko, G.; et al. Rheological study of chitosan acetate solutions containing chitin nanofibrils. *Carbohydr Polym.* **2014**, *112*, 753–757.
18. Morganti, P.; Fabrizi, G.; Bruno, C. Protective effects of oral antioxadants on skin and eye function. *Skinmed.* **2004**, *3*(6), 310–316.
19. Biagini, G.; Zizzi, A.; Tucci, G.; et al. Chitin nanofibrils linked to chitosan glycolate as spray, gel and gauze preparations for wound repair. *J Bioact Compat Polym.* **2007**, *22*, 525–538.
20. Mezzana, P. Clinical efficacy of a new chitin-nanofibrils based gel in wound healing. *Acta Chirurgiae Plasticae.* **2008**, *50*(3), 81–84.
21. Biagini, G.; Zizzi, A.; Giantomassi, F.; et al. Cutaneous absorption of nanostructured chitin associated with natural synergistic molecules (lutein). *J Appl Cosmetol.* **2008**, *26*, 69–80.

22. Morganti, P.; Fabrizi, G.; Palombo, P.; et al. Chitin-nanofibrils: a new active cosmetic carrier. *J Appl Cosmetol.* **2008**, *26*, 105–120.
23. Morganti, P.; Morganti, G.; Fabrizi, G.; et al. A new sun to rejuvenate the skin. *J Appl Cosmetol.* **2008**, *26*, 159–166.
24. Morganti, P. Chitin nanofibrils for advanced cosmeceuticals. Clinics in Dermatology **2008**, *26*, 334–340.
25. Morganti, P.; Fabrizi, G.; Ruocco, E.; et al. Chitin nanofibrils improved photoprotection. *Cosmetics and Toiletries* **2009**, *124*(9), 66–73.
26. Fraser, J.R.; Laurent, T.C.; Laurent, U.B. Hyaluronan: its nature, distribution, functions and turnover. *J Intern Med.* **1997**, *242*(1), 27–33.
27. Jin Huang. Organic Chemicals from Bioprocesses in China. Advances in Biochemical Engineering/Biotechnology **2010**, *122*, 43–71.
28. Frampton, J.E.; Hylan G.-F. 20 single-injection formulation. *Drugs Aging.* **2010**, *27*(1), 77–85.
29. Rutjes, A.W.; Juni, P.; da Costa. B.R.; et al. Viscosupplementation for osteoarthritis of the knee: a systematic review and meta-analysis. *Ann Intern Med.* **2012**, *157*(3), 180–191.
30. Josefsson, A.; Adamo, H.; Hammarsten, P. Prostate cancer increases hyaluronan in surrounding nonmalignant stroma, and this response is associated with tumor growth and an unfavorable outcome. *Am J Pathol.* **2011**, *179*(4), 1961–1968.
31. Gritsenko, P.G.; Ilina, O.; Friedi, P. Interstitial guidance of cancer invasion. *J Pathol.* **2012**, *226*(2), 185–199.
32. Shu, X.Z.; Ghosh. K.; Liu, Y. Attachment and spreading of fibroblasts on an RGD peptide-modified injectable hyaluronan hydrogel. *J Biomed Mater Res A.* **2004**, *68*(2), 365–375.
33. Morganti, P.; Palombo, P.; Fabrizi, G.; et al. Biweekly in-office injectable treatment of striae distensae versus a long-term daily use of topical vitamin C. *J Appl Cosmetol.* **2001**, *19*, 107–112.
34. Marenkov, L.N.; Steinert, P.M. Ceramides are bound to structural proteins of the human foreskin epidermal cornified envelope. *J Biol Chem.* **1998**, *273*, 17763–17770.
35. Morganti, P.; Randazzo, S.D.; Giardina, A.; et al. Effects of phosphatidylcholine linoleic acid-rich and glicoli acid in acne vulgaris. *J Appl Cosmetol.* **1997**, *15*, 21–41.
36. Morganti, P.; Agostini, A.; Bruno, C. ; et al. Role of topical glicolic acid and phosphatidylcholine linoleic acid in the pathogenesis of acne. Linoleic acid versus squalene. *J Appl Cosmetol.* **1997**, *15*, 33–41.
37. Morganti, P.; Morganti, G.; Fabrizi, G. ; et al. A new sun to rejuvenate the skin. *J Appl Cosmetol.* **2008**, *26*, 159–168.
38. Morganti, P. Chitin nanofibrils for cosmetic delivery. *Cosmet Toil.* **2010**, *125*, 36–39.
39. Morganti, P.; Fabrizi, G.; Palombo, M.; et al. New chitin complexes and their anti-aging activity from inside out. *J Nutr Health Aging.* **2012**, *16*(3), 242–245.
40. Morganti, P.; Palombo, M.; Palombo, P.; et al. Cosmetic science in skin aging: achieving the efficacy by the chitin nano-strucured chrystallites. *SOFW J.* **2010**, *136*, 14–24.

Chitin Nanocomposite Scaffolds for Advanced Medications

Xue-Gang Xu, Xing-Hua Gao, Hong-Duo Chen and Pierfancesco Morganti

Abstract: The rapid development of nanotechnology, especially in the biological material sciences, has stimulated the demand for new hybrid materials and biocomposites for wide range of applications. Chitin, the second most abundant polysaccharide after cellulose, is a naturally occurring biopolymer found in yeast, fungi and exoskeletal structures of numerous invertebrates. Chitin meets the several desirable properties of biomaterial, which includes mechanical strength, chemical and thermal stability, biocompatibility, etc. Both chitin and its deacetylated product chitosan are considered of great economic values because of their versatile biological activities which made them proper to various applications in personal care products, pharmaceutical, medicine, food, agriculture and environmental sectors. This chapter gives an overview of the extensive research and recent developments on chitin/chitosan based nanocomposite scaffolds for application as advanced medicines. Several kinds of nanocomposite scaffolds have been applicable or under investigations, such as chitin-chitosan-gelatin scaffolds, pectin-chitin/$CaCO_3$ nanocomposite scaffold, et al. Here we highlight the most recent research on different aspects of chitin based nanocomposite scaffolds, including their preparations, properties and applications, especially the applications of chitin nanocomposite scaffolds in tissue engineering, stem cell technologies, and vaccine preparations.

1. Introduction

There has been rapid development of nanotechnology recently, especially in the biological material sciences, which stimulates the demand for new hybrid materials and biocomposites of specific and defined properties for a wide range of applications. Nanomaterials exhibit novel characteristics compared to bulk materials, such as high mechanical strength, high surface-area-to-volume ratio, and extremely miniaturized size [1].

The interest in chitin as a raw material began in the 1920s owing to market pressure for low cost fibers, which promoted research into artificial silks. In particularly, chitin's biocompatibility and wound-healing properties made it attractive for different biomedical applications including tissue engineering scaffolds, wound dressing and sutures, and biocompatible devices. Chitin is the second most abundant polysaccharide after cellulose. It is a natural biopolymer found in fungi, yeast and the exoskeletal structures of numerous invertebrates including insects, sponges, worms and aquatic arthropods [2].

The industrial processing of shellfish such as crab, shrimp and krill for human food generates a huge amount of waste and it is said that 50–60% of the total weight is chitin. Annually, worldwide production is about 1.44 million metric tons dry weight [3]. These abundant and renewable marine wastes can be commercially exploited for the extraction of chitin [3]. However, traditional chitin extraction processes employ harsh chemicals and produce other waste-water or solvent. Advances in the biological extraction of this natural polymer, which reduces energy cost and waste water—producing valuable by-products—will certainly have high economic and environmental impacts [3]. Thus, the bio-extraction of chitin from crustacean shell waste has been increasingly studied in order to develop eco-friendly, cleaner, and economical processes.

Due to their molecular and supramolecular structures characterized for their intrinsic antimicrobial and wound healing properties, chitin and derived compounds (chitosan and chito-oligosaccharides) have been identified as suitable bio platforms to make specialized polymers functionalized for the advanced medicine.

What is chitin? As a derivative of glucose, it is a long-chain natural occurring polymer of *N*-acetyl-D-glucosamine with a molecular length varying from 5000 to 8000 *N*-acetyl-D-glucosamine units found in crabs, to up to only 100 units contained in yeast.

Chitosan, a chitin-derived compound, is usually produced by chitin deacetylation with concentrated alkali solutions at elevated temperatures. This process of deacetylation, which does not proceed to completion, implies that chitosan, obtained from commercial sources, is in reality chitin with a low degree of acetylation. When the number of the units is higher than 50%, the polymer is termed chitin, when less than 50% it is termed chitosan.

In conclusion, chitin and chitosan may be considered as two points of a continuum material that share the same basic structure differing in their acetylation degree. As to the function of its cell structural support and defense against environmental aggressions, chitin may be compared to human keratin, which supports skin, hair and nails [4].

It has therefore proven useful for several medical and industrial purposes from tissue engineering to making advanced medications for wounded and burned skin [5], to producing smart and innovative colored dressings, imitating the iridescent colors used by birds and butterflies in nature. Birds' plumage and butterfly wing scales, in fact, are often organized into stacks of nano-sticks or nano-layers made of chitin nano-crystals, which produce various iridescent colors thin-film interference (Figure 1) [6,7].

Figure 1. The different colors and iridescences are due to chitin nanocrystals. Source: Morganti 2012 [6].

Chitin is insoluble in aqueous solution and relatively difficult to process. To improve its processability, many different approaches have been studied for the successful employment of this polymer in the areas of advanced medicine including stem cells, tissue engineering etc.

Chitosan displays interesting physicochemical properties, differing in its properties and organized structure. In a solid state, relatively rigid crystals are formed due to its regularly arranged hydroxyl and amino groups, while in solution the hydrogen bondings drive the formation of microfibrils, depending on the chitosan concentration. Furthermore, chitosan is easy to develop into various designs, i.e., films, sponges, scaffolds and hydrogels, which result in the ability to make various kinds of tissue-engineering materials and wound dressings [7]. Due to their unique structural, physico-chemical and functional properties, both chitin and chitosan are good candidates for the preparation of scaffolds and dressing materials for tissue regeneration [8].

In recent years, considerable attention has been attracted to ameliorating the functionality of these biopolymers for improving their properties by increasing their solubility or introducing selective active functions or making blends with other intrinsically bioactive polymers [9].

This chapter tries to give an overview of the recently reported chitin and chitosan nanocomposite scaffolds, emphasizing their characteristics and applications in tissue engineering and other applications.

2. Characteristics of Potential Chitin/Chitosan Nanocomposite Scaffolds

2.1. Nanohydroxyapatite/Gelatin/Carboxymethyl Chitin Composite Scaffold

Sagar et al. developed a novel 3D scaffold from the unique combination of nanohydroxyapatite/gelatin/carboxymethyl chitin (*n*-HA/gel/CMC) for bone tissue engineering by using the solvent-casting method combined with vapor-phase cross-linking and freeze-drying [10]. An optimized (composition and processing parameters) ratio of *n*-HA:gel:CMC (1:2:1) exhibited an ideal porous structure with regular interconnected pores (75–250 μm) and a higher mechanical strength.

Their results suggested that the divalent (Ca^{2+}), carboxyl (COO^-), amino (NH_4^+), and phosphate (PO_4^{3-}) groups created favorable ionic interactions which facilitated structural stability and integrity of the composite scaffold. Further, hemocompatibility and biocompatibility with MG-63 osteoblast cells indicated that the structural and dimensional stability of a composite scaffold provided an optimal mechanosensory environment for the enhancement of cell adhesion, proliferation, and network formation [10].

2.2. Pectin-Chitin/$CaCO_3$ Nanocomposite Scaffold

Kumar et al. developed a nanocomposite scaffold using a mixture of chitin, pectin and nano $CaCO_3$ through the technique of lyophilisation, with specific biomedical applications for tissue engineering and drug delivery [2].

The developed composite scaffold showed controlled swelling and degradation, compared to the control scaffold. Moreover, cells attached onto the scaffolds started to proliferate after 48 h of incubation and demonstrated negligible toxicity towards cells. Drug delivery through the scaffold was confirmed using a bisphosphonate called Fosamax. The results suggested that the developed composite scaffold possessed the essential requisites for their applications in the fields of drug delivery and tissue engineering.

2.3. α-Chitin Hydrogel/Nano Hydroxyapatite Composite Scaffold

In another paper, Kumar et al. synthesized α-chitin hydrogel/nano hydroxyapatite (*n*HAp) composite scaffold using a freeze-drying approach [11] with nHAp and α-chitin hydrogel. Hydroxyapatite [(HAp), $Ca_{10}(PO_4)_6(OH)_2$] is a major inorganic ceramic material and an essential component of bone. nHAp is used for various applications like dental filling material, bone tissue engineering, etc. Because it offers a high surface area to volume ratio, a small concentration is sufficient to enhance its bioactivity and osseointegration. Hap nanoparticles were mixed with α-chitin hydrogel at concentrations of 0.5% and 1% (w/w), stirred for 30 min, frozen at −20 °C and lyophilized to get a microporous nanocomposite scaffold [11]. The prepared composite scaffolds were characterized using Fourier transform infrared spectroscopy (FTIR), X-ray diffraction (XRD), Thermogravimetric analysis and differential thermal analysis (TG-DTA) and scanning electron microscopy (SEM). Porosity, swelling ability, protein adsorption, in vitro biodegradation and biomineralization of the scaffolds were evaluated. These supports showed interconnected micro-pores, sufficient swelling ratio of 15–20, good protein adsorption, controlled degradation and wonder biomineralization ability.

2.4. Chitin-Chitosan/Nano ZrO_2 Composite Scaffolds

According to de Moraes et al. [11], innovative biomaterials can provide a promising new direction for the treatment of bone defects, stimulating a proper repair process with no damage to the adjacent tissues [12]. Collagen, in fact, is one of the most used biomaterials due to its biocompatibility and bioactivity [13]. Thus, Soares et al. [14], evaluating the odontogenic potential of human dental pulp cells in contact with a porous system of chitosan-collagen mineralized with calcium-aluminate, concluded that this scaffold seems to be an interesting candidate for in vivo applications to exposed pulp tissue.

Owing to its good mechanical strength and biocompatibility, zirconia is also considered to be one of the most used materials after titanium over a period of about 20 years, especially in dentistry. Cultured osteoblasts proliferate and differentiate in zirconia with no adverse reaction. Jayakumar R et al. tried incorporation of nano ZrO_2 onto the chitin-chitosan scaffold to enhance osteogenesis [15]. They fabricated a nanocomposite scaffold using a lyophilisation technique with chitin-chitosan and nano ZrO_2. The prepared nanocomposite scaffolds were characterized by FTIR, SEM, XRD and TGA. The swelling, degradation, cell attachment, cell viability and biomineralization of the composite scaffolds were also detected. The results showed better swelling and controlled degradation compared to the control scaffold [15]. Cytocompatibility studies proved the non-toxic nature of chitin-chitosan/nano ZrO_2 scaffolds against MG-63, L-929 and hMSCs [15]. Additionally, cell attachment studies showed the nanocomposite scaffold significantly increased the cell attachment when compared to control scaffolds. All these results suggested that the developed nancomposite scaffolds possess the prerequisites for tissue engineering scaffolds and could be used for bone tissue engineering and other bio-engineering purposes of human interest.

3. Chitosan/Gelatin/nSiO$_2$ Composite Scaffold

Kavya et al. [16] fabricated a 3D nanocomposite scaffold of chitosan, gelatin and nano-silica by lyophilisation to produce a better candidate for bone tissue engineering compared to pure chitosan and chitosan/gelatin scaffolds.

To prepare composite scaffolds with better biological compatibility and hydrophilicity, they added gelatin to enhance the performance of chitosan. Unique biocomposites of amorphous silica, chitin, and crystalline aragonite have been discovered in marine sponges. Silica is believed to be essential in skeletal development so that critical amounts of silicon ions are found to up-regulate genes like collagen type-, BSP, osteocalcin and osteopontin in osteoblasts.

To this purpose, these authors developed a 3D composite scaffold of chitosan/ chondroitin sulfate/$nSiO_2$ to bring out the combined properties of chitosan, gelatine and $nSiO_2$ to facilitate bone regeneration. Porosity, swelling, density, mechanical integrity, degradation, biomineralization and protein adsorption studies, favored it in comparison to the conventional chitosan and chitosan/gelatin scaffolds. In vitro cyto-compatability, cell attachment-proliferation and ALP activity studies by using MG-63 cells, advocated its remarkable performance. These cumulative results indicate the chitosan/gelatin/nSiO$_2$ nanocomposite scaffold as a suitable candidate for bone tissue engineering.

Chitosan-Graphene Oxide Network Structure Scaffold

Graphene—a single layer of sp2 bonded carbon atoms in a two-dimensional hexagonal lattice—has attracted considerable attention as a potential biomaterial because of its physic-chemical properties such as a large surface area, high hydrophilicity and dispersibility. Chitosan-graphene network structure scaffolds were synthesized by covalent linkage of the carboxyl groups of grapheme oxide with the amine groups of chitosan [10].

The covalent incorporation of graphene oxidant into a CS network favorably modulated the biological response of osteoblasts, such that cell attachment and growth were significantly enhanced. Thus, related to a combination of a number of physic-chemical factors, including a large surface area, nanoscale roughness, the presence of pendant groups, a hydrophilic nature and high water retention ability, this network is believed to be a promising material for tissue engineering applications in regenerative medicine (Figure 2) [17,18].

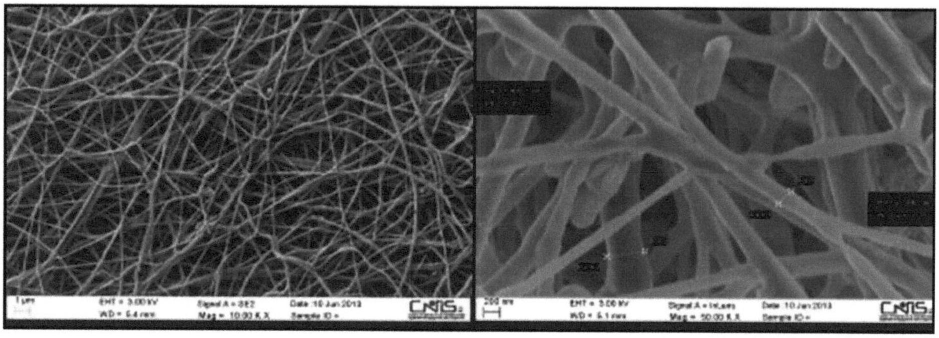

Figure 2. Engineered non-woven tissue made by chitin nanofibrils with scanning electron microscopy (SEM). Source: Morganti et al. [19].

4. Applications of Nanocomposite Scaffolds for Advanced Medications

Tissue Engineering

Extensive research is in progress to develop biosmart materials in the field of tissue engineering, being considered of prime importance for biomedical and microbiological applications.

Chitin, chitosan and its derivatives have received much attention because of their biocompatibility and other advantages. Chitin composite scaffolds have sufficient mechanical strength. And due to their viscoelasticity in wet conditions, they can be cut, deformed, and fitted according to bone defects. This allows us to use them in cases of bone cysts and large defects with smaller openings.

These scaffolds can spontaneously grow by consuming the calcium and phosphate ions from the surrounding fluid which render them osteoinductive structures, analogous to living bone.

Besides being a matrix of a composite material, chitin has also been confirmed to promote osteogenesis in mesenchymal stem cells, accelerate wound healing, enhance cell migration, and form granulation tissue with angiogenesis [19–22]. These effects might be mediated by the production of cytokines and growth factors by fibroblasts that come into contact with the chitin material. For tissue engineering applications, the ideal situation would be gradual but complete scaffold degradation concomitant to tissue remodeling, while foreign material might elicit an adverse host tissue response in the long term, so the degradability of polymeric scaffolds should be of primary concern.

Chitin is susceptible to lysozyme and chitotriosidases, which are ubiquitously present in humans. In this regard, it is interesting to underline that this natural polymer has to be considered as a pathogen-associated molecular pattern (PAMP) regulating the macrophage function and inflammation, depending on its size [23].

Moreover, it has been shown that chitin nanofibrils (240 × 7 × 5 nm) stimulate the defensive production of defensine-2, contextually reducing the production of metallo-proteinases and inflammation [24,25]. This is another advantage for chitin in the application of tissue engineering. Combinations of chitin/chitosan with other polymers have been fabricated to nanocomposite scaffolds for tissue engineering as we mentioned above. Recent work established that a chitin-hydroxyapatite composite, loaded with mesenchymal stem cell-induced osteoblasts, was able to support bone regeneration when implanted into bone defects in a rabbit femur [26].

Composite scaffolds of chitin with nanohydroxyapatite and nanotitania, obtained by dispersing the particles in a chitin hydrogel cast from a $CaCl_2$/methanol solvent system and freeze-drying of the mixture, supported apatite deposition and adhesion of a variety of cell lines [11,15]. Composite scaffolds, prepared with *n*-HA/gel/CMC, have sufficient mechanical strength and, along with their viscoelasticity in wet conditions, they can be cut, deformed, and fitted according to bone defects which allows them to be used in cases of bone cysts and large defects with smaller openings [10].

Another composite scaffold, based on α-chitin hydrogel/nano hydroxyapatite composite scaffold, showed that its viability in the presence of scaffold leachables and *n*HAp did not affect its cyto-compatibility. It was also found that the scaffolds were cytocompatible and cells were well attached and distributed throughout these structures [11]. Moreover, scaffolds made by chitin nanofibrils developed antibacterial and anti-inflammatory activity and increased the production of defensine-2, speeding up the skin repairing process in wounded and burned skin [17,24–27].

All these results suggested that these scaffolds can be used for bone and wound tissue engineering also because of their effectiveness and safeness. However, biopolymers offer a highly effective flexibility to design porous matrices by chitosan, chitin, collagen, alginate and other natural compounds. Moreover, when at the nanodimension, they have to be considered the best candidates for drug delivery applications due to their controllable pore sizes, high surface area with favorable properties.

5. Stem Cell Technologies

Because of the fast development in technologies, these years, stem cell had bright prospects in regenerative medicine and organ transplantations. Chitin and other biomaterials can help to realize the vast potential of stem cells in regenerative medicine by: (1) playing a role in providing substrates that support stem cell self-renewal, while maintaining stem cell pluri- or multipotency; (2) favoring the provision of a matrix permissive to stem cell differentiation; (3) representing a matrix of composite material as physical support of cells to regenerate organs; and (4) guiding their differentiation when appropriate signals are provided [28]. To this purpose, the advantages of a chitin matrix in supporting stem cell proliferation and differentiation has extended its applications in vivo, where chitin has been shown to be effective as a carrier material for mesenchymal stem cells in the treatment of large physical defects.

Mesenchymal stem cells seeded or encapsulated in water-soluble chitin-alginate fibrous scaffolds have been differentiated into chondrogenic and osteogenic lineages by immersion in the respective differentiation media [29]. Nanocomposite scaffolds of chitin or chitosan would definitely have more broad application prospects.

6. Vaccine Preparation

Vaccination is one of the major keys to maintaining a good public health and wellbeing status of society. It induces specific adaptive immune responses and memory responses against infections, tumors, etc. [30]. Unfortunately, not all vaccines are as effective, often showing low efficiency of antibody production with a weak host T cell response and T cell memory, which require repeated booster injections to obtain longer host memory immune responses.

To solve this problem, adjuvants [31] are often used to augment the effects of a vaccine by stimulating the immune system to respond to the vaccine more vigorously, thus providing increased immunity to a particular disease. There are many adjuvants, such as aluminium salts, virosomes, etc. Aluminium—the only adjuvant approved by the Food and Drug Administration (FDA) for clinical use—stimulates B cell response for antibody production but is not very effective at inducing host T cell responses and does not work well with all antigens.

Chitin nanocomposite scaffolds are a biodegradable polymer, which not only has immune adjuvant effects on its own [23,24] but is also able to release incorporated cytokines and antigens in a controlled manner, therebby synergistically boosting adjuvant effects. Chitin also accelerates macrophage migration and fibroblast proliferation with a particular role in vascularisation [32].

For clinical applications, chitin/chitosan is frequently cited to possess low immunogenicity as an advantage, and at the same time, the ability to act as an immunoadjuvant. Glycolic acid-g-chitosan-gold nanoflower nanocomposite scaffolds, for example, were confirmed to be used for the sustained delivery of drug [33]. This property would also help chitin/chitosan nanocomposite scaffolds to be used in vaccine preparation as an adjuvant.

7. Conclusions

Recent progress in chitin and chitosan nano-composite scaffolds highlights great potential usage in wound-healing and tissue engineering due to their unique structural, functional, physical and chemical properties. Another area of particular interest is also gene-therapy, referring to methods aimed at influencing gene expression in living organisms through the delivery of integrating exogenous DNA or RNA to treat or prevent disease [34]. For these reasons, many innovative approaches are emerging in recent years with the use of novel materials and technologies. However, biomedical modifications of these biopolymers and their preparation in different designs have been reported extensively, but until now with limited commercial impact. Further studies on the clinical applications of these chitin and chitosan nanocomposite scaffolds would be the most important issue in this field.

Acknowledgments: The work was sponsored by the Project for Construction of Major Discipline Platform at the Universities of Liaoning Province.

Conflicts of Interest: The authors declare no conflict of interests.

References

1. Shukla, S.K.; Mishra, A.K.; Arotiba, O.A.; Mamba, B.B. Chitosan-based nanomaterials: A state-of-the-art review. *Int. J. Biol. Macromol.* **2013**, *59*, 46–58. [CrossRef] [PubMed]
2. Kumar, P.T.S.; Ramya, C.; Jayakumar, R.; Nair, S.K.V.; Lakshmanan, V.K. Drug delivery and tissue engineering applications of biocompatible pectin-chitin/nano $CaCO_3$ composite scaffolds. *Colloid Surface B* **2013**, *106*, 109–116. [CrossRef] [PubMed]
3. Kaur, S.; Dhillon, G.S. Recent trends in biological extraction of chitin from marine shell wastes: A review. *Crit. Rev. Biotechnol.* **2013**, *35*, 44–61. [CrossRef] [PubMed]
4. Morganti, P.; Chen, H.-D.; Gao, X.-H. Chitin Nanofibril: A Natural Eco-Friendly and immunoadjuvant active carrier for medical use. *J. Appl. Cosmetol.* **2016**, *34*, 141–154.
5. Morganti, P.; Febo, P.; Cardillo, A.; Cardillo, M. Chitina e Lignina: Polimeri di supporto per la rigenerazione dei tessuti. *Natural* **2016**, *140*, 72–79.
6. Morganti, P. Nanoparticles and Nanostructures Man-Made or Naturally Recovered: The Biomimetic Activity of Chitin Nanofibrils. *J. Nanomater. Mol. Nanotechnol.* **2012**, *1*, 2. [CrossRef]
7. Morganti, P.; Palombo, M.; Chen, H.D.; Gao, X.H. Medical Textile and Nanotechnology. *Cosmet. Sci. Technol.* **2013**, 128–138.
8. Francesko, A.; Tzanov, T. Chitin, chitosan and derivatives for wound healing and tissue engineering. *Adv. Biochem. Eng. Biotechnol.* **2011**, *125*, 1–27. [PubMed]
9. Ifuku, S. Chitin and Chitosan Nanofibers: Preparation and chemical modifications. *Molecules* **2014**, *19*, 18367–18380. [CrossRef] [PubMed]
10. Sagar, N.; Soni, V.P.; Bellare, J.R. Influence of carboxymethyl chitin on stability and biocompatibility of 3D nanohydroxyapatite/gelatin/carboxymethyl chitin composite for bone tissue engineering. *J. Biomed. Mater. Res. B Appl. Biomater.* **2012**, *100*, 624–636. [CrossRef] [PubMed]
11. Kumar, P.T.; Srinivasan, S.; Lakshmanan, V.K.; Tamura, H.; Nair, S.V.; Jayakumar, R. Synthesis, characterization and cytocompatibility studies of alpha-chitin hydrogel/nano hydroxyapatite composite scaffolds. *Int. J. Biol. Macromol.* **2011**, *49*, 20–31. [CrossRef] [PubMed]

12. Moraes, P.C.; de Souza Marques, I.C.; Basso, F.G.; Rosseto, H.L.; Pires de Souza, P.; de Carvalho, F.; de Souza Cidta, C.A.; de Fonseca, R.C. Repair of Bone Defects with Chitosan-collagen Biomembrane and Scaffold Containing Calcium Aluminate Cement. *Braz. Dent. J.* **2017**, *28*, 287–295. [CrossRef]
13. Soares, G.D.; Rosseto, H.L.; Scheffel, D.S.; Basso, F.G.; Huck, C.; Hebling, J.; de Souza Cista, C.A. Ontogenic different ustion potential of human dental pulp cells cultured on a Calcium-Aluminate enriched Chitosan-collagen scaffold. *Clin. Oral. Investig.* **2017**. [CrossRef] [PubMed]
14. Soares, D.G.; Rosseto, H.L.; Bassso, F.G.; Hebling, J.; Costa, C.A. Chitosan-collagen biomembrane embedded with calcium-aluminate enhances dentinogenic potential of pulp cells. *Braz. Oral Res.* **2016**, *30*, e54. [CrossRef] [PubMed]
15. Jayakumar, R.; Ramachandran, R.; Sudheesh Kumar, P.T.; Divyarani, V.V.; Srinivasan, S.; Chennazhi, K.P.; Tamura, H.; Nair, S.V. Fabrication of chitin-chitosan/nano ZrO(2) composite scaffolds for tissue engineering applications. *Int. J. Biol. Macromol.* **2011**, *49*, 274–280. [CrossRef] [PubMed]
16. Kavya, K.C.; Jayakumar, R.; Nair, S.; Chennazhi, K.P. Fabrication and characterization of chitosan/gelatin/*n*SiO2 composite scaffold for bone tissue engineering. *Int. J. Biol. Macromol.* **2013**, *59*, 255–263. [CrossRef] [PubMed]
17. Depan, D.; Girase, B.; Shah, J.S.; Misra, R.D. Structure-process-property relationship of the polar graphene oxide-mediated cellular response and stimulated growth of osteoblasts on hybrid chitosan network structure nanocomposite scaffolds. *Acta Biomater.* **2011**, *7*, 3432–3445. [CrossRef] [PubMed]
18. Kumar Dutta, P.G. *Chitin and Chitosan for Regenerative Medicine*; Springer: New Delhi, India, 2016.
19. Morganti, P.; Carezzi, F.; Del Ciotto, P.; Morganti, G.; Nunziata, M.L.; Gao, X.H.; Chen, H.D.; Galina Tishenko, G.; Yudin, V.E. Chitin Nanofibrils: A Natural Multifunctional Polymer. In *Physicochemical Characteristics, Effectiveness and Safeness*; Phenix, D.A., Ahmed, W., Eds.; One Central Press Ltd.: Altrincham, UK, 2014; pp. 1–37.
20. Udenni Gunathilake, T.M.S.; Ching, Y.C.M.; Ching, K.Y.; Chuah, C.H.; Abdullah, L.C. Biomedical and Microbiological Applications of Bio-Based Porous Materials: A Review. *Polymers* **2017**, *9*, 160. [CrossRef]
21. Lieder, R.; Thormodsson, F.; Ng, C.H.; Einarsson, J.M.; Gislason, J.; Petersen, P.H.; Sigurjonsson, O.E. Chitosan and Chitin Hexamers affect expansion and differentiation of mesenchymal stem cells differently. *Int. J. Biol. Macromol.* **2012**, *51*, 675–680. [CrossRef] [PubMed]
22. Muzzarelli, R.A.; Mattioli-Belmonte, M.; Pugnaloni, A.; Biagini, G. Biochemistry, histology and clinical uses of chitins and chitosans in wound healing. *EXS* **1999**, *87*, 251–264. [PubMed]
23. Da Silva, C.A.; Chalouni, C.; Williams, A.; Hartl, D.; Lee, C.G.; Elias, J.A. Chitin is a size-dependent regulator of macrophage TNF and IL-10 production. *J. Immunol.* **2009**, *182*, 3573–3582. [CrossRef] [PubMed]

24. Morganti, P.; Fusco, A.; Paoletti, I.; Perfetto, B.; Del Ciotto, P.; Palombo, M.; Chianese, A.; Baroni, A.; Donnarumma, G. Anti-Inflammatory, Immunomodulatory, and Tissue Repair Activity on Human Keratinocytes by Green Innovative Nanocomposites. *Materials* **2017**, *10*, 843. [CrossRef] [PubMed]
25. Morganti, P.; Febo, P.; Cardillo, M.; Donnarumma, G.; Baroni, A. Chitin Nanofibril and Nanolignin: Natural Polymers of Biomedical Interest. *J. Clin. Cosmet. Dermatol.* **2017**, *1*. [CrossRef]
26. Morganti, P.; Del Ciotto, P.; Stoller, M.; Chianese, A. Antibacterial and anti-inflammatory Green Nanocomposites. *Chem. Eng. Trans.* **2016**, *47*, 61–66.
27. Ge, Z.; Baguenard, S.; Lim, L.Y.; Wee, A.; Khor, E. Hydroxyapatite-chitin materials as potential tissue engineered bone substitutes. *Biomaterials* **2004**, *25*, 1049–1058. [CrossRef]
28. Wan, A.C.; Tai, B.C. CHITIN—A promising biomaterial for tissue engineering and stem cell technologies. *Biotechnol. Adv.* **2013**, *31*, 1776–1785. [CrossRef] [PubMed]
29. Yim, E.K.; Wan, A.C.; Le Visage, C.; Liao, I.C.; Leong, K.W. Proliferation and differentiation of human mesenchymal stem cell encapsulated in polyelectrolyte complexation fibrous scaffold. *Biomaterials* **2006**, *27*, 6111–6122. [CrossRef] [PubMed]
30. Andre, F.E.; Booy, R.; Bock, H.L.; Clemens, J.; Datta, S.K.; John, T.J.; Lee, B.W.; Lolekha, S.; Peltola, H.; Ruff, T.A.; et al. Vaccination greatly reduces disease disability deaths and inequity worldwide. *WHO Bull.* **2008**, *8612*, 81–160. [CrossRef]
31. Di Pasquale, A.; Preiss, S.; Da Silva, T.; Garcon, N. Vaccine Adjuvants: From 1920 to 2015 and beyond. *Vaccines* **2015**, *3*, 320–343. [CrossRef] [PubMed]
32. Fischer, T.H.; Bode, A.P.; Demcheva, M.; Vournakis, J.N. Hemostatic properties of glucosamine-based materials. *J. Biomed. Mater. Res. A.* **2007**, *80*, 167–174. [CrossRef] [PubMed]
33. Kumari, S.; Singh, R.P. Glycolic acid-g-chitosan-gold nanoflower nanocomposite scaffolds for drug delivery and tissue engineering. *Int. J. Biol. Macromol.* **2012**, *50*, 878–883. [CrossRef] [PubMed]
34. Reley, M.K., II; Vermeris, W. Recent advances in nanomaterials for gene delivery. A review. *Nanomaterials* **2017**, *7*, 94. [CrossRef] [PubMed]

Flexible Food Packaging Using Polymers from Biomass

Maria-Beatrice Coltelli, Vito Gigante, Patrizia Cinelli and Andrea Lazzeri

Abstract: The use of polymers from biomass in the production of flexible packaging is an important challenge to give an answer to both the reduction of oil-derived materials and the increase of waste production. Oil-derived materials are now employed in recyclable packaging, but, although the management of waste recycling is improving, it cannot allow the complete mechanical recycling of the plastic fraction. It would be important to optimize the system by replacing packaging difficult to be recyclable, such as multi-layer flexible packaging mainly based on the use of polyolefines, with alternative packaging consisting of biodegradable materials, thus managing its disposal by considering composting. In the present paper, a classification of polymeric materials available from biomass for flexible packaging is reported. Moreover, the biodegradable and renewable materials investigated or yet employed in this sector are described. In addition, some insights on the actual production of flexible packaging is given, to describe also what technical specification the polymers from biomass should have to replace commodities. Finally, the perspectives in the field of flexible packaging for polymers from biomass are discussed.

---

1. Polymeric Materials from Biomass

As material scientists, we often need to study and select the best material for a specific application. In many cases, especially for applications requiring the combination of lightness and resistance, polymeric materials are quite interesting, as well as blends or composites obtained by using them blended with other polymers or with inorganic or organic fillers.

In fact, polymers, with respect to other materials, show a higher versatility due to the possibility of modulating their macromolecular structure by controlling the industrial synthesis in order to achieve specific final properties.

The structural materials selected by nature in plants and animals are based on polymers too. The main structural polymer for plants is cellulose, a polyether, which is the polymer of D-glucose having each repeating unit linked to the following one by a β(1-4)-linkage. The main structural polymers for animals are proteins, characterized by a polyamide structure.

In general, polymers can be natural, artificial or synthetic (Table 1). An example of a natural polymer is cellulose, contained in paper or cotton fabric; among natural polymers are also the polymers produced by microorganisms, such as poly(hydroxyalcanoates) (PHAs). An example of an artificial polymer, obtained by chemical modification of natural polymers, is cellulose acetate, a plastic material, or cellulose xanthate, a fiber for textile applications, or nitro-cellulose, which can be an explosive, but with a low degree of nitration is employed in varnishes and enamels in the cosmetic sector [1].

The natural and artificial polymers show the advantage of being renewable; that is, they are obtained not from petrochemical resources but from natural sources.

Examples of synthetic polymers are given, in the packaging sector, by the so-called commodities, consisting in polyethylene (PE), polypropylene (PP), poly(ethylene terephthalate) (PET) or polystyrene (PS).

Currently biosynthetic polymers are distinguished from synthetic polymers because they were industrially produced by traditional polymerization methods, but starting from monomers available from natural sources [2]. Hence they are renewable. Poly(lactic acid) (PLA) is an example of biosynthetic polymer. The advantage of biosynthetic polymers with respect to natural and artificial polymers is the possibility of controlling the primary structure of the polymer and consequently its processability and its properties.

Table 1. Classification of polymers with respect to renewability and biodegradability.

Class	Renewability	Biodegradability
natural	YES	YES [a]
artificial	YES [b]	POSSIBLE
synthetic	NO	NO [c]
bio-synthetic	YES	YES/NO

[a] although some evaluations about it were made [3], natural rubber cannot be considered biodegradable; [b] the renewability of the chemicals used for modification should be considered as well; [c] some petrochemical-based polymer are biodegradable.

From the point of view of biodegradability [4,5] usually natural polymers are biodegradable where biodegradability is the capability of resulting in the complete oxidation in the environment giving CH_4, CO_2 and other simple compounds. This is typical of all polymers, but for synthetic polymers very long times—centuries—are required (Figure 1). Artificial polymers can be biodegradable, but usually it depends much on the degree of chemical modification reached. For instance, cellulose acetate, that can have an acetylation degree between 0 and 3, is reported to be biodegradable with an acetylation degree below 2.5 [6]. Biosynthetic polymers are usually both bio-based and biodegradable, such as PLA, but also polymers defined as synthetic, such as PE, traditionally obtained by petrochemicals, can now be produced from natural sources. These polymers are not biodegradable.

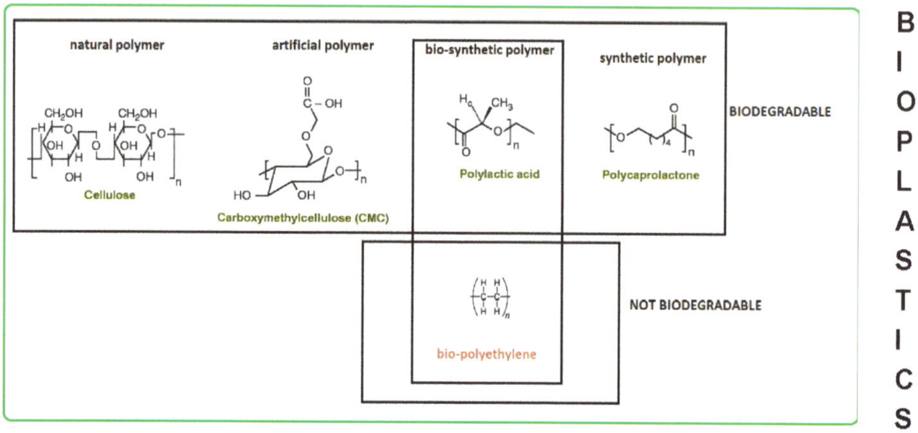

Figure 1. Scheme related to bioplastics with examples of specific polymers.

In addition, in bioplastics also the polymers currently not yet synthesized by biotechnology but from not biobased monomers, such as poly(butylene succinate) (PBS), polycaprolactone (PCL) and poly(butylene adipate-co-terephthalate) (PBAT) et cetera, are included, because they are biodegradable. The bioplastics family thus regroups polymers coming from renewable and/or biodegradable sources [7].

In the field of packaging, the most interesting options for polymers in applications depend much on the end-life option [8,9]. While in durable applications biodegradability is not an interesting property, in the food-packaging sector it could be so, especially if the perspective of composting is selected for the waste packaging.

More important than biodegradability is compostability. The latter is related to the capability of a material to effectively be disintegrated because of microorganism activity in the environment in standard conditions. Biodegradable and compostable products should be certified according to EN 13432/14995 standards [10,11], defining procedures for testing the effective compostability, granting the correct behavior of the material in the composting plants.

The other possible option for packaging management is recycling, which requires the possibility of separating the different polymers, avoiding contamination as well as good durability of the plastic material, especially in re-processing conditions. Thus, it is an option suitable for mono-material packages.

2. Commercially Available Polymeric Materials from Biomass

In considering bioplastics that can be employed in flexible packaging preparation, polymers produced on an industrial scale and yet available in the market and validated in processes commonly used in plastic industry have to be taken mainly into account.

Two different types of basic formulations were developed on a mature technological level: starch-based materials and poly(lactic acid) (PLA)-based materials.

Starch is a natural polymer, consisting of amylose and amylopectin, with the former being linear and water soluble, and the latter branched and water insoluble. The processing of starch was possible some decades ago by controlling the plasticizing effect of water [12], culminating in the thermoplastic processing of starch at approximately its natural water content (about 15%) at a temperature of about 100 °C. Amorphous thermoplastic starch (TPS)-based polymers were thus obtained. An important characteristic of thermoplastic starch formation is the thermal and mechanical (shear-based) destructuring of the starch granules to form a homogeneous melt, the formation being accompanied by swelling. In particular, the gelatinization of starch occurs [13] thanks to the processing, in which the starch granules become swollen and destructured and lose amylose by diffusion. This process, having a typical temperature dependent on the water content, results in the destruction of amylopectin crystallites and molecular order in the granules. The gelatinization represents an undesired state. Hence the range of temperature for processing is superiorly limited by the gelatization point. From a rheological point of view TPSs show the possibility of being processed only in a restricted screw speed range, but in that range the behavior is shear thinning, similar to the one of low density polyethylene, with apparent viscosity decreasing with the increase of screw speed [14]. Destructured starch behaves as a thermoplastic polymer and can be processed as a traditional plastic; when alone, however, its sensitivity to humidity makes it unsuitable for most of the applications. The main use of destructured starch alone is in soluble compostable foams such as loose-fillers, and other expanded items as a replacement for polystyrene.

The attaining of processable starch-based formulations suitable for flexible film production was possible by blending starch with thermoplastic hydrophilic synthetic polymers such as poly(caprolactone) [15] and poly(ethylene-vinyl alcohol) [16]. Usually plasticizers such as glycerol [17] or polyethylene glycols [12] are used for optimizing TPS processing. Cassava, also known as manioc or yucca, is a plant producing tuberous roots, typical of South America. Ezehoa et al. [18] reported the preparation of cassava starch-based formulations employed for preparing films by blown film extrusion. The films were obtained by adding poly(vinyl alcohol) (PVA), which is the matrix in the system. Ali et al. [19] also reported the preparation of films for packaging containing starch, but these films consisted mainly of polyethylene. However, in general it is not possible to employ starch for packaging films without blending it with plasticizers and polymers. The granules available on the market and widely employed for soft packaging, especially for pouches, were developed by considering this approach. NOVAMONT, which is the main producer of MATER-BI material, also follows this approach, but using renewable and biodegradable self-produced additives. These intermediates are produced by vegetable oils and are defined polymeric complexing agents [20]. In fact, they interact with the starch, incorporating it. Hence the processing and mechanical properties were easily modulated and the starch is protected by environmental humidity by the barrier properties of the host polymers/additives.

Another promising material for packaging application, because of its cost, which is now not much higher than the one of PS (at about 2 €/kg in Europe [21]), is poly(lactic acid) (PLA), which is renewable and biodegradable. Because of its rigidity it is not suitable alone for flexible packaging applications. However, commercial granules suitable for flexible packaging applications are available. In fact, in the last decade, many studies were carried out about PLA blends with other biodegradable polyesters or plasticizers. Plasticized and nano-filled films were prepared by flat die extrusion by Scatto et al. [22]. The rheological measurements carried out by capillary viscometer suggested that the processability of plasticized and nano-filled PLA is suitable for the industrial production of cast films. Plasticization reduces the glass transition temperature of the plastic material, making it ductile at room temperature. A good miscibility of the plasticizer is important, to avoid demixing and loss in transparency of films. Moreover, the plasticizer should not migrate out of the film. The production of film based on PLA, biodegradable plasticizers and nano clays was investigated in the running EC project DIBBIOPACK "Development of injection and extrusion blow molded biodegradable and multifunctional packages by nanotechnologies" [23].

Blending with other commercial polymers is also another possibility for increasing PLA flexibility. Blends of PLA and poly(butylene adipate-co-terephthalate) (PBAT) show a good flexibility and ductility in a proper composition range [24,25]. The addition of a plasticizer to PLA/PBAT blends can also provide a further alternative for properties modulation [26].

The use of chain extenders [27] or peroxides [28] can be also important for modulating rheological properties introducing an increase in melt viscosity, some branching and hence a shear thinning behavior.

Blends of PLA and PBS were also studied [29,30] for application in flexible film preparation, thanks to the good compatibility between these polyesters.

Many companies are developing new PLA-based formulations of granules suitable for flexible packaging applications and many different granules can be found in the market. The formulations are often very complex, as they contain fillers, such as calcium carbonate or talc, or nucleating agents, to control crystallization of PLA during processing by increasing resistance without compromising the transparency of films. Our research unit has recently developed novel copolymers consisting of PLA, organic plasticizers with epoxy functions and biodegradable elastomeric polyesters, which allows the production of transparent PLA-based films, maintaining compostability [24,31] and activities for exploiting this know-how producing commercial granules are ongoing.

3. Production of Flexible Packaging by Using Biobased Polymers

The production of flexible packaging is based on the preparation of films. Plastic films and multilayer systems can be manufactured using different converting processes such as blown film extrusion, flat die extrusion, extrusion coating, extrusion laminating and co-extrusion. These processes have advantages and disadvantages depending on the material type in use, the width and thickness of film and the required film properties. The use of starch or biopolyester in such processes introduced some modification with respect to the use of polyolefins (high density poly(ethylene)—HDPE and low density poly(ethylene)—LDPE) and PET. Biopolyesters showed a processability similar to traditional plastics. The control of humidity of biopolyesters is significant for limiting the increase of melt fluidity due to chain scission, hence industrial plants must be equipped with proper drying sections. In this equipment usually biopolyester pellets are heated in presence of a dried air flow, with a low dew point, maintaining them stirred [32] to avoid undesired agglomeration, before extrusion.

One of the most commonly employed methods for preparing plastic film is blown film extrusion [30,33]. The produced film is tubular, hence this process is usually employed for the production of pouches, industrial bags or packaging films for shrink wrapping. The necessary industrial equipment consists of an extruder equipped with an annular die. The blown film process involves the biaxial stretching of annular extrudate to make a suitable bubble according to the product requirements. During this film-blowing process, the molten polymer from the annular die is pulling upward applying the take-up force; air is introduced at the bottom of the die to inflate the bubble and an air ring is used to cool the extrudate. The nip rolls are used to provide the axial tension needed to pull and flatten the film into the winder (Figure 2). The speed of the nip rolls and the air pressure inside the bubble are adjusted to maintain the process and product requirements. At a certain height from the die exit, molten polymer is solidified due to the effect of cooling followed by crystallization, and a freeze line height (FLH) can be observed. Above this point the bubble diameter is assumed to be constant.

Bubble size is maintained by controlling the air through a hole in the die face. Addition of air inside the bubble will expand it to a larger diameter and vice versa. This inflation process will stretch the bubble in the circumferential direction (CD). The ratio of this expanded bubble diameter and the die diameter is defined as blow-up ratio (BUR). To pull the extrudate in the upward direction, an axial force is applied by means of nip rollers and hence another stretching in the axial or machine direction (MD) occurs. Draw-down ratio (DDR), which is another important process variable is defined as the ratio of the linear speed of the film at the nip rolls and the average melt velocity at the die exit. Melt rheological properties are important in the blown film extrusion to determine the processability, shape and stability of the film bubble and the onset of surface roughness.

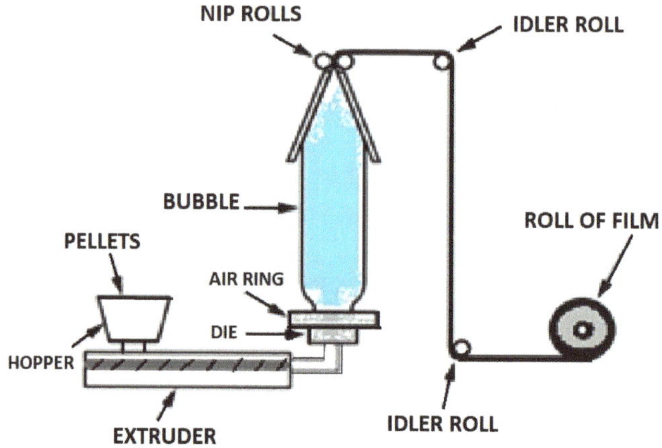

Figure 2. Schematic diagram of the blown film process.

Usually the polymer employed in this process for film production is low-density polyethylene (LDPE) or linear low-density polyethylene (LLDPE). The degree of branching and the length of the branching are particularly important for the success of the process. Generally, a high short chain branching of LDPE influences much the capability of crystallizing of the polymer, much influencing the mechanical and optical properties of the film. A polymer having a more irregular structure is thus more transparent. Long chain branching (LCB) is usually necessary because, considering the rheological behavior, it is correlated to shear thinning (viscosity decreasing by increasing shear rate in a stationary flow) and tension thickening (viscosity increasing by increasing the extensional flow) [34].

PLA-based materials show generally a melt fluidity too high to be processed in these plants. Usually chain extenders are used to improve the viscosity of the melt and the melt strength to allow the production of films, and commercial ones are available on the market [35]. Lazzeri et al. developed a patent to produce transparent PLA-based blends containing an epoxidized molecule acting both as plasticizer, compatibilizer and fluidity regulator [24], thus allowing a good control of processability and final properties. With respect to traditional polyolefins, the tearing strength of these blends should be improved. Recently it was evidenced that by using the epoxidized molecule in combination with a nucleating agent it was possible to improve the tearing strength of the blends [36] reaching values comparable to those of PP.

Flat die extrusion consists of an extrusion through a linear die of adjustable thickness (die gap) usually between 3 and 1.4 mm. This technology allows the production of polymeric sheets and films (with thicknesses ranging from 50 microns to a millimeter) and consists of the extrusion of the molten polymer through a die of rectangular geometry. The geometry generally used for the flat die distribution channel is the coat hanger die. The material comes out from the die in the form of a molten plate that is immediately in contact with a thermostatic roller to allow cooling and solidification. Due to the motion of the roller, the film undergoes elongation with a consequent reduction in thickness. The film then passes through a second roller and to a measuring, cutting and winding station.

The operating parameters to be controlled during flat die extrusion are the extrusion and windup rolls temperatures, distance between die and the first roll and the draw ratio (ratio between windup roll speed and polymer speed at the die exit).

All of these parameters have a big influence on the final product characteristics (morphologically and mechanically). (Figure 3).

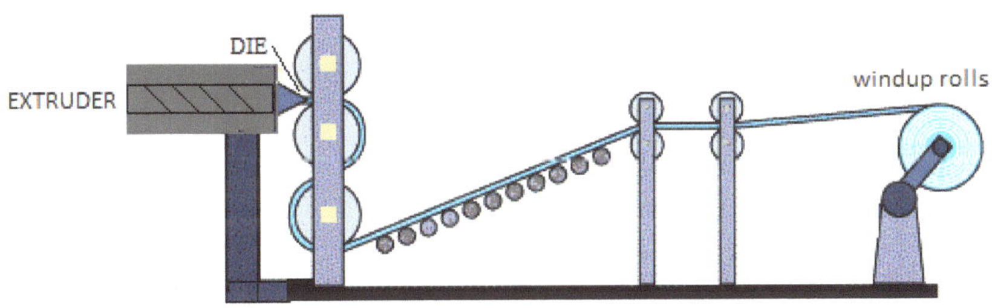

Figure 3. Schematic diagram of the flat die extrusion.

In flat die extrusion, as well as in blown extrusion, the necessity of increasing the melt viscosity of PLA-based blends is often an issue, but usually melt fluidity values higher than those adopted in the blown extrusion are suitable for this processing technique.

Films produced through this technique (Figure 3) can be used in flexible packaging and they can be used also in multilayer systems. Diaper top-sheets are generally made by using polyolefins with a specific patterning of holes on the surface allowing the controlled passage of liquids through the film. Currently the ongoing POLYBIOSKIN European project [37] is active in developing a diaper top-sheet based on biobased formulations to replace traditional ones. The final objective is developing diapers completely compostable and with improved compatibility with skin. Hence PLA-based granules are selected and processed by flat die extrusion to obtain flexible films (Figures 3 and 4).

Figure 4. Flat die extrusion of PLA (**left**) and flat die extruded films based on PLA/PBAT blends (**right**).

Extrusion coating consists of the extrusion of a film of plastic onto a preformed film made of plastic, cellulose or aluminum, hence it is a technique suitable for multilayer packages. The current process of extrusion coating involves extruding resin from a slot at temperatures up to 300 °C directly onto the moving substrate. The control of velocity at which the substrate runs allows to control the coating thickness. High extensional flows are encountered at non-isothermal conditions, especially in industrial conditions requiring higher and higher output speeds. Currently it is very difficult for biopolyester-based materials to have the suitable processing properties in terms of fluidity and melts strength typical of polyolefins to perform extrusion coating in industrial plant currently used for producing PE/aluminum bilayer typical of the most employed multilayer packages such as TETRAPACK [38].

Coating of plastic films with water-based solution containing whey was studied in the framework of the WHEYLAYER European project [39]. The project results showed that the layer made of whey, coming from the waste cheese industry, a protein in between other polymeric films (for instance PET and PE films), allowed separation of different materials in a water bath containing enzymatic detergents [40]. Hence a multilayer package allowing the full recovery and recycling of all the polymeric components was developed in the framework of this project. Moreover, the whey layer shows interesting barrier properties towards oxygen.

Another technique for producing multilayer packaging is extrusion laminating. It is similar to the extrusion coating process except that the extruded melt polymer acts as the adhesive to a second film of material (Figure 5). Specific scientific papers about this technique could not be found. In any case, in this sector, the development of biodegradable glues is an interesting topic of current research [41].

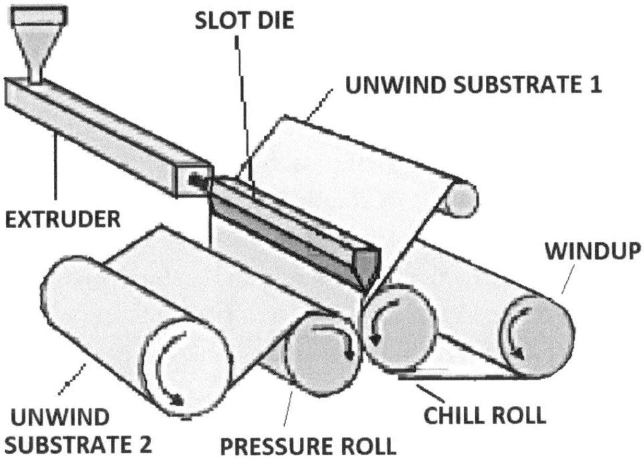

Figure 5. Schematic diagram of extrusion laminating.

Co-extrusion is a similar process to extrusion coating but with two or more extruders coupled to a single die head in which the individually extruded melts are brought together and finally extruded as a multi-layer film. In this case, the viscosity differences between the different layers are of critical importance to avoid distortions of the produced multilayer system. Regarding this process applied to bioplastics, only a patent reported the co-extrusion of PLA with poly(ethylene-co-vinyl acetate) (EVA), a not biodegradable polymer, to obtain multilayer films [42]. Certainly the study of processes allowing to obtain completely biodegradable systems would be useful.

In general, for the processing of biobased polymers, structural and rheological properties of extruded polymers are fundamental for optimizing industrial processes. In general, as stated above, usually for extrusion coating, less viscous polymer melts are required than for blown film extrusion, but a polymer having proper shear thinning and tension thickening behavior must be selected to avoid instabilities and defects thus avoiding the breakage of the film, leading to the undesired interruption of the industrial production [43].

Regarding the processing of starch-based materials, the method based on blown film extrusion is quite common. In fact, the commercial starch-based materials are used for producing pouches and shopping bags. The use of these bags to replace polyolefin ones is supported by EU regulation [44]. Regarding the processing of PLA, films are produced by flat die extrusion [37] or blown film extrusion [36].

4. Innovative Plastic Formulations for Flexible Packaging

Innovative biopolymer-based materials are developed by academies and companies all over the world. Many studies are focused on improving the properties of starch or PLA-based formulations, but also in fostering the use of other biopolymers by making easier their preparation and finding new applications or also by trying to develop new materials employing natural abundant polymers such as polysaccharides (cellulose or chitin) or proteins. In both cases the studies are focused onto waste coming from agricultural or food industries as starting materials.

Innovative materials based on starch are currently developed by preparing blends with other biodegradable polymers to improve processability and mechanical properties. Recently Hwang et al. [45] prepared PLA/starch blends by melt blending by using maleic anhydride in the presence of a peroxide in order to obtain in one processing step both functionalization of PLA and reactive compatibilization between starch and PLA. Although some improvements in compatibility were observed, the tensile properties were very similar to the one of pure PLA. Reasonably, the combination of increased compatibility with biopolyesters and a better knowledge about the effect of plasticizers coming from renewable sources, as more and more of them will be obtained by new biorefineries [46], will lead towards the formulation of materials with improved properties.

PLA-based materials for flexible packaging applications will be improved by deepening the knowledge about nano-composite preparation and properties and selecting suitable and performing plasticizers. PLA nano-composites are promising because the presence of nano-fillers can improve some functional properties (barrier properties, thermo-mechanical properties, anti-microbial properties, etc.) without affecting the density of material, because generally only a few percent of nano-fillers is needed to improve physical-chemical and mechanical properties. These properties are in fact very important for flexible packaging applications. The addition to PLA of layered nano-filler was widely studied [47] as these kinds of fillers affect barrier properties and so are very interesting in the attempt to develop packaging films. The barrier properties of PLA to O_2, CO_2 and water vapor are reported by Auras et al. [48] and summarized on Table 2. It can be noticed that in general the permeability properties are slightly lower than the ones of PET, with the exception of water vapor permeability, where the PLA is slightly better.

Table 2. Barrier properties of PET and PLA on the basis of Auras paper [48].

	CO_2 Permeability	Oxygen Permeability	Water Vapor Permeability Coefficient (WVPC)
PLA	1.99^{a}–2.77^{b} × 10^{-14} g·m/(m^2·s·Pa) at 0% RH from 25 °C	1.39^{a}–1.21^{b} × 10^{-15} g·m/(m^2·s·Pa) at 25 °C and 70% RH	1.61^{a}–1.65^{b} × 10^{-11} g·m/(m^2·s·Pa) at 30 °C in the 40% to 90% RH range
PET	1.73 × 10^{-15} g·m/(m^2·s·Pa) at 0% RH from 25 °C	1.88 × 10^{-16} g·m/(m^2·s·Pa) measured at 25 °C and 70% RH	1.1 × 10^{-12} g·m/(m^2·s·Pa) at 25 °C in the 40% to 90% RH range

[a] PLA 4040-D commercial product; [b] PLA 4030-D commercial product.

On the basis of these data, PET and PLA are both hydrophobic films that absorb very low amounts of water and show similar barrier property behavior. Hence the replacement of PET in flexible packaging applications would be possible.

As stated above, many researchers tried to improve the barrier properties of PLA by preparing nano-composites. The dispersion of phyllosilicates into polymers or blends at the nanometer scale [49] allowed improvement of the properties of the polymer matrix such as thermal stability. PLA nano-composites that contain phyllosilicates are considered superior for providing improved gas barrier properties, due mainly to the strong effect of confinement as the result of a high surface:volume ratio (i.e., reducing chain mobility and permeability [49]), as well as to the enhancement of tortuosity [50] of the path required for small molecules to permeate through a polymer film due to the presence of silicate lamellae.

New synthetic and processing routes were investigated by Castiello et al. for the production of PLA-based nano-composites containing a modified clay [51]. The direct synthesis of PLA by ring opening polymerization in the clays was compared in terms of structure and property features of nano-composites with a method based on melt processing in a laboratory batch mixer adding clay or modified clay to PLA. Although the in situ polymerization could be promising, the latter method showed evident advantages in terms of time and cost. Successively, the processing was successfully scaled up in a semi-industrial extruder in order to prepare extruded films for flexible packaging [22,23]. However, since the addition of plasticizer was necessary for attaining the suitable flexibility, the barrier property improvements due to the clay addition was counterbalanced by its reduction due to plasticization, making the material in the amorphous state at room temperature and so more suitable for gases and vapors diffusion.

Nano-cellulose, which represents the crystalline part of cellulose fibers, is also very interesting as a filler to be dispersed in PLA in an attempt to improve the barrier properties of PLA to oxygen, as evidenced by several authors [52,53].

Chitin is another interesting source, as it is very abundant because it represents a waste of the food industry being present in the exoskeletons of crabs, krills, lobsters and shrimps [54]. From chitin it is possible to obtain on an industrial scale chitin nano-fibrils [55] up to now employed in the sector of cosmesis and biotextiles [56]. The addition of chitin nano-fibrils to PLA is very innovative and plasticized, and nano-filled PLA showed composition-dependent tensile properties [57] suitable for flexible packaging. As an antimicrobial and antifungal behavior was reported for chitin [58] and chitin nano-fibrils [59], the possibility of exploiting this approach for producing nano-reinforced antimicrobial films is quite interesting. This represented one of the objectives of the European project N-CHITOPACK [60]. One of the aims of the project was to develop a biodegradable nanostructured film containing chitin nano-fibrils. The observed transparency of the films was in agreement with a nano-scaled dispersion of chitin nano-fibrils inside the material (Figure 6). This result was achieved by studying a suitable method for adding chitin nano-fibrils to polymers during extrusion avoiding nano-fibril agglomeration. These films were biodegradable, as evidenced by Cinelli et al. [61].

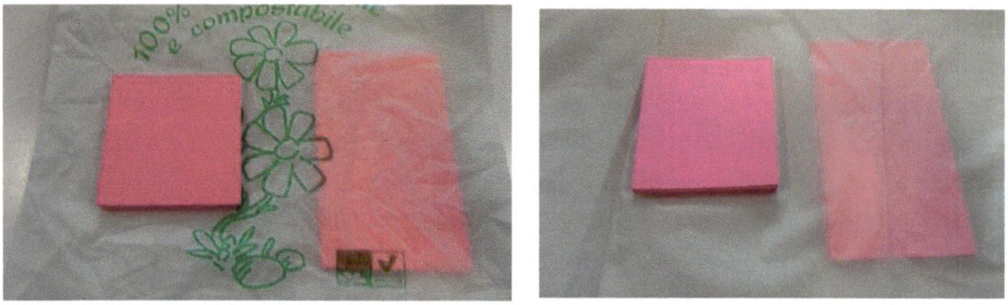

Figure 6. Commercial compostable biobased films (**left**) based on BioComp® BF 3051; films (**Right**) obtained with the same material but with addition of chitin nano-fibrils.

In the field of polyesters, polyhydroxyalcanoates (PHAs) should also be considered [62], as they are produced by microorganisms. Since the biotechnology for obtaining them is not yet at a mature level, the price of these polymers is quite high (4–5 €/kg [63]). Several researchers are targeting optimization of production and extraction process and the use of waste material-based substrates for the growth of microorganism producing PHAs, as developed in the EC project Oli-PHA "A novel and efficient method for the production of PHA polymer-based packaging from olive oil waste water" under the scientific coordination of our research unit [64]. Their processability changes as a function of the molecular weight and also primary structure of the PHAs. However, in general these materials are brittle and very viscous when melted. Hence the use of a proper plasticizer is fundamental for processing. The necessity of controlling the migration of plasticizer from PHAs to avoid stickiness of the material and loss of ductility was evidenced by Farris et al. [65].

The other important class of biomaterials is represented by proteins, widely available as by-products or waste of the agricultural and horticultural industries and the industry of food or leather.

Hence different kinds of proteins are also studied as potential plastic materials for producing films. As a result, proteins from plants (wheat gluten, soy, sunflower, and corn) and animals (gelatin, keratin, casein, and whey) were employed in plastic formulation [66–68].

However many studies have been carried out using casting and compression molding techniques. In fact, the processing of proteins is still difficult to control with a conventional extrusion machine, as evidenced by Verbeek et al. [69].

In particular the employment of flat die extrusion and calendering is quite difficult. In this case, a successful production of protein-based laminates was reported by blending with non-biodegradable polymers such as poly(ethylene-co-vinyl acetate) (EVA) [70]. Moreover, new fully biodegradable composites based on whey/PBS blends were developed in the framework of the BIOBOARD European project [38,70] but it was possible to prepare films with a higher thickness than conventional films used in packaging. On the other hand, to exploit the barrier properties of whey, the application of layer coating made of protein onto a polymeric seemed easier, as evidenced by Cinelli et al. [36]. If techniques based on casting from water solutions are considered, also chitosan, the polymer obtained industrially by the deacetilation of chitin, can be a very interesting material for the production of coatings [71,72].

Chitosan is a biodegradable polymer that may be used to elaborate edible films or coatings to enhance shelf life of foods. Its water vapor permeability is about 20 times higher with respect to PS, so it is too high with respect to conventional packaging. However, it was demonstrated to have an anti-microbial activity that can be interesting for the production of biopolymer-based biodegradable packaging materials with additional bioactive functions [60].

The necessity of improving the chitosan barrier properties was evidenced by Morreno-Osorio et al. [73], who added to chitosan a natural compound acting as cross-linker, thus improving both mechanical and barrier properties. More frequently nano-fillers were used to improve chitosan barrier properties. Azeredo et al. [74] demonstrated that cellulose nano-fibers (CNF) can improve the mechanical and water vapor barrier properties of chitosan films. A nano-composite film with 15% CNF and plasticized with 18% glycerol was comparable to some synthetic polymers in terms of strength and stiffness, but with poorer elongation and water vapor barrier, indicating that they can be used for applications that do not require high flexibility and/or water vapor barrier. The more important advantage of such films when compared to synthetic polymer films is their environmentally friendly properties.

However up to now, despite of the production of cast films based on chitosan being successfully set-up [75], there are not yet suitable continuous machines for preparing chitosan-based films. In addition, the packaging of wet food cannot be done, as the material can dissolve in water and has a too much high water permeability. However anti-microbial properties are reported for chitosan films and biotextiles [76]. Hence the preparation of multilayer systems including one layer of chitosan should be very interesting.

5. Multi-Layer Systems: The Most Environmental Friendly Application in Food Packaging

The production of multilayer packaging films is very important in the field of packaging because it offers the opportunity of protecting the content of packaging from the interaction with different gases. As an example, the case of traditional multilayer sheets based on poly(ethylene terephthalate) (PET) and poly(ethylene) (PE) can be reported.

The recycling of these multilayer packages is not easy because the different polymeric, cellulosic or metal-based materials must be separated. Natural polymers are much used in flexible packaging, since cellulose is employed in paper-based packaging. Usually this kind of packaging consists of two layers for dried food, as the paper does not have the suitable barrier properties to water (but it has good barrier properties towards oxygen). Hence, the paper is present in a multilayer system in which also a polyethylene layer, granting the suitable water barrier properties, is present. For non-dried food, creams or liquids, also an aluminum foil layer is present to grant a total barrier effect for the packaging. For cellulose-based materials the option of recycling is possible also by considering multilayer systems. The recycling of cellulose is actually carried out in paper plants, by repulping the grinded recovered packaging material. A residual fraction (about the 30% by weight of the material) consists of aluminum and polyethylene, which can be employed in the injection-molding sector. The recycling process should be optimized, for example by separating the aluminum from the PE or replacing the PE with a biodegradable polymer and recovering Al after composting. These possible options for multilayer paper-based packaging are the object of the BIOBOARD European project [38] research activities. In the perspective of having a composting option for future packaging, the possible future scenario can be the one described in Figure 7. The monolayer plastic-based, the multilayer plastic-based and the multilayer paper-based flexible packaging systems need to be modified by replacing the layer with compostable ones. Overall, the packaging must grant the same properties in terms of content protection as the synthetic polymer-based options.

Extensive research has been devoted to the set-up of new synthetic routes for compostable polymer synthesis [77,78], and to the recovery process of materials from biomass in order to make them available for employment in materials production [79–81]. Many papers have been devoted to the blending of currently available polymers with processing aids in order to allow them to have the same properties of commodities [82,83] and others (less numerous) have been trying to find out new industrial technologies suitable for biomass derived materials [84]. All this research work can contribute to the aim of replacing flexible packaging with compostable and renewable ones in the future, but it should be stressed that the last point is fundamental. In fact, often the new materials cannot be processed exactly as commodities. Given that the plastic industrial sector mainly consists of numerous and small enterprises, plastics converters requiring new biobased plastics that have the same characteristics and processability of conventional plastic-based counterparts, as the conversion of industrial lines to new ones, could require huge investments for them.

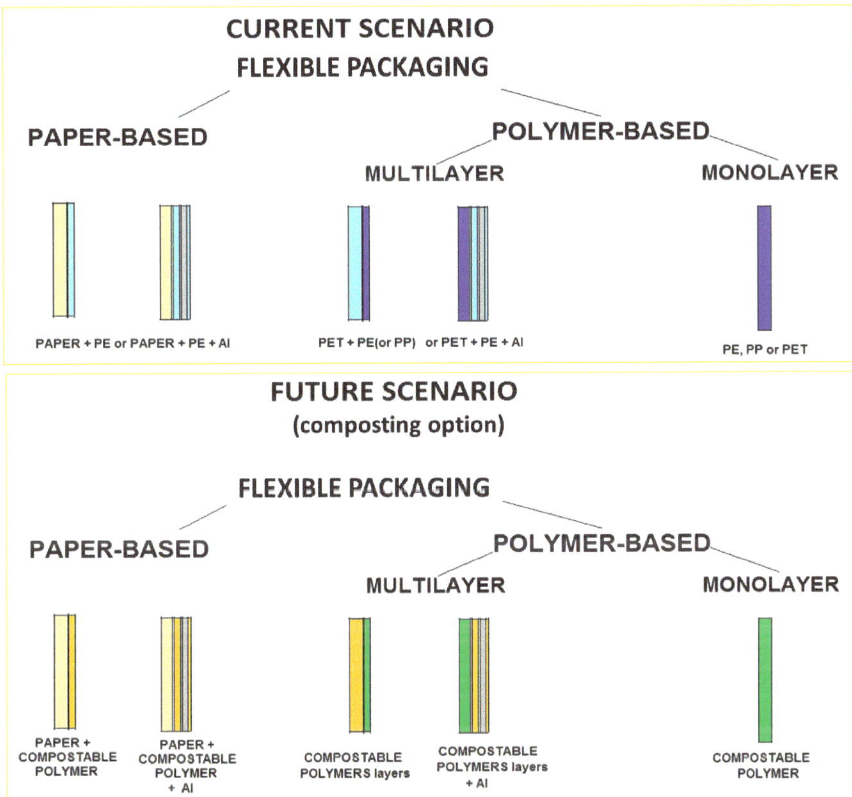

Figure 7. Polymeric materials currently employed in packaging and future perspectives in the hypothesis of composting option for flexible packaging.

The preparation of nano-composites can be effective in reducing gas permeability, hence a system in which the different layers are nanostructured can allow an optimized set of properties for a specific packaging.

Multilayer systems should be prepared preferentially by co-extrusion or lamination. Hence the material employed for making the different layers must be processable by these techniques. It is fundamental for controlling the structure of the polymers and especially their rheological behavior. However recently some researchers published a work about a new technique for preparing a peculiar multilayer system by using a wet method [85]. A PLA film was immersed in a aqueous solution (Figure 8) of chitosan and successively in a aqueous suspension of sodium montmorillonite (MMT).

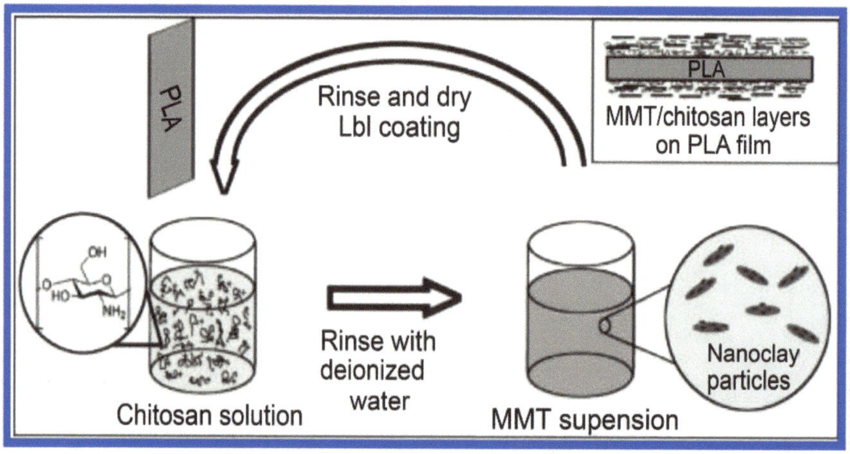

Figure 8. Illustration about LBL technique for depositing a chitosan/montmorillonite nano-structured coating on the surface of a PLA film. Reprinted with permission from [80], Copyright 2012, American Chemical Society.

By making 40 successive immersion steps, 40 alternate layers of chitosan (having a positive charge) and MMT platelets (having a negative charge) were deposited on the PLA films, thus producing a nanostructured coating. The peculiar technique was called LBL (layer by layer). Successively Laufer et al. [86] employed again the technique for preparing multilayer nano-coatings made with three food contact-approved components (chitosan (CH), poly(acrylic acid) (PAA), and montmorillonite (MMT) clay). They deposited the layers onto PET and polylactic acid (PLA) substrates. At 38 °C and 90% relative humidity (RH), the oxygen transmission rate (OTR) of 500 μm PLA was reduced from 50 to 4.6 cc/(m^2 day atm), which is lower than 179 μm PET film under the same conditions. This good gas barrier is believed to be due to the nano-brick wall structure present in this thin film, where clay platelets act as bricks held together by polymeric mortar. These assembled thin films are also very transparent, which combined with ambient processing and the use of renewable and food contact approved ingredients, makes this a promising foil replacement technology. This technology is quite interesting even if it is not easy to apply it on an industrial scale.

The coating of polymeric films made with biodegradable polymers with a film made with chitosan or another hydro-suspendable biopolymer-based formulation can be obtained by continuously spraying a film at the end of a flat die extrusion plant. In this way a bilayer system can be obtained continuously without introducing enormous changes in the plant.

These papers evidence that the preparation of a multilayer containing biopolyesters and chitosan, also with the presence of nano-filler, is a quite interesting scientific and technological challenge.

The preparation of multilayer packaging containing paper layers and biobased polymers can be very interesting for producing fully biodegradable soft packaging with optimal barrier properties. Coltelli et al. reviewed the use of proteins from many natural sources in paper and paperboard coatings, making an extensive comparison in mechanical and barrier properties between proteins from several sources and having different primary structures [87].

In multilayer systems the formulation of biodegradable glues for lamination or biodegradable hot melt for extrusion coating is an interesting topic for current research [41]. For binding, protein or starch-based adhesives can be used, whereas for hot melt glues, PLA or PCL can be used. Sodergard et al. [88,89] studied blends of lactic acid-based hot melt adhesives with oxidized potato starch and poly(ethylene glycol). Thanks to this approach the authors increased the disintegration rate of the glue, making it suitable for the production of compostable multilayer packaging.

6. Challenges and Perspectives

The use of biobased plastics in flexible packaging is possible by employing different products. The main commercial products are based on starch or PLA. In both cases biopolymers, plasticizers, chain extenders and fillers are added to the material to modulate their properties. Some commercial products are available on the market, but research is in progress to better understand the correlation between formulation (new biobased additives are continuously introduced in the market especially thanks to the ongoing developments of biorefineries [46]), processability and final properties of such bioplastic materials.

In the field of starch-based materials, systematic research aimed at developing a knowledge about the integrated effect of biopolyesters, plasticizers and fillers in the starch-based material could be useful. PLA-based nanostructured composites, containing nano-fillers (such as phillosilicates or cellulose or chitin nano-fibrils) can be promising to improve the barrier properties of biodegradable films without modifying their optical properties (transparency).

A bioplastic material for flexible packaging applications should have rheological and mechanical behavior similar to polyolefins. In particular, these conditions are difficult to respect for materials based on natural polymers, such as polysaccharides or proteins. In this case, the application to flexible packaging requires further efforts. In the case of PLA the melt viscosity is low, and usually for developing the blown film extrusion or flat die extrusion a chain extension approach, also leading to branching, can be followed. On the other hand, in the case of proteins, their tendency to give extensive crosslinking by heating during processing can be counterbalanced by the addition of plasticizers or reactive molecules, which results in the decrease in melt viscosity.

In addition, the preparation of multilayer systems by employing biopolymer sheets can be a promising technique for flexible packaging. The replacement of current multilayers packaging with fully biodegradable ones, by keeping into account the scheme of Figure 7, will require a good adhesion between the different polymers and paper sheets. In the case of starch-based materials the good adhesion is granted by the high chemical affinity between starch and cellulose. In some cases the necessity of improving the adhesion by the use of proper biodegradable glues, by the application of surficial treatments (mechanical, plasma or corona) or by peculiar thermal and mechanical treatments to films could be useful to better design multilayer packages taking into account their end-life management.

Acknowledgments: EU is gratefully acknowledged for supporting the authors research activities thanks to the funding of several projects, all cited in the present chapter: Nano-chitopack (GA 315233), Dibbiopack (GA 280676), WheyLayer (218340), OliPHA (GA 280604), Bioboard (GA 315313) and the ongoing H2020 BBI project PolyBioSkin (GA 745839).

Author Contributions: Maria-Beatrice Coltelli made the outline and mainly wrote the chapter; Vito Gigante contributed to describing the processing of biopolymers; Patrizia Cinelli wrote the parts related to starch and PHB and revised thoroughly the chapter; Andrea Lazzeri revised the chapter.

Conflicts of Interest: The authors declare no conflict of interest.

References

1. Fernandez de la Ossa, M.A.; Torre, M.; Garcia-Ruiz, C. Nitrocellulose in propellants: Characteristics and thermal properties. In *Advances in Materials Science Research*; Wythers, M.C., Ed.; Nova Science Publishers: Hauppauge, NY, USA, 2012; Volume 7, pp. 201–220.
2. Harmsen, P.F.H.; Hackmann, M.M.; Bos, H.L. Green building blocks for bio-based plastics. *Biofuels Bioprod. Bioref.* **2014**, *8*, 306–324. [CrossRef]
3. Linos, A.; Steinbüchel, A. Biodegradation of Natural and Synthetic Rubbers. In *Biopolymers Online*; Steinbuchel, A., Ed.; Wiley: Hoboken, NJ, USA, 2005; Volume 2, pp. 321–335.
4. Babu, R.P.; O'Connor, K.; Seeram, R. Current progress on bio-based polymers and their future trends. *Prog. Biomater.* **2013**, *2*, 81–116. [CrossRef]
5. Tokiwa, Y.; Calabia, B.P.; Ugwu, C.U.; Aiba, S. Biodegradability of Plastics. *Int. J. Mol. Sci.* **2009**, *68*, 3722–3742. [CrossRef] [PubMed]
6. Cantor, P.A.; Mechalas, B.J. Biological degradation of cellulose acetate reverse-osmosis membranes. *J. Polym. Sci. Polym. Symp.* **1969**, *28*, 225–241. [CrossRef]

7. What are Bioplastics? European Bioplastics Fact Sheet. Available online: http://docs.european-bioplastics.org/2016/publications/fs/EUBP_fs_what_are_bioplastics.pdf (accessed on 25 November 2017).
8. Raschka, A.; Carus, M.; Piotrowski, S. Renewable raw materials and feedstock for bioplastics. In *Bio-Based Plastics: Materials and Applications*; Kabasci, S., Ed.; Wiley: Hoboken, NJ, USA, 2013; pp. 331–334.
9. Malinconico, M.; Cerruti, P.; Santagata, G.; Immirzi, B. Natural Polymers and Additives in Commodity and Specialty Applications: A Challenge for the Chemistry of Future. *Macromol. Symp.* **2014**, *337*, 124–133. [CrossRef]
10. EN13432: Packaging. Requirements for packaging recoverable through composting and biodegradation. Test scheme and evaluation criteria for the final acceptance of packaging. 2000.
11. EN14995: Plastics—Evaluation of compostability—Test Scheme and Specifications, 2006.
12. Stepto, R.F.T. Development and Processing of Thermoplastic Starch Materials. In *Modification and Blending of Synthetic and Natural Macromolecules*; NATO Science Series; Ciardelli, F., Penczek, L., Eds.; Kluwer Academic Publishers: Dordrecht, The Netherland, 2004; pp. 219–240.
13. Pereira, A.G.B.; Gouveia, R.F.; de Carvalho, G.M.; Rubira, A.F.; Muniz, E.C. Polymer blends based on PEO and starch: Miscibility and spherulite growth rate evaluated through DSC and optical microscopy. *Mater. Sci. Eng. C* **2009**, *29*, 499–504. [CrossRef]
14. Willet, J.L.; Jasberg, B.K.; Swanson, C.L. *Polymers from Agricultural Coproducts*; ACS Symposium Series 575; Fishman, M.L., Friedman, R.B., Haag, S.J., Eds.; American Chemical Society: Washington, DC, USA, 1994; pp. 51–68.
15. Kalambur, S.; Rizvi, S.S.H. Biodegradable and Functionally Superior Starch–Polyester Nanocomposites from Reactive Extrusion. *J. Appl. Polym. Sci.* **2005**, *96*, 1072–1082. [CrossRef]
16. Contat-Rodrigo, L.; Ribes-Greus, A. Thermal and viscoelastic properties of some commercial starch products. *J. Appl. Polym. Sci.* **2003**, *88*, 1242–1251. [CrossRef]
17. Rahman, W.A.W.A.; Sin, L.T.; Rahmat, A.R.; Samad, A.A. Thermal behaviour and interactions of cassava starch filled with glycerol plasticized polyvinyl alcohol blends. *Carbohyd. Polym.* **2010**, *81*, 805–810. [CrossRef]
18. Ezeoha, S.L.; Ezenwanne, J.N. Production of Biodegradable Plastic Packaging Film from Cassava Starch. *IOSR J. Eng.* **2013**, *3*, 14–20. [CrossRef]
19. Ali, R.R.; Rahman, W.A.W.A.; Kasmani, R.M.; Ibrahim, N. Starch Based Biofilms for Green Packaging. In *Developments in Sustainable Chemical and Bioprocess Technology*; Pogaku, R., Ed.; Springer: New York, NY, USA, 2013; pp. 347–354.
20. Bastioli, C. Properties and applications of Mater-Bi starch-based materials. *Polym. Degrad. Stab.* **1998**, *59*, 263–272. [CrossRef]
21. Platt, D. Polymer Demand Picks up as Holiday Season Comes on, Plastics News Europe. Available online: http://www.plasticsnewseurope.com/article/20171011/PNE/171019982/polymer-demand-picks-up-as-holiday-season-comes-to-end (accessed on 11 October 2017).

22. Scatto, M.; Salmini, E.; Castiello, S.; Coltelli, M.B.; Conzatti, L.; Stagnaro, P.; Andreotti, L.; Bronco, S. Plasticized and Nanofilled Poly(lactic acid)-Based Cast Films: Effect of Plasticizer and Organoclay on Processability and Final Properties. *J. Appl. Polym. Sci.* **2013**, *127*, 4947–4956. [CrossRef]
23. DIBBIOPACK. *Development of Injection and Extrusion blow Moulded Biodegradable and Multifunctional Packages by Nanotechnologies*; GA 280676; European Commission: Brussels, Belgium, 2012.
24. Lazzeri, A.; Thanh Phuong, V.; Cinelli, P. Copolymers Based on Reactive Polyesters and Plasticisers for the Manufacture of Transparent, Biodegradable Packaging Film. WO2013164743, 7 November 2013.
25. Coltelli, M.B.; Toncelli, C.; Ciardelli, F.; Bronco, S. Compatible blends of biorelated polyesters through catalytic transesterification in the melt. *Polym. Degrad. Stab.* **2011**, *96*, 982–990. [CrossRef]
26. Coltelli, M.B.; Della Maggiore, I.; Bertoldo, M.; Bronco, S.; Signori, F.; Ciardelli, F. Poly(lactic acid) (PLA) properties as a consequence of poly(butylene adipate-co-terephtahlate) (PBAT) blending and acetyl tributyl citrate (ATBC) plasticization. *J. Appl. Polym. Sci.* **2008**, *110*, 1250–1262. [CrossRef]
27. Dong, W.; Zou, B.; Yan, Y.; Ma, P.; Chen, M. Effect of Chain-Extenders on the Properties and Hydrolytic Degradation Behavior of the Poly(lactide)/Poly(butylene adipate-co-terephthalate) Blends. *Int. J. Mol. Sci.* **2013**, *14*, 20189–20203. [CrossRef] [PubMed]
28. Coltelli, M.B.; Bronco, S.; Chinea, C. The effect of free radical reactions on structure and properties of poly(lactic acid) (PLA) based blends. *Polym. Degrad. Stab.* **2010**, *95*, 332–341. [CrossRef]
29. Park, J.W.; Soon Im, S. Phase behavior and morphology in blends of poly(L-lactic acid) and poly(butylene succinate). *J. Appl. Polym. Sci.* **2002**, *86*, 647–655. [CrossRef]
30. Pucci, A.; Ruggeri, G.; Bronco, S.; Signori, F.; Donati, F.; Bernabò, M.; Ciardelli, F. Colour responsive smart polymers and biopolymers films through nanodispersion of organic chromophores and metal particles. *Prog. Org. Coat.* **2011**, *72*, 21–25. [CrossRef]
31. Fehri, S.M.K.; Cinelli, P.; Coltelli, M.B.; Anguillesi, I.; Lazzeri, A. Thermal Properties of Plasticized Poly (Lactic Acid)(PLA) Containing Nucleating Agent. *Int. J. Chem. Eng. Appl.* **2016**, *7*, 85.
32. Mujumdar, A.S. classification and selection of industrial dryers. In *Handbook of Industrial Drying*; Mujumdar, A.S., Ed.; CRC Press: Boca Raton, FL, USA, 2006.
33. Majumbder, K.K. Blown Film Extrusion: Experimental, Modelling and Numerical Study. Ph.D. Thesis, School of Civil, Environmental and Chemical Engineering Science, Engineering and Technology Portfolio, RMIT University, Melbourne, VIC, Australia, 2008.
34. Dartora, P.C.; Campomanes Santana, R.M.; Fontes Moreira, A.C. The influence of long chain branches of LLDPE on processability and physical properties. *Polímeros* **2015**, *25*, 531–539. [CrossRef]

35. Al-Itry, R.; Lamnawar, K.; Maazouz, A. Improvement of thermal stability, rheological and mechanical properties of PLA, PBAT and their blends by reactive extrusion with functionalized epoxy. *Polym. Degrad. Stab.* **2012**, *97*, 1898–1914. [CrossRef]
36. Mallegni, N.; Phuong, T.V.; Coltelli, M.B.; Cinelli, P.; Lazzeri, A. Poly(lactic acid) (PLA) based tear resistant and biodegradable flexible films by blown film extrusion. *Materials* **2018**, in press.
37. POLYBIOSKIN. *High Performance Functional Bio-Based Polymers for Skin-Contact Products in Biomedical, Cosmetic and Sanitary Industry*; GA 745839; European Commission: Brussels, Belgium, 2017.
38. BIOBOARD. *Development of Sustainable Protein-Based Paper and Paperboard Coating Systems to Increase the Recyclability of Food and Beverage Packaging Materials*; GA 315313; European Commission: Brussels, Belgium, 2012.
39. WHEYLAYER. *Whey Protein-Coated Plastic Films to Replace Expensive Polymers and Increase Recyclability*; GA 218340; European Commission: Brussels, Belgium, 2011.
40. Cinelli, P.; Schmid, M.; Bugnicourt, E.; Coltelli, M.B.; Lazzeri, A. Recyclability of PET/WPI/PE Multilayer Films by Removal of Whey Protein Isolate-Based Coatings with Enzymatic Detergents. *Materials* **2016**, *9*, 473. [CrossRef] [PubMed]
41. Jennyangel, S.; Dhandapani, R. Eco-friendly biopolymers as adhesives—An overview. *Int. J. Pharm. Bio Sci.* **2013**, *4*, 524–533.
42. Gullick, G.; Isonhood, S.; Sanders, E. Coextruded Film with Polylactic Acid (PLA) and Ethylene Vinyl Acetate (EVA). U.S. 20080026171 A1, 31 January 2008.
43. Shi, B.; Topolkaraev, V.; Wang, J. Biopolymers, Processing, and Biodegradation. In *Renewable and Sustainable Polymers*; Chapter 8; ACS Symposium Series; American Chemical Society: Washington, DC, USA, 2011; Volume 1063, pp. 117–132.
44. Directive (EU) 2015/720 of the European Parliament and of the Council of 29 April 2015 Amending Directive 94/62/EC as Regards Reducing the Consumption of Lightweight Plastic Carrier Bags. Available online: https://publications.europa.eu/en/publication-detail/-/publication/58d93aee-f3bc-11e4-a3bf-01aa75ed71a1/language-en (accessed on 25 November2017).
45. Hwang, S.W.; Shim, J.K.; Selke, S.; Soto-Valdez, H.; Rubino, M.; Auras, R. Effect of Maleic-Anhydride Grafting on the Physical and Mechanical Properties of Poly(L-lactic acid)/Starch Blends. *Macromol. Mater. Eng.* **2013**, *298*, 624–633. [CrossRef]
46. Clark, J.H. Green Chemistry for the second generation biorefinery—Sustainable chemical manufacturing based on biomass. *J. Chem. Technol. Biotechnol.* **2007**, *82*, 603–609. [CrossRef]
47. Coltelli, M.B.; Coiai, S.; Bronco, S.; Passaglia, E. Nanocomposites Based on Phyllosilicates: From Petrochemicals to Renewable Thermoplastic Matrices. In *Advanced Nanomaterials I*; Geckeler, K.E., Nishide, H., Eds.; Wiley: New York, NY, USA, 2010; pp. 403–458.
48. Auras, R.A.; Harte, B.; Selke, S.; Hernandez, R.J. Mechanical, physical and barrier properties of Poly(lactic acid) films. *J. Plast. Film Sheet* **2003**, *19*, 123–135. [CrossRef]
49. Pavlidou, S.; Papaspyrides, C.D. A review on polymer–layered silicate nanocomposites. *Prog. Polym. Sci.* **2008**, *33*, 1190–1198. [CrossRef]

50. Neilsen, L.E. Models for the permeability of filled polymer systems. *J. Macromol. Sci.* **1967**, *1*, 929–942. [CrossRef]
51. Castiello, S.; Coltelli, M.B.; Conzatti, L.; Bronco, S. Comparative Study About Preparation of Poly(lactide)/Organophilic Montmorillonites Nanocomposites Through Melt Blending or Ring Opening Polymerization Methods. *J. Appl. Polym. Sci.* **2012**, *125*, E413–E428. [CrossRef]
52. Beck, S.; Bouchard, J.; Berry, R. Dispersibility in water of dried nanocrystalline cellulose. *Biomacromolecules* **2012**, *13*, 1486–1494. [CrossRef] [PubMed]
53. Espino-Perez, E.; Bras, J.; Ducruet, V.; Guinault, A.; Dufresne, A.; Domenek, S. Influence of chemical surface modification of cellulose nanowhiskers on thermal, mechanical and barrier properties of poly(lactide) based bionanocomposites. *Eur. Polym. J.* **2013**, *49*, 3144–3154. [CrossRef]
54. Mincea, M.; Negrulescu, A.; Ostafe, V. Preparation, modification and application of chitin nano-whiskers: A review. *Rev. Adv. Mater. Sci.* **2012**, *30*, 225–242.
55. Muzzarelli, C.; Morganti, P. Preparation of Chitin and Derivatives Thereof for Cosmetic and Therapeutic Use. U.S. WO2006048829 A2, 11 May 2006.
56. Morganti, P.; Morganti, G.; Morganti, A. Transforming nanostructured chitin from crustacean waste into beneficial health products: A must for our society. *Nanotechnol. Sci. Appl.* **2011**, *4*, 123–129. [CrossRef] [PubMed]
57. Coltelli, M.B.; Cinelli, P.; Anguillesi, I.; Salvadori, S.; Lazzeri, A. Preparation of biopolyester composites containing nano-chitin from sea food waste. In Proceedings of the BIOPOLPACK, Parma, Italy, 29–30 May 2014.
58. Singh, R.; Chacharkar, M.P.; Mathur, A.K. Chitin membrane for wound dressing application—Preparation, characterisation and toxicological evaluation. *Int. Wound J.* **2008**, *5*, 665–673. [CrossRef] [PubMed]
59. Salaberria, A.M.; Fernandes, S.C.M.; Herrera Diaz, R.; Labidi, J. Processing of α-chitin nanofibers by dynamic high pressure homogenization: Characterization and antifungal activity against A. niger. *Carbohydr. Polym.* **2015**, *116*, 286–291. [CrossRef] [PubMed]
60. N-CHITOPACK. *Sustainable Technologies for the Production of Biodegradable Materials Based on Natural Chitin-Nanofibrils Derived by Waste of Fish Industry, to Produce Food Grade Packaging*; GA 315233; European Commission: Brussels, Belgium, 2012.
61. Cinelli, P.; Coltelli, M.B.; Mallegni, N.; Morganti, P.; Lazzeri, A. Degradability and sustainability of nanocomposites based on polylactic acid and chitin nanofibrils. *Chem. Eng. Trans.* **2017**, *60*, 115–120.
62. Reddy, C.S.K.; Ghai, R.; Rashmi; Kalia, V.C. Polyhydroxyalkanoates: An overview. *Bioresour. Technol.* **2003**, *87*, 137–146. [CrossRef]
63. Kootstra, M.; Elissen, H.; Huurman, S. *PHA's (Polyhydroxyalkanoates): General Information on Structure and Raw Materials for Their Production*; Report 727; ACRRES: Wageningen, The Netherlands, 2017.
64. Oli-PHA. *A Novel and Efficient Method for the Production of Polyhydroxyalkanoate (PHA) Polymer-Based Packaging from Olive Oil Waste Water*; GA 280604; European Commission: Brussels, Belgium, 1 June 2012.

65. Farris, G.; Cinelli, P.; Anguillesi, I.; Salvadori, S.; Coltelli, M.B.; Lazzeri, A. Effect of Ageing Time on Mechanical Properties of Plasticized Poly(hydroxybutyrate) (PHB). *AIP Conf. Proc.* **2014**, *1599*, 294–297.
66. Hatti-Kaul, R.; Tornvall, U.; Gustafsson, L.; Borjesson, P. Industrial biotechnology for production of bio-based chemicals—A cradle to grave perspective. *Trends Biotechnol.* **2007**, *25*, 119–124. [CrossRef] [PubMed]
67. Hernandez-Izquierdo, V.M.; Krochta, J.M. Thermoplastic processing of proteins for film formation—A review. *J. Food Sci.* **2008**, *73*, 30–39. [CrossRef] [PubMed]
68. Barone, J.R.; Arikan, O. Composting and biodegradation of thermally processed feather keratin polymer. *Polym. Degrad. Stab.* **2007**, *92*, 859–867. [CrossRef]
69. Verbeek, C.J.R.; Van de Berg, L.E. Extrusion Processing and Properties of Protein-Based Thermoplastics. *Macromol. Mater. Eng.* **2010**, *295*, 10–21. [CrossRef]
70. Schmid, M.; Herbst, C.; Müller, K.; Stäbler, A.; Schlemmer, D.; Coltelli, M.B.; Lazzeri, A. Effect of potato pulp filler on the mechanical properties and water vapor transmission rate of thermoplastic WPI/PBS blends. *Polym. Plast. Technol. Eng.* **2016**, *55*, 510–517. [CrossRef]
71. Elsabee, M.Z.; Abdou, E.S. chitosan based edible films and coating: A review. *Mater. Sci. Eng. C* **2013**, *33*, 1819–1841. [CrossRef] [PubMed]
72. Rhim, J.W.; Hong, S.I.; Park, H.M.; Perry, N.G. Preparation and Characterization of Chitosan-Based Nanocomposite Films with Antimicrobial Activity. *J. Agric. Food Chem.* **2006**, *54*, 5814–5822. [CrossRef] [PubMed]
73. Moreno-Osorio, L.; Garcia, M.; Villalobos-Carvajal, R. Effect of Polygodial on Mechanical, Optical and, Barrier Properties of Chitosan Films. *J. Food Proc. Preserv.* **2010**, *34*, 219–234. [CrossRef]
74. Azeredo, H.M.C.; Mattoso, L.H.C.; Avena-Bustillos, R.J.; Filho, G.C.; Munford, M.L.; Wood, D.; McHugh, T.H. Nanocellulose Reinforced Chitosan Composite Films as Affected by Nanofiller Loading and Plasticizer Content. *J. Food Sci.* **2010**, *75*, N1–N7. [CrossRef] [PubMed]
75. Mikešováa, J.; Hašeka, J.; Tishchenko, G.; Morganti, P. Rheological study of chitosan acetate solutions containing chitin nanofibrils. *Carbohyd. Polym.* **2014**, *112*, 753–757. [CrossRef] [PubMed]
76. Naseri, N.; Algan, C.; Jacobs, V.; John, M.; Oksman, K.; Mathew, A.P. Electrospun chitosan-based nanocomposite mats reinforced with chitin nanocrystals for wound dressing. *Carbohydr. Polym.* **2014**, *109*, 7–15. [CrossRef] [PubMed]
77. Alves, M.H.; Sfeir, H.; Tranchant, J.F.; Gombart, E.; Sagorin, G.; Caillol, S.; Billon, L.; Save, M. Terpene and Dextran Renewable Resources for the Synthesis of Amphiphilic Biopolymers. *Biomacromolecules* **2014**, *15*, 242–251. [CrossRef] [PubMed]
78. Oh, J.K. Polylactide (PLA)-based amphiphilic block copolymers: Synthesis, self-assembly, and biomedical applications. *Soft Matter* **2011**, *7*, 5096–5108. [CrossRef]
79. Liu, Z. Preparation of biopolymers from plant oils in green media. *BioEnergy Res.* **2013**, *6*, 1230–1236. [CrossRef]

80. Mozejko, J.; Ciesielski, S. Saponified waste palm oil as an attractive renewable resource for mcl-polyhydroxyalkanoate synthesis. *J. Biosci. Bioeng.* **2013**, *116*, 485–492. [CrossRef] [PubMed]
81. Bernardini, J.; Cinelli, P.; Coltelli, M.B.; Lazzeri, A. Flexible polyurethane foams green production employing lignin or oxypropylated lignin. *Eur. Polym. J.* **2014**, *64*, 147–156. [CrossRef]
82. Gurgel, A.V.M.; Altenhofen da Silva, M.; Oliveira dos Santos, L.; Masumi Beppu, M. Natural-based plasticizers and biopolymer films: A review. *Eur. Polym. J.* **2011**, *47*, 254–263.
83. Fehri, M.K.; Mugoni, C.; Cinelli, P.; Anguillesi, I.; Coltelli, M.B.; Fiori, S.; Montorsi, M.; Lazzeri, A. Composition dependence of the synergistic effect of nucleating agent and plasticizer in poly(lactic acid): A mixture design study. *Express Polym. Lett.* **2016**, *10*, 274–288. [CrossRef]
84. Loppinet-Serani, A.; Aymonier, C.; Cansell, F. Supercritical water for environmental technologies. *J. Chem. Technol. Biotechnol.* **2010**, *85*, 583–589. [CrossRef]
85. Svagan, A.J.; Åkesson, A.; Cárdenas, M.; Bulut, S.; Knudsen, J.C.; Risbo, J.; Plackett, D. Transparent Films Based on PLA and Montmorillonite with Tunable Oxygen Barrier Properties. *Biomacromolecules* **2012**, *13*, 397–405. [CrossRef] [PubMed]
86. Laufer, G.; Priolo, M.A.; Kirkland, C.; Grunlan, J.C. High oxygen barrier, clay and chitosan-based multilayer thin films: An environmentally friendly foil replacement. *Green Mater.* **2013**, *1*, 4–10. [CrossRef]
87. Coltelli, M.B.; Wild, F.; Bugnicourt, E.; Cinelli, P.; Lindner, M.; Schmid, M.; Weckel, V.; Müller, K.; Rodriguez, P.; Staebler, A.; et al. State of the Art in the Development and Properties of Protein-Based Films and Coatings and Their Applicability to Cellulose Based Products: An Extensive Review. *Coatings* **2016**, *6*, 1. [CrossRef]
88. Viljanmaa, M.; Södergård, A.; Törmälä, P. The use of lactic acid-based hot melt adhesive in the industrial lamination process. *Int. J. Adhes. Adhes.* **2003**, *23*, 151–154. [CrossRef]
89. Inkinen, S.; Stolt, M.; Södergård, A. Stability studies on blends of a lactic acid-based hot melt adhesive and starch. *J. Appl. Polym. Sci.* **2008**, *110*, 2467–2474. [CrossRef]

Naturally-Made Hard Containers for Food Packaging: Actual and Future Perspectives

Patrizia Cinelli, Maria-Beatrice Coltelli and Andrea Lazzeri

Abstract: The huge amount of plastic used in packaging application and the related problems of disposal of the packaging after its use have gained research and industrial interest in both bio-based polymers that are not biodegradable but produced by renewable resources and bio-based compostable polymers, to be used for production of packaging. The production of bio-polymers is more costly than that of conventional fossil fuel-based polymers. Moreover, the range of applications of polymers derived from natural sources was limited due to difficulties in processing natural materials, moisture sensitivity, incompatibility at the interface between natural fillers and polymeric matrices, possible toxicity related to natural material degradation, poor mechanical properties, etc. Mass transfer properties are very important for packaging application, and at present, bio-based packaging generally lacks in maintaining barrier properties compared to traditional petro-derived polymers. The combination of traditional polymers and biodegradable ones in multilayer systems allowed obtaining a good balance of mass transfer properties, but the presence of the non-biodegradable layer negatively affects, and in some cases compromises, the composting of the final packaging. Recent advances in technology are reducing the cost of manufacturing bio-based plastics and are producing materials with an expanded range of properties that has made them suitable for low cost and high demanding applications such as packaging. Chemical modifications of the bio-based polymers, as well as blending with biodegradable additives or polymers are the preferred solutions to improve and control the properties of these materials. In this work, several materials derived from biomasses such as polyesters (poly lactic acid, poly hydroxyl alkanoates), polysaccharides (starch), vegetal and animal proteins, etc., which can be used in hard packaging applications are reviewed. Modification and processing of bio-based products with additives, polymers and natural fibers (cellulose, wood fibers, etc.) are discussed as well. The suitability of these materials for the industrial processing required for the production of hard food packaging is reported critically in order to evidence the challenges and perspectives for polymers derived from renewable resources to be applied in this sector.

---

1. Introduction

Plastic materials applied in packaging can match almost all requirements, perfectly meeting customers' demands, offering important advantages such as low weight, protection of the packaged good (and consequent reduction of food waste), suitable mass transfer properties and low cost. Hard packaging comprises containers, bottles, canisters, jars, cups, buckets, trays, clamshells, blisters and the like.

In 2012 in the European Union, the post-consumer plastic waste ending up in the waste upstream was 25.2 million tons; 62% of this waste was recovered through energy recovery and recycling processes, while 38% still was sent to landfill [1].

The recent European Commission (EC) regulation promotes the bio-recycling of plastic waste through composting or anaerobic digestion, as well as the utilization of renewable resources for the production of plastics. Most recently, the development of environmental-friendly bio-based and or biodegradable polymeric materials from renewable sources has attracted extensive interest. This is due to the waste accumulation, but also to the limitation of crude oil or gas supply. Crude oil is the starting point of any conventional plastic, and it will only last for a few more decades, since it is used mainly to serve the energy demand of the population on a worldwide scale. An early conversion to renewable sources is thus important for the plastics industry.

Although only 4% of the global oil consumption is used to produce plastics (and a further 4% is used to produce the energy for the production of plastics), the new technologies required to process renewable sources thus preparing for the "post-oil era" require a sufficient time to be developed. About 46 million tons of plastics were used in the European Union in 2012 [1], about 40% for packaging production, and ca. 12 million tons of this plastic were used for rigid packaging.

In a recent fact sheet [2], European Bioplastic reports that the bioplastic producers are engaged to life cycle thinking and further improving their products, increasing the yield of their processes, thus producing more with less and ensuring sustainable resource supply to decrease the impact on the environment. Various life cycle analyses are in agreement with the better performance of bio-based rigid packaging, in particular in the impact categories of global warming potential (often referred to as a carbon footprint) and in the category of the consumption of fossil resources.

Some bio-based plastics are commercialized or are announced to be ready for the market such as polylactic acid (PLA), polybutylene succinate (PBS), polyhydroxy alkanoates (PHA), polyethylene furanoate (PEF) and also bio-based commodity plastics such as polyethylene (PE), polyethylene terephthalate (PET) or, in the very near future, polypropylene (PP). These polymers can be used for the production of rigid packaging.

The traditional market of rigid containers for food and beverages consists essentially of two different categories: the first one, usually defined as mono-material packaging, is based on unique plastic materials; the second is based on two or more materials (multi-material).

In the first category, the most used polymers are poly(ethylene terephthalate) (PET), high density poly(ethylene) (HDPE), poly(propylene) (PP) and polystyrene (PS). PET and HDPE are employed for the production of bottles, as they can be processed by blow molding, whereas PP and PS are much employed in the production of rigid food trays or jar as they can be extruded in sheets and thermoformed. At room temperature, all plastic materials are rigid, but HDPE and PP are above their glass transition temperature. Hence, the rigidity is provided by the presence of the crystalline phase domains, embedded in the main amorphous matrix. As many molecular motions are possible above the glass transition, this peculiar morphology can explain the performance impact properties of these polymers.

Multilayer rigid packaging can consist of thermoplastic polymers or of different materials, such as a multilayer system consisting of plastic, board and aluminum, with board representing the most abundant material. The former is processed usually by thermoforming starting from extruded sheets, whereas the latter is processed by properly creasing and sealing the board-based multilayer sheet.

On the market, rigid packaging based on bio-based and biodegradable polymers is already present, and some examples are reported in Figure 1.

Figure 1. Examples of rigid packaging produced with bio-based polymers.

In all cases, as the market of products is huge, the containers, which are designed purposely for preserving the specific properties of food or beverages, can consist of many different polymeric materials. The polymers cited above are the most employed, because of their wide-scale production and low price. However, many other polymers with specific properties can be addressed to allow the achievement of the desired requirement, and some examples are reported in Table 1.

Table 1. Main application of bio-based polymers in hard packaging.

Hard Packaging	Materials					
	PLA	PHA	Starch	PBS	BioPET	BioPE
Bottles	x	-	-	-	x	x
Trays	x	-	x	-	x	x
Containers	x	x	x	x	x	x
Caps	x	x	x	x	-	x
Blisters	x	-	x	x	x	-
Foamed packaging	x	-	x	x	-	-
Cutlery	x	-	x	x	-	x

PLA = poly lactic acid, PHA = poly hydroxyl alkanoate, PE = poly ethylene, PET = poly ethylene terephthalate. x = used and present on the market, - = not commonly used.

The applications of these packaging have still a huge potential to increase, and at present, most of the rigid packaging from renewable resources is represented by cardboard and its laminates. This is due to difficulties in processing of bio-based biodegradable polymers for the production of hard packaging and the cost being higher than that of petro-derived polymers.

Thus, an overview of processing technologies for the production of hard packaging is reported below with particular focus on issues related to processing of bio-based-biodegradable polymers.

2. Processing for Production of Hard Packaging

2.1. *Injection Molding*

The production of plastic rigid containers is carried out by different processes. Among those, injection molding is a processing method applied for polymeric thermoplastic materials. In this process, the polymer is melted in a heated barrel and injected in a mold to produce pieces with a defined shape. The automation of the process allows for series production. This method can be employed for example for producing caps, thick jars for cosmetics, cutlery and coffee capsules. The process consists of three main phases: the injection, in which the partial filling of the mold with the molten polymer is achieved; the holding pressure and plastification, in which the mold is completely filled by the molten polymer and kept in the mold at a defined pressure and temperature; and the ejection, in which the solid piece is extracted from the mold (Figure 2).

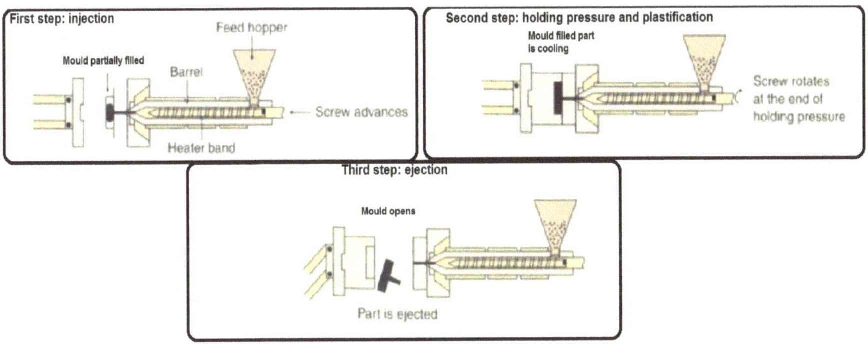

Figure 2. Three main steps of the injection molding process.

The selection and control of the temperature is fundamental in the first step of the process, since, when processing polymers with this technique, a very low viscosity at a high shear rate is required in order to grant a rapid and perfect filling of the mold. The knowledge and control of the rheological behavior of the molten polymer is thus very important to allow a good processing. In particular, melt flow index (MFI) determination at the same temperature of injection molding can be useful, and generally, thermoplastic polymers having MFI above about 10 g/min are suitable for injection molding. In some cases, the dimensions of the filling channel in the mold should be well dimensioned as a function of the melt rheology of the thermoplastic material employed.

In the second step, the temperature of the mold, the holding pressure and the holding time are the parameters that are important to be controlled. Some polymers can crystallize during the holding step; hence, the temperature of the mold and the holding time are quite important to allow the material to reach the desired crystalline morphology, as the amount and distribution of crystals in the material influence its final properties. This is particularly the case for bio-polyesters such as PLA and PHAs.

A specific selection of parameters is necessary also for the further steps. In fact, the ejection step should be done when the piece is solid and resistant enough, in order to avoid the breakage of the piece inside the open mold. The thermal properties such as the glass transition temperature of the material and/or its crystallization kinetics can greatly influence this step. Hence, the determination of the heat deflection temperature (HDT) usually allows one to know the maximum temperature of the mold suitable to allow the ejection without deformation of the prepared item.

This point is critical as the deformation or breakage of the item would require interrupting the processing cycle, thus wasting time, material and energy, but at the same time, in the least time possible, to avoid the waste of time and energy as well.

2.2. *Injection Stretch Blow Molding Process*

The injection stretch blow molding process, which is the most advanced method for producing bottles, consists of the preliminary injection molding of a pre-form, having the shape of a test tube with a threaded neck. Then, this piece is transferred to a mold where it is blown with an air jet in order to obtain a container having the shape of the mold. The container wall is thus bi-oriented, that is stretched both in the direction of air-flow and in the radial direction. By using this method, bottles of many different shapes and dimensions can be obtained. Transparent bottles available for water or fizzy beverages are made of PET, whereas the opaque ones, used for milk or liquid detergents, are made of HDPE. PLA is also suitable for this process, and most recently, bottles produced from PLA have been proposed; one example on the market is reported in Figure 3 [3–5].

Figure 3. Plastic bottle based on PLA.

2.3. *Thermoforming*

Arrays, plastic cups, blisters and jars are produced by another important processing method, thermoforming. Whereas in the injection molding or injection stretch blow molding, granules are fed into the equipment, in this case, the polymeric material must be fed in sheets having a thickness in the range of 50–300 microns for packaging production. The process can be applied also to thicker sheets, but in this case, the applications are in the automotive or electric and electronic fields, for the production of bodies or shells. Hence, a preliminary flat die extrusion step is necessary for producing suitable sheets for thermoforming. In the packaging field, PS or PP is usually employed, but positive results and interesting products were produced also with PLA and PLA filled with natural fibers. In Figure 4 are reported examples of rigid packaging trays and egg containers produced by thermoforming of sheets based on PLA and wood fibers, produced by the authors in the activity of the EC project FORBIOPLAST "Forest Resource Sustainability through Bio-Based-Composite Development". GA 212239 [6].

Figure 4. Trays and egg containers based on PLA and wood fibers.

The method consists of heating the material above its glass transition temperature, but below its melting point, thus obtaining a softened sheet usually by using infrared heaters. Then, a mold is inserted (or a vacuum is applied), which gives the softened sheet the desired shape (Figure 5).

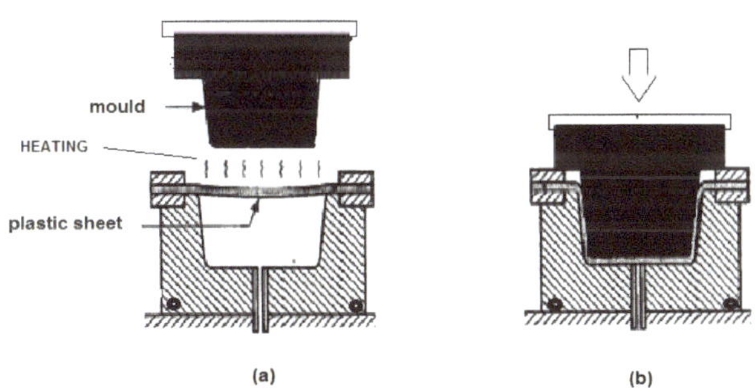

Figure 5. Thermoforming process.

2.4. Multi-Material Packaging

The methods listed up to now are suitable for packages consisting of one single material. However, usually, fresh food requires packaging with enhanced barrier properties, and for these reasons, packaging consisting of two or more layers is necessary. An interesting example is the packaging consisting of one layer of PET and one layer of poly(ethylene) (PE). These bi-layer sheets can be obtained by lamination of PE and PET or also by co-extrusion, usually employing compatibilizers consisting of ethylene copolymers, such as poly(ethylene-co-vinyl acetate) (EVA) in between the two layers to enhance adhesion and barrier properties. The multi-layer sheets can be thus thermoformed to obtain arrays or blisters, especially employed to pack fresh foods in supermarkets because of their structural, barrier and optical properties (high transparency).

The production of rigid packaging for liquids can be made also by using board-based materials. The usual rigid multilayer system consists of different layers (Figure 6) of different thicknesses, with a layer of PE in contact with the food. In dependence of the perishability of the liquid, a total barrier layer of aluminum may be necessary. This is the case of rigid containers for milk. However, also in the most complex multilayer sheet, the content of cellulose is at least 70% by weight, as the board is the main structural material of the packaging.

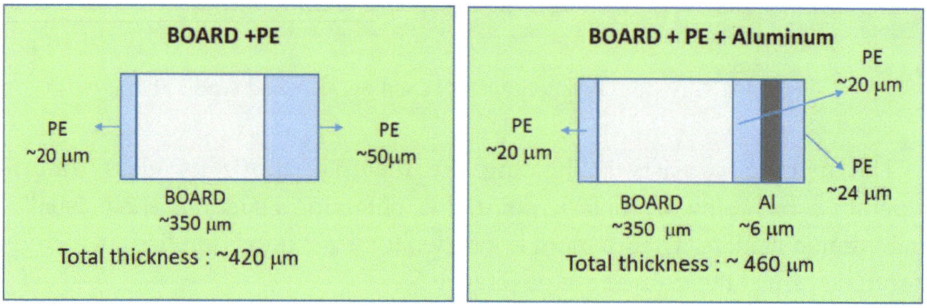

Figure 6. Structure of typical board-based multilayer packaging.

The sheets can be obtained by extrusion coating of PE on Al foil and co-extrusion of PE and board. To obtain the final packaging, the sheets must be properly creased and folded (Figure 7) [7].

Usually, the multilayer sheet is produced by worldwide producers, then adapted to the final users by the conversion plant, where also the printing is realized, whereas the folding is usually carried out in the packaging plant.

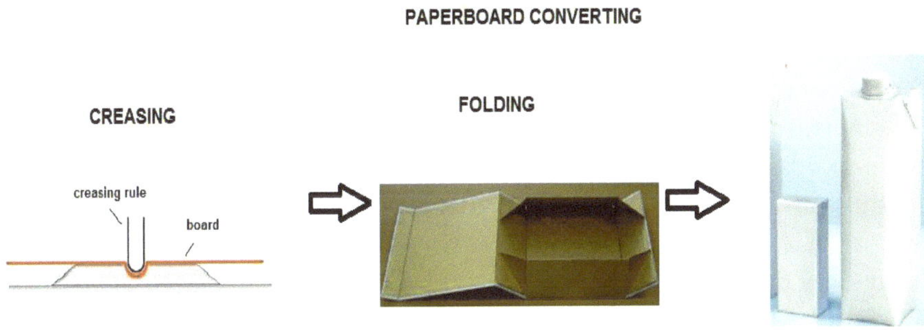

Figure 7. Scheme of paperboard-based packaging conversion.

In Table 2, a summary of the different processes used for the different types of rigid packaging is reported.

Table 2. Examples of processes used for the production of rigid packaging.

Rigid Packages or Products	Process	Polymers
Caps, coffee capsules, jars, etc.	Injection molding	HDPE, PP, PLA; PHA
Bottles	Injection stretch blow molding	PET, HDPE, PLA
Arrays, cups, jar, trays, blisters, etc.	Thermoforming	PS, PP, PET, PET/PE, PLA
Bricks for beverages or cups, etc.	Extrusion coating/laminating, creasing, folding	Cellulose/PE(/Al)

3. Adapting Biobased Polymers to Process for Rigid Packaging

3.1. Poly Lactic Acid

PLA is an aliphatic, thermoplastic polyester derived from renewable resources such as starch and appears as one of the best sustainable alternatives to petrochemical-derived products for applications in packaging [8].

PLA has been found to have good stiffness and strength, and it can be processed with conventional plastic processing machinery (extrusion, extrusion blow molding, injection molding, vacuum forming) and is being used in several applications for hard packaging such as boxes, jars, trays, water and milk bottles.

Products made from PLA degrade in the environment, fragmenting into small molecules, and can be compostable, as, depending on their thickness, they can fully disappear in less than 30 days in ideal conditions [9,10].

Lim et al. [11] explained that the processing of PLA requires a preliminary drying step in order to reduce the detrimental effect of water on the PLA properties [12]. Thus, during the processing in the melt, water can result in the chain scission of PLA macromolecules because of hydrolysis. The hydrolysis is due to the nucleophilic attack of the water molecules on the carbonyl ester groups of PLA. All the authors that processed PLA for producing blends [13,14] adopted this strategy or took into account the effect of hydrolysis on the final properties of their material. Lim [11] evidenced that commercial-grade PLA pellets are usually crystallized. This permits drying at higher temperatures to reduce the required drying time. In fact, if amorphous pellets are available, they must be dried below the Tg (~60 °C) to avoid pellets sticking together. In this case, drying under vacuum is suggested.

Industrial desiccators are available to control the water content before processing. Drying of PLA is usually attained using a closed-loop dual-bed regenerative desiccant-type dryer, where the dew point of the drying air should be at least −40 °C or lower.

The extrusion of PLA is the processing step allowing granules to be obtained. To ensure the melting of all the crystalline phases and to achieve an optimized melt viscosity for processing, the heater set point is usually set at 200–210 °C. Commercial-grade PLA can be processed in a conventional extruder equipped with a general purpose screw of an l/d ratio in the range of 24–30. Screws for extruding PET, which are low-shear to grant mild mixing (to minimize chain scission and acetaldehyde generation), are also suitable for processing PLA [10]. The compression ratio (the ratio of the flight depth in the feed section to the flight depth in the metering section) for PLA extrusion is in the range of 2–3 [15].

In injection molding, the cycle time is an important parameter, which is often minimized to maximize the production throughput. The filling, holding and cooling events that occur during injection molding have an important implication on the shrinkage of the injection molded articles, which must be controlled to avoid undesired piece deformation. In general, injection molded PLA pieces are relatively brittle, and this brittleness is attributed to the rapid physical aging of the polymer, as the ambient temperature is only about 25 °C below the glass transition temperature. Cai et al. [16] showed that endothermic enthalpy relaxation measured in the glass transition region by DSC increased with increasing aging time. Moreover, it was observed that when molded specimens were aged at room temperature for 3–8 h, they became very brittle [17]. This occurrence was attributed to the reduction of the free volume of the polymer due to rapid relaxation to the equilibrium amorphous state.

Increasing the crystallinity of the polymer can reduce the aging effect [18]. Several authors investigated and are still investigating the effect of nucleating agents on PLA [19–21] in order to make compatible effective crystallization, mechanical stability during time and short injection molding cycles enabling series production. In fact, the formed crystallites act like physical crosslinks to retard the polymer chain mobility. In the framework of the DIBBIOPACK European project *"Development of Injection and Extrusion Blow Moulded Biodegradable and Multifunctional Packages by Nanotechnologies"* GA 280676 [22], some nucleating agents were studied in combination with a plasticizer, to control the evolution of crystallinity in injection molded specimens [23]. To reduce the brittleness, the plasticization [24–26] and the blending with biodegradable elastomeric polymers [27,28] was also investigated and successfully applied, in some cases following reactive blending [29] or reactive extrusion [30,31] approaches.

Interestingly, by considering the injection stretch blow molding process, it was observed that the molecular orientation induced by this process limits the effect of aging by stabilizing the polymer free volume [17]. Aging is also reduced by the crystallites produced during strain-induced crystallization acting as physical crosslinks to stabilize the amorphous fraction, thus reducing the brittleness. Similarly to PET, PLA exhibits strain-hardening when stretched to high strain. This is desirable for blow molding of pre-forms to minimize wall thickness variation. As strain-hardening occurs only when the PLA is highly stretched, the pre-form must be designed as a function of mold dimension such that optimal stretch ratios are achieved during blow molding. It was noticed that the crystallinity after stretching decreases by decreasing the stereo isomeric fraction of the polymer [22]. Thus, the optimal stretch ratios depend on the grade of employed PLA.

The thermoforming of PLA can be generally made with aluminum molds in the range of 80–110 °C. Molds, trim tools and ovens designed for thermoforming PET and PS can be used for forming PLA containers. On the other hand, molds for thermoforming of PP cannot be used likewise for PLA, since PP shrinks significantly more than PLA during cooling [11].

Many processing methods are thus available to obtain rigid packages from PLA, especially if its properties are modulated by using proper additives, such as other polymers, fillers or low molecular weight additives such as stabilizers and chain extenders.

3.2. Poly Hydroxy Alkanoate

PHAs are gaining attention among biodegradable polymers to be used for packaging production due to their characteristic properties (such as high biodegradability in different environments such as soil and marine water, not just in composting plants) coupled with high barrier properties to oxygen and moisture and relatively high thermal stability [32–35]. These properties differentiate PHAs from PLA and are particularly interesting because of addressing the issue of plastic waste accumulation in the oceans.

The brittleness of polyhydroxy Butyrate (PHB) and polyhydroxy butyrate-co-valerate (PHB/V) is due to the secondary crystallization of the amorphous phase that occurs during storage at room temperature; in fact, the glass transition temperature (Tg) of PHB is close to room temperature. The crystallization kinetics of PHB usually starts from homogeneous nuclei, unless specific nucleating agents are added, since this polymer is free of heterogeneities. In fact, PHB does not contain catalyst residues or other impurities that can act as heterogeneous nuclei [36–40]. To achieve high elongation at break and a higher flexibility for modified/formulated PHB, the glass transition temperature must reach a lower value than the testing temperature. By adding plasticizers, the molecular mobility is improved and the glass transition temperature is lowered, as well as the melting temperature. For processing, it must be considered also that PHB thermally decomposes at temperatures just above its melting point. Short exposure of PHB to temperatures near 180 °C could induce severe degradation. The main reaction consists of a random chain scission that involves a cis-elimination reaction of β-CH and a six-member ring transition, which results in a rapid decrease in molecular weight accompanied by production of the degraded products of olefinic and carboxylic acid compounds, such as crotonic acid and various oligomers [41–44].

Several blends among PHB and other biodegradable polymers and several types of plasticizers have been investigated [45,46]. Materials of relatively low cost, biodegradable and possibly produced by renewable resources are preferred as plasticizers. Examples are: oxypropylated glycerin (or laprol), glycerol, glycerol triacetate, 4-nonylphenol, 4,40-dihydroxydiphenylmethane, acetyl tributyl citrate, salicylic ester, acetylsalicylic acid ester, soybean oil, epoxidized soybean oil, dibutyl phthalate, triethyl citrate, dioctyl phthalate, dioctyl sebacate, acetyl tributyl citrate, di-2-ethylhexylphthalate, tri(ethylene glycol)-bis(2-ethylhexanoate), triacetin, fatty alcohols with or without glycerol fatty esters, polyethylene glycol (PEG), as well as low molecular weight polyhydroxybutyrate since PHAs with a medium chain length are elastomers with a low melting point and a relatively lower degree of crystallinity [47,48]. In our research unit, we have recently investigated the performances of different degrees of polyethylene glycol (PEG) as plasticizers for PHB-based blends compared to tributyl citrate and the effect of the kinetic of crystallization by the use of these plasticizers [49]. PEG400 and tributyl citrate are very efficient plasticizers for PHB, and PEG400 resulted in being an efficient lubricating agent for the production of composite based on PHB and wood fibers, allowing better processing of the viscous melt containing up to 30% by weight of wood fibers [50]. Blends based on either PHB or PHA and PEG400 have been used for the production of hard packaging jars intended for cosmetic applications, in the research activity of the EC project Oli-PHA [51], as reported in Figure 8.

Although plasticization was the object of many studies, it must be noticed that the improvement in mechanical properties is often limited. For instance, for PHBV, Martino et al. [52], in the framework of the ECOBIOCAP project "Ecoefficient Biodegradable Composite Advanced Packaging" GA No. 265669 [53], obtained an improvement of the elongation at break of 6% starting from 2% from the pure PHBV. Hence, the improvement in properties of PHA polymers is still an open issue in current bioplastics research.

Figure 8. Cosmetic jars produced with PHB.

3.3. Starch

Polysaccharides such as starch are considered for the production of rigid packaging due to relatively low cost and high degradability, in compost, soil and marine water. Thus, as previously addressed, the long-term impact of plastic waste in the marine environment is a primary issue since birds, mammals and fish may become entangled in plastic films or ingest plastic particles [54].

Non-durable plastic hard packaging includes plates, bowls and cups, as well as peanuts used for packaging fragile materials. Polystyrene (PS) is the plastic mostly used in plastic food service items (79%), and a low recycling rate is reported for the packaging waste made of PS also due to difficulties in collection and cleaning of the packaging after use [55]. Polymers such as PS are especially persistent and difficult to degrade in the environment, and for foamed items, their light weight promotes the dispersion in the environment by wind or storm drains [56]. Native starch is not a thermoplastic material, and it thermally degrades before its glass transition temperature (Tg) when its melting temperature (230–240 °C) is reached [57,58].

However, as evidenced in the previous section, when starch is heated in the presence of plasticizers such as water or polyols, the semi-crystalline structure of the starch granule is disrupted. Then, the glass transition temperature decreases below the thermal degradation temperature, and the starch actually behaves as a thermoplastic, it is then denominated thermoplastic starch (TPS) [56]. In extrusion processing, the TPS melts at much lower moisture content (10–20%) than that used for conventional cooking methods [59,60].

Several companies have introduced starch-based products on the market, Novamont (Novara, Italy), Cereplast (Seymour, IN, USA), BASF (Suffolk, VA, USA), Biotec GmbH (Gütersloh, Germany), Plantic (Altona, VIC, Australia) and Biolice

(Ennezat, France), among others. Some images of starch-based hard packaging (peanuts and clams shell) are reported in Figure 9.

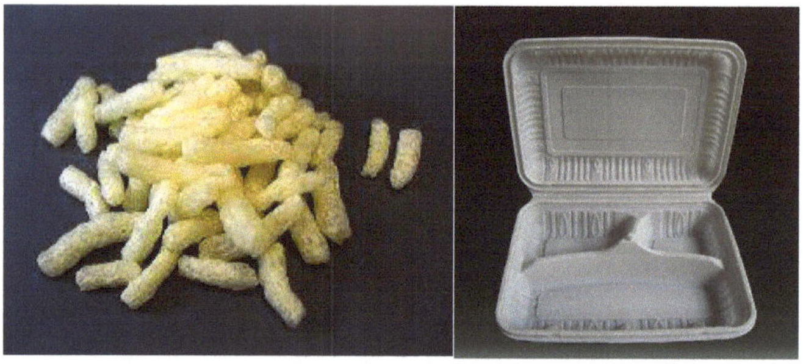

Figure 9. Image of hard packaging made by starch.

An alternative procedure to produce foamed starch-based packaging is based on a baking technology that was firstly developed for the food industry and was successively adapted for making starch-based foam food service products with a process similar to that used to produce waffles [61–64]. A starch dough is first prepared containing gelatinized or pre-gelatinized starch, native starch, water, fiber, fillers and other additives and then mixed for about 10 min. A predetermined amount of aqueous starch dough is placed into a preheated (150–200 °C) mold cavity. The dough rapidly heats, and the starch component is gelatinized, forming a melt that fills the mold cavity. A skin forms on the upper and lower surfaces where the dough contacts the mold surfaces. The steam formed during the process acts as a blowing agent and allows the formation of a foam structure in the core region of the product. Steam is then allowed to vent from the mold, and within about 45–60 s, the product dries and solidifies into the desired shape. The starch-based products are very similar to PS foam and are marketed by Biopack (Graz, Austria), Apack-IBEK Verpackungshandel GmbH (Markt Erlbach, Germany), Earthshell (Santa Barbara, CA, USA) and Biosphere Industries (Carpinteria, CA, USA).

Packages based on starch are sensitive to water; hence, some researchers studied the possible cross-linking of starch to decrease its solubility in water [65,66] with the aim of using starch for application as paper coating.

3.4. Proteins

Proteins (casein, collagen, gelatine, corn, soy, wheat, etc.), can be obtained from a variety of agricultural commodities and/or wastes and food products. Proteins can be processed by casting or by melt extrusion in the presence of plasticizers or other polymers [67–79]. The mechanical and barrier properties offered by protein-based films are generally superior to those offered by polysaccharide-based films due to, and contrary to polysaccharides, which are mainly homopolymers, the specific structure (based on 20 different monomers), which confers functional properties and high intermolecular binding. High molecular weight proteins are insoluble, or only partially soluble, in water and, thus, present themselves as very interesting film-forming molecules for the formation of water-resistant films, particularly after crosslinking [80]. As proteins are hydrophilic, they adhere very well to polar surfaces such as paper, thus acting as barriers to oxygen and carbon dioxide. Materials with protein coatings are expected to show good barrier properties offering an alternative to nonrenewable polymers such as ethylene vinyl alcohol (EVOH) or silica-based coating. Achievements in this application were gained by the EC project WHEYLAYER "Whey Protein-Coated Plastic Films to Replace Expensive Polymers and Increase Recyclability" and the following demo action WHEYLAYER2 "Barrier Properties for Sustainable Packaging" [81]. An image of the WHEYLAYER approach and of the products are reported in Figure 10.

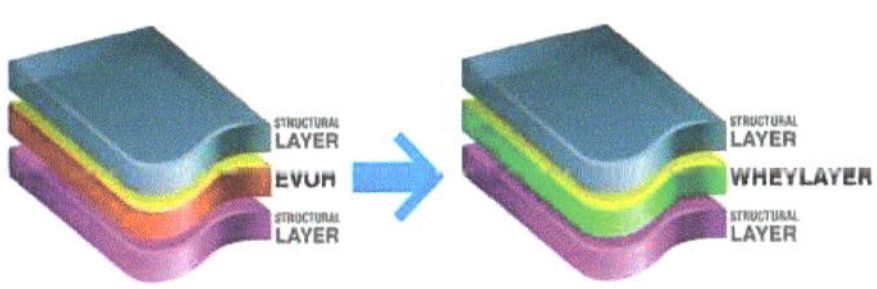

(a)

(b)

Figure 10. (**a**) WHEYLAYER approach and products; (**b**) EVOH, ethylene vinyl alcohol.

In the WHEYLAYER-based packaging, a whey-based coating is applied on a plastic film substrate, such as PET, and the coated film is used for the production of multi-layer packaging.

It is possible to achieve a separation of the different layers of the packaging by removing the whey protein layer by washing with water containing a protease enzymatic detergent [82,83]. This procedure promotes the recyclability of the material that constitutes the packaging. When WHEYLAYER is applied on a compostable substrate such as PLA, it is possible to produce a completely biodegradable multi-layer material with improved barrier properties [84].

Furthermore, in rigid paperboard packages, the main innovation can be the introduction of a new bio-based layer that can replace, at least partially, polyethylene, thus increasing the bio-content of the packaging and allowing a better management of the package end life.

More research activity is running on this very promising approach of the application of protein in packaging. The most recent project BIOBOARD [85] aims at developing a food waste-based protein-based film by flat die extrusion. In this project, both whey protein from the dairy industry and potato pulp from the starch industry are used [86,87].

The project LEGUVAL"Valorisation of Legumes Co-Products and by-Products for Package Application and Energy Production from Biomass" GA No. 15241 [88] aims at the use and valorization of co-products and by-products of processed grain legumes thanks to the extraction of their proteinaceous fraction, which can be used as a raw material in packaging.

4. Conclusions

This overview of processing technologies for the production of hard packaging from bio-based polymers and the strategies for maximum valorization of bio-based polymer in this application outlines the potentiality and the growing attention not just of researcher, but also of producers and consumers towards bio-based materials. The renewable origin, as well as the biodegradability and compostibility of several products based on bio-based and biodegradable polymers represent a benefit for the environment and human life.

The research will continue to improve the properties of packaging based on these materials, as well as optimize their processing, with also the aim to lower the cost of bio-based products, which at present represents the main limit to the wide spreading of their applications. With the growing awareness of the need for environment and resource preservation, the market for bio-based products applied in single use applications is forced to increase. Thus, we expect more and more achievements and innovative bio-based products suitable for use in the production of hard packaging.

Author Contributions: Conceptualization, P.C. and M.B.C.; Writing – Original Draft Preparation, P.C. and M.B.C.; Writing – Review & Editing, P.C. and A.L.; Supervision, A.L.

Acknowledgments: The authors acknowledge the financial support from the European Community's Seventh Framework Programme under: Grant agreement (GA) No. 280604 Oli-PHA, GA No. 280676 DIBBIOPACK, GA No. 315313 BIOBOARD, N-Chitopack GA 315233, WHEYLAYER GA No. 218340-2, LEGUVAL GA No. 315241.

Conflicts of Interest: The authors declare no conflict of interest.

References

1. Plastic Europe. *Plastics—The Facts 2014–2015, An Analysis of European Plastics, Production, Demand and Waste Data*; Plastic Europe: Brussels, Belgium, 2015.
2. Benefits of Biobased Rigid Packaging. 2014. Available online: http://en.european-bioplastics.org (accessed on 9 August 2018).
3. Sant'Anna. Available online: www.santanna.it (accessed on 9 August 2018).

4. Acqua e Bevande. Available online: www.sks-bottle.com/PLAValueAdded.html (accessed on 9 August 2018).
5. Roos, G. 100% Bio-Plastic Water Bottles Trickle into Marketplace, 2010. Environmental Leader. Available online: http://www.environmentalleader.com/2010/03/08/bio-plastic-water-bottles-trickle-into-marketplace/ (accessed on 9 August 2018).
6. Lazzeri, A. FORBIOPLAST: Bio-Composites Based on Forest Derived Materials and Biodegradable Polymers. Presented at the PLASTiCE Conference Europe for Sustainable Plastics, Bologna, Italy, 24–25 October 2011; Available online: http://www.plastice.org/fileadmin/files/FORBIOPLAST_Plastice2011.pdf; www.forbioplast.eu (accessed on 9 August 2018).
7. Cavlin, S.I. The Unique convertibility of paperboard. *Packag. Technol. Sci.* **1988**, *1*, 77–92. [CrossRef]
8. Liu, H.; Zhang, J. Research progress in toughening modification of poly(lactic acid). *J. Polym. Sci. Part B Polym. Phys.* **2011**, *49*, 1051–1083. [CrossRef]
9. Sarasa, J.; Jose, M.G.; Javierre, C. Study of the biodisintegration of a bioplastic material waste. *Bioresour. Technol.* **2009**, *100*, 3764–3768. [CrossRef] [PubMed]
10. Rudeekit, Y.; Numnoi, J.; Tajan, M.; Chaiwutthinan, P.; Leejarkpai, T. Determining Biodegradability of Polylactic Acid under Different Environments. *J. Met. Mater. Miner.* **2008**, *18*, 83–87.
11. Lim, L.T.; Auras, R.; Rubino, M. Processing technologies for poly(lactic acid). *Prog. Polym. Sci.* **2008**, *33*, 820–852. [CrossRef]
12. Henton, D.E.; Gruber, P.; Lunt, J.; Randall, J. Polylactic acid technology. In *Natural Fibers, Biopolymers, and Biocomposites*; Mohanty, A.K., Misra, M., Drzal, L.T., Eds.; Taylor & Francis: Boca Raton, FL, USA, 2005; pp. 527–577.
13. Ke, T.; Sun, X. Effects of moisture content and heat treatment on the physical properties of starch and poly(lactic acid) blends. *J. Appl. Polym. Sci.* **2001**, *81*, 3069–3082. [CrossRef]
14. Signori, F.; Coltelli, M.B.; Bronco, S. Thermal degradation of poly(lactic acid) (PLA) and poly(butylene adipate-*co*-terephthalate) (PBAT) and their blends upon melt processing. *Polym. Degrad. Stab.* **2009**, *94*, 74–82. [CrossRef]
15. Natureworks. *Natureworks Polymer Technical Data Sheets—PLA Polymer 7032D*; Natureworks LLC: Minnetonka, MN, USA, 2005.
16. Cai, H.; Dave, V.; Gross, R.A.; McCarthy, P. Effects of physical aging, crystallinity, and orientation on the enzymatic degradation of poly(lactic acid). *J. Polym. Sci.* **1996**, *34*, 2701–2708. [CrossRef]
17. Witzke, D.R. Introduction to Properties, Engineering, and Prospects of Polylactide Polymers. Ph.D. Thesis, Michigan State University, East Lansing, MI, USA, 1997.
18. Saeidlou, S.; Huneault, M.A.; Li, H.; Park, C.B. Poly(lactic acid) crystallization. *Prog. Polym. Sci.* **2012**, *37*, 1657–1677. [CrossRef]
19. Harris, A.M.; Lee, E.C. Improving Mechanical performance of injection molded PLA by controlling Crystallinity. *J. Appl. Polym. Sci.* **2008**, *107*, 2246–2255. [CrossRef]

20. Tábi, T.; Sajó, I.E.; Szabó, F.; Luyt, A.S.; Kovács, J.G. Crystalline structure of annealed polylactic acid and its relation to processing. *Express Polym. Lett.* **2010**, *4*, 659–668. [CrossRef]
21. Xuetao, S.; Zhang, G.; Phuong, T.V.; Lazzeri, A. Synergistic Effects of Nucleating Agents and Plasticizers on the Crystallization Behavior of Poly(lactic acid). *Molecules* **2015**, *20*, 1579–1593.
22. Dibbiopack. *Development of Injection and Extrusion Blow Moulded Biodegradable and Multifunctional Packages by Nanotechnologies*; FP7-NMP.2011.LARGE.5, GA 280676; Dibbiopack: Zaragoza, Spain, 2011.
23. Kazem Fehri, M.; Cinelli, P.; Phuong, T.V.; Anguillesi, I.; Salvadori, S.; Montorsi, M.; Mugoni, C.; Fiori, S.; Lazzeri, A. New heat-resistant PLA blends. *Bioplast. Mag.* **2014**, *8*, 26–27.
24. Li, H.; Huneault, M.A. Effect of nucleation and plasticization on the crystallization of poly(lactic acid). *Polymer* **2007**, *48*, 6855–6866. [CrossRef]
25. Kulinski, Z.; Piorkowska, E. Crystallization, structure and properties of plasticized poly(L-lactide). *Polymer* **2005**, *46*, 10290–10300. [CrossRef]
26. Coltelli, M.B.; Della Maggiore, I.; Bertoldo, M.; Bronco, S.; Signori, F.; Ciardelli, F. Poly(lactic acid) (PLA) properties as a consequence of poly(butylene adipate-*co*-terephtahlate) (PBAT) blending and acetyl tributyl citrate (ATBC) plasticization. *J. Appl. Polym. Sci.* **2008**, *110*, 1250–1262. [CrossRef]
27. Farsetti, S.; Cioni, B.; Lazzeri, A. Physico-mechanical properties of biodegradable rubber toughened polymers. *Macromol. Symp.* **2011**, *301*, 82–89. [CrossRef]
28. Lazzeri, A.; Phuong, T.V.; Cinelli, P. Copolymers Based on Reactive Polyesters and Plasticisers for the Manufacture of Transparent, Biodegradable Packaging Film. WO 2013/164743A1, 7 November 2013.
29. Coltelli, M.B.; Bronco, S.; Chinea, C. The effect of free radical reactions on structure and properties of poly(lactic acid) (PLA) based blends. *Polym. Degrad. Stab.* **2010**, *95*, 332–341. [CrossRef]
30. Penco, M.; Lazzeri, A.; Phuong, T.V.; Cinelli, P. Copolymers Based on Polyester and Aromatic Polycarbonate. WO 2012/025907A1, 1 March 2012.
31. Drumright, R.E.; Gruber, P.R.; Henton, D.E. Polylactic acid technology. *Adv. Mater.* **2000**, *12*, 1841–1846. [CrossRef]
32. Tokiwa, Y.; Calabia, B.P.; Ugwu, C.U.; Aiba, S. Biodegradability of Plastics. *Int. J. Mol. Sci.* **2009**, *10*, 3722–3742. [CrossRef] [PubMed]
33. Boopathy, R. Factor limiting bioremediation technologies. *Bioresour. Technol.* **2000**, *74*, 63–67. [CrossRef]
34. Lee, S.Y. Bacterial polyhydroxyalkanoates. *Biotechnol. Bioeng.* **1996**, *49*, 1–14. [CrossRef]
35. Bugnicourt, E.; Cinelli, P.; Lazzeri, A.; Alvarez, V. Polyhydroxyalkanoate (PHA): Review of synthesis, characteristics, processing and potential applications in packaging. *Express Polym. Lett.* **2014**, *8*, 791–808.

36. El-Hadi, A.; Schnabel, R.; Straube, E.; Muller, G.; Henning, S. Correlation between degree of crystallinity, morphology, glass temperature, mechanical properties and biodegradation of poly (3-hydroxyalkanoate) PHAs and their blends. *Polym. Test.* **2002**, *21*, 665–674.
37. Holmes, P.A. Applications of PHB—A microbially produced biodegradable thermoplastic. *Phys. Technol.* **1985**, *16*, 32–36. [CrossRef]
38. Di Lorenzo, M.L.; Raimo, M.; Cascone, E.; Martuscelli, E. Poly(3-hydroxybutyrate)-based copolymers and blends: Influence of a second component on crystallization and thermal behaviour. *J. Macromol. Sci. Part B Phys.* **2001**, *40*, 639–667. [CrossRef]
39. De Konig, G.J.M.; Lamstra, P.J. Crystallization phenomena in bacterial poly[(R)-3-hydroxybutyrate]: 2. Embrittlement and rejuvenation. *Polymer* **1993**, *34*, 4089–4094. [CrossRef]
40. Doi, Y.; Mukai, K.; Kasuya, K.; Yamada, K. Biodegradation of biosynthetic and chemosynthetic polyhydroxyalkanoates. In *Biodegradable Plastics and Polymers*; Doi, Y., Fukuda, K., Eds.; Elsevier: Amsterdam, The Netherlands, 1994; pp. 39–51.
41. Grassie, N.; Marray, E.J.; Holmes, P.A. The thermal degradation of poly(-(D)-b-hydroxybutyric acid): Part 1—Identification and quantitative analysis of products. *Polym. Degrad. Stab.* **1984**, *6*, 47–61. [CrossRef]
42. Grassie, N.; Marray, E.J.; Holmes, P.A. The thermal degradation of poly(-(D)-b-hydroxybutyric acid): Part 2—Changes in molecular weight. *Polym. Degrad. Stab.* **1984**, *6*, 95–103. [CrossRef]
43. Grassie, N.; Murray, E.J. The thermal degradation of poly(-(D)-bhydroxybutyric acid), part 3—The reaction mechanism. *Polym. Degrad. Stab.* **1984**, *6*, 127–134. [CrossRef]
44. Mohanty, A.K.; Misra, M.; Drzal, L.T. Sustainable Bio-Composites from Renewable Resources: Opportunities and Challenges in the Green Materials World. *J. Polym. Environ.* **2002**, *10*, 19–26. [CrossRef]
45. Innocenti-Mei, L.H.; Bartoli, J.; Baltieri, R.C. Mechanical and thermal properties of poly(3-hydroxybutyrate) blends with starch and starch derivates. *Macromol. Symp.* **2003**, *197*, 77–87. [CrossRef]
46. Shen, L.; Haufe, J.; Patel, M. *Product Overview and Market Projection of Emerging Bio-Based Plastics*; PRO-BIP: Utrecht, The Netherlands, 2009.
47. Shinn-Gwo, H.; Heng-Wei, H.; Min-Tzung, Y. Thermal properties and applications of low molecular weight Polyhydroxybutyrate. *J. Therm Anal. Calorim.* **2013**, *111*, 1243–1250.
48. Parra, D.F.; Fusaro, J.; Gaboardi, F.; Rosa, D.S. Influence of poly (ethylene glycol) on the thermal, mechanical, morphological, physical chemical and biodegradation properties of poly (3-hydroxybutyrate). *Polym. Degrad. Stab.* **2006**, *91*, 1954–1959. [CrossRef]
49. Farris, G.; Cinelli, P.; Anguillesi, I.; Salvadori, S.; Coltelli, M.B.; Lazzeri, A. Effect of agening time on mechanical properties of plasticized poly(hydroxybutyrate) (PHB). *AIP Conf. Proc.* **2014**. [CrossRef]
50. Seggiani, M.; Cinelli, P.; Verstichel, S.; Puccini, M.; Vitolo, S.; Anguillesi, I.; Lazzeri, A. Development of fibres-reinforced biodegradable composites. *Chem. Eng. Trans.* **2015**, in press.

51. European Community's Seventh Framework Programme (FP7/2007–2013) NMP Grant Agreement No. 280604 Oli-PHA. A Novel and Efficient Method for the Production of Polyhydroxyalkanoate (PHA) Polymer-Based Packaging from Olive Oil Waste Water. Available online: http//www.olipha.eu/ (accessed on 9 August 2018).
52. Martino, L.; Berthet, M.A.; Angellier-Coussy, H.; Gontard, N. Understanding External Plasticization of Melt Extruded PHBV—Wheat Straw Fibers Biodegradable Composites for Food Packaging. *J. Appl. Polym. Sci.* **2015**. [CrossRef]
53. European Commission. Ecoefficient Biodegradable Composite Advanced Packaging (EcoBioCAP); Project Report on Grant Agreement No. 265669, European Community's 7th Framework Programme (FP7/2007-2013): France, 2015. Available online: https://cordis.europa.eu/project/rcn/97870_de.html (accessed on 9 August 2018).
54. Boerger, C.M.; Lattin, G.L.; Moore, S.L.; Moore, C.J. Plastic ingestion by planktivorous fishes in the North Pacific Central Gyre. *Mar. Pollut. Bull.* **2010**, *60*, 2275–2278. [CrossRef] [PubMed]
55. Glenn, G.M.; Orts, W.; Imam, S.; Chiou, B.S.; Wood, D.F. *Starch Plastic Packaging and Agriculture Applications*; Chapter 15, Paper 1459; USDA-ARS/UNL Faculty: Lincoln, NE, USA, 2014.
56. Gorden, M. *Ban on Expanded Polystyrene Foam Takeout Food Packaging*; California Water Action: Oakland, CA, USA, 2006; Available online: http://www.cleanwateraction.org/ca (accessed on 9 August 2018).
57. Dufresne, A.; Dupeyre, D.; Vignon, M.R. Cellulose microfibrils from potato tuber cells: Processing and characterization of starch–cellulose microfibril composites. *J. Appl. Polym. Sci.* **2000**, *76*, 2080–2092. [CrossRef]
58. Shogren, R.L. Effect of moisture content on the melting and subsequent physical aging of cornstarch. *Carbohydr. Polym.* **1992**, *19*, 83–90. [CrossRef]
59. Kokini, J.L. The effect of processing history on chemical changes in single- and twin-screw extruders. *Trends Food Sci. Technol.* **1993**, *4*, 324–329. [CrossRef]
60. Wiedmann, W. Extrusion cooking of starches for semi-products. *Starch-Starke* **1987**, *39*, 352–357. [CrossRef]
61. Hass, F.; Hass, J.; Tiefenbacher, K. Process of Manufacturing Rottable Thin-Walled Starch-Based Shaped Elements. U.S. Patent No. 5,576,049, 19 November 1996.
62. Tiefenbacher, K.F. Starch-based foamed materials-use and degradation properties. *J. Macromol. Sci. Part A Pure Appl. Chem.* **1993**, *30*, 727–731. [CrossRef]
63. Cinelli, P.; Chiellini, E.; Lawton, J.; Imam, S.H. Foamed articles based on potato starch, corn fibers and poly(vinyl alcohol). *Polym. Degrad. Stab.* **2006**, *91*, 1147–1155. [CrossRef]
64. Chiellini, E.; Cinelli, P.; Ilieva, V.I.; Imam, S.H.; Lawton, J.W. Environmentally Compatibile Foamed Articles Based on Potato Starch, Corn Fiber, and Poly (Vinyl Alcohol). *J. Cell. Plast.* **2009**, *345*, 17–32. [CrossRef]
65. Menzel, C. Starch Structures and Their Usefulness in the Production of Packaging Materials. Ph.D. Thesis, Swedish University of Agricultural Sciences, Uppsala, Sweden, 2014.

66. Olsson, E.; Hedenqvist, M.S.; Johansson, C.; Järnström, L. Influence of citric acid and curing on moisture sorption, diffusion and permeability of starch films. *Carbohydr. Polym.* **2013**, *94*, 765–772. [CrossRef] [PubMed]
67. Tang, C.H.; Xiao, M.L.; Chen, Z.; Yang, X.Q.; Yin, S.W. Properties of cast films of vicilin-rich protein isolates from Phaseolus legumes: Influence of heat curing. *LWT Food Sci. Technol.* **2009**, *42*, 1659–1666. [CrossRef]
68. Tang, C.; Xiao, M.L.; Chen, Z.; Yang, X.Q. Influence of Succinylation on the Properties of Cast Films from Red Bean Protein Isolate at Various Plasticizer Levels. *J. Appl. Polym. Sci.* **2011**, *120*, 1934–1941. [CrossRef]
69. Saremnezhad, S.; Azizi, M.H.; Barzegar, M.; Abbasi, S.; Ahmadi, E. Properties of a New Edible Film Made of Faba Bean Protein Isolate. *J. Agric. Sci. Technol.* **2011**, *13*, 181–192.
70. Viroben, G.; Barbot, J.; Mouloungui, Z.; Gueguen, J. Preparation and characterization of films from pea protein. *J. Agric. Food Chem.* **2000**, *48*, 1064–1069. [CrossRef] [PubMed]
71. Brandenburg, A.H.; Weller, C.L.; Testin, R.F. Edible Films and Coatings from Soy Protein. *J. Food Sci.* **1993**, *58*, 1086–1089. [CrossRef]
72. Bamdad, F.; Goli, A.H.; Kadivar, M. Preparation and characterization of proteinous film from lentil (Lens culinaris) Edible film from lentil (Lens culinaris). *Food Res. Int.* **2006**, *39*, 106–111. [CrossRef]
73. Farris, S.; Schaich, K.M.; Liu, L.S.; Piergiovanni, L.; Yam, K.L. Development of polyion-complex hydrogels as an alternative approach for the production of bio-based polymers for food packaging applications: A review. *Trends Food Sci. Technol.* **2009**, *20*, 316–332. [CrossRef]
74. Cuq, B.; Gontard, N.; Guilbert, S. Proteins as agricultural polymers for packaging production. *Cereal Chem.* **1998**, *75*, 1–9. [CrossRef]
75. Chiellini, E.; Cinelli, P.; Grillo Fernandes, E.; Kenawy, E.R.; Lazzeri, A. Gelatin based Blends and Composites. Morphological and Thermal Mechanical Characterization. *Biomacromolecules* **2001**, *2*, 806–811. [CrossRef] [PubMed]
76. Gennadios, A.; Brandenburg, A.H.; Weller, C.L.; Testin, R.F. Effect of pH on properties of wheat gluten and soy protein isolate films. *J. Agric. Food Chem.* **1993**, *41*, 1835–1839. [CrossRef]
77. Kowalczyk, D.; Baraniak, B. Effects of plasticizers, pH and heating of film-forming solution on the properties of pea protein isolate films. *J. Food Eng.* **2011**, *105*, 295–305. [CrossRef]
78. Kenawy, E.R.; Cinelli, P.; Corti, A.; Miertus, S.; Chiellini, E. Biodegradable Composite Films Based on Waste Gelatin. *Macromol. Symp.* **1999**, *144*, 351–364. [CrossRef]
79. Chiellini, E.; Cinelli, P.; Corti, A.; Kenawy, E.R. Composite Films Based on Waste Gelatin Thermal-Mechanical Properties and Biodegradation Testing. *Polym. Degrad. Stab.* **2001**, *73*, 549–555. [CrossRef]
80. Cutter, C.N.; Sumner, S.S. Application of edible coatings on muscle foods. In *Protein-Based Films and Coatings*; Gennadios, A., Ed.; CRC Press: Boca Raton, FL, USA, 2002; pp. 467–484.

81. European Community's Seventh Framework Programme (FP7/2007-2013) Grant agreement No. 218340-2 WHEYLAYER "Whey Protein-Coated Plastic Films to Replace Expensive Polymers and Increase Recyclability" and WHYELAYER 2 "Barrier Properties for Sustainable Packaging" Seventh Framework Programme (FP7/2007-2013) Grant Agreement No. 315743. Available online: http://www.wheylayer.eu/ (accessed on 9 August 2018).
82. Bugnicourt, E.; Schmid, M.; Mc Nerney, O.; Wildner, J.; Smykala, L.; Lazzeri, A.; Cinelli, P. Processing and validation of whey protein coated films and laminates at semi-industrial scale as novel recyclable food packaging materials with excellent barrier properties. *Adv. Mater. Sci. Eng.* **2013**, *2013*, 496207. [CrossRef]
83. Schmid, M.; Müller, K.; Sängerlaub, S.; Stäbler, A.; Starck, V.; Ecker, F.; Noller, K. Mechanical and barrier properties of thermoplastic whey protein isolate/ethylene vinyl acetate blends. *J. Appl. Polym. Sci.* **2014**, *131*. [CrossRef]
84. Cinelli, P.; Schmid, M.; Bugnicourt, E.; Wildner, J.; Bazzichi, A.; Anguillesi, I.; Lazzeri, A. Whey protein layer applied on biodegradable packaging film to improve barrier properties while maintaining biodegradability. *Polym. Degrad. Stab.* **2014**, *108*, 151–157. [CrossRef]
85. Bioboard.eu. Consortium of Bioboard Project: Seventh Framework Programme (FP7/2007-2013) under Grant Agreement No. 315313 (Updated May 2014). Available online: http://www.bioboard.eu/ (accessed on 9 August 2018).
86. Schmid, M.; Herbst, C.; Müller, K.; Stäbler, A.; Schlemmer, D.; Coltelli, M.B.; Lazzeri, A. Effect of Potato Pulp Filler on the Mechanical Properties and Water Vapor Transmission Rate of Thermoplastic WPI/PBS Blends. *Polym.-Plast. Technol. Eng.* **2016**, *5*, 510–517. [CrossRef]
87. Coltelli, M.B.; Wild, F.; Bugnicourt, E.; Cinelli, P.; Lindner, M.; Schmid, M.; Weckel, V.; Müller, K.; Rodriguez, P.; Staebler, A.; et al. State of the Art in the Development and Properties of Protein-Based Films and Coatings and Their Applicability to Cellulose Based Products: An Extensive Review. *Coatings* **2016**, *6*, 1. [CrossRef]
88. European Commission. LEGUVAL, Valorisation of Legumes Co-Products and by-Products for Package Application and Energy Production from Biomass; Project Report on Grant Agreement No. 15241, European Community's 7th Framework Programme (FP7/2007-2013): France, 2015. Available online: https://cordis.europa.eu/project/rcn/111096_en.html (accessed on 9 August 2018).

PART V

Biological Activity, Safety & Patent Application & Legal Considerations

Biological Activity of Innovative Polymeric Nanoparticles and Non–Woven Tissue

Giovanna Donnarumma, Brunella Perfetto, Adone Baroni, Iole Paoletti, Maria Antonietta Tufano, Paola Del Ciotto and Pierfrancesco Morganti

Abstract: In the last few years many studies have focused their attention on different potential biomedical applications for biocompatibile polymers, especially nanopolymers, in shape of nanotubes, nanofibers, and nanoparticles. Polymer nanocomposites arising from different chemistries and constructions include aliphatic polyester such as polylactide (PLA) and poly (DL-lactic-cd-glycolic acid) (PLGA), poly (ε-caprolactone) (PCL), poly(*p*-dioxanone) (PPDO), poly(butylenes succinate) (PBS), poly (hydrdroxyalkanoate), and natural biopolymers, such as starch, cellulose, chitin, chitosan, lignin, and protein. In medicine, they contribute to applications in surgery, dentistry, and pharmachology for scaffolds for tissue regeneration, tissue engineering, drug delivery devices, and gene transfection. Bionanopolymers can be used in many other fields including food, cosmetics and agriculture. The research efforts are focused on the study of new polymeric nano-constructs that exploit the body's natural biological response and are environment-friendly.

1. Introduction

For thousands of years humans made tools and devices from naturally available compounds, and later artificial compounds were created to make them.

The interdisciplinary field of biotechnology and particularly nanobiotechnology, which combines biology, chemistry, engineering, nanopharmacology, and nanomedicine, is revolutionizing the development of drug delivery systems and tissue engineering, such as tissue engineered for vascular grafts and wound healing. Research in this area has provided unlimited potential to improve human health [1–3]. A large variety of materials can be used and are classified as synthetic, natural, or hybrid. The synthetic materials can be further classified in degradable and non-degradable materials. The synthetics include polymers, such as (poly(ethyleneglycol), *N*-(2-hydroxypropyl) and methacrylamide co-polymers, Natural polymers can be classified as those obtained from natural sources such as animal, microbial, and vegetable sources. They are usually natural proteins or polysaccharides, such as chitin, chitosan, (dextran (α-1,6 polyglucose), dextrin (α-1,4 polyglucose), lignin, collagen, gelatin, and hyaluronic acid. Interestingly, in the interaction of nanopolymers with the human immune system, is that chitin and hyaluronic acid show the same backbone (Figures 1 and 2).

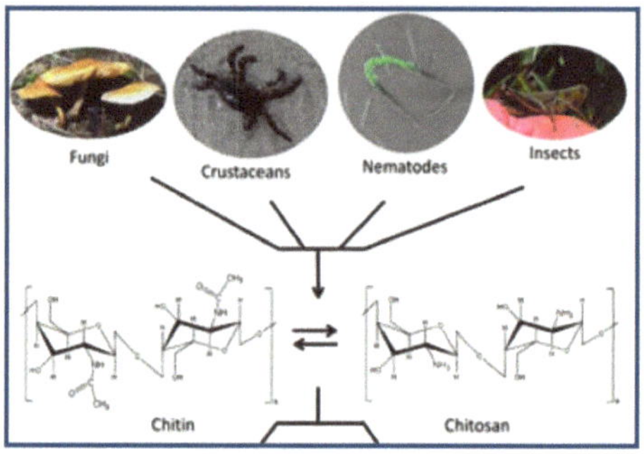

Figure 1. Chemical structures of Chitin and Chitosan from natural sources such as crustacean exoskeleton and fungi wall. Adapted from Morganti, P. et al. [4].

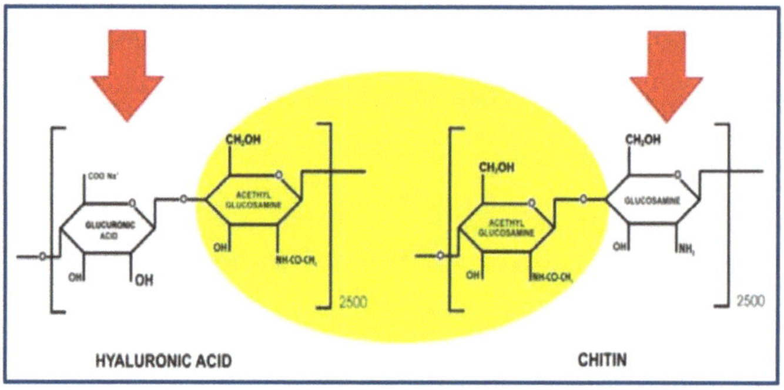

Figure 2. Chitin and hydrauronic acid structures. Adapted from Morganti, P. et al. [4].

Others include linear polyamidoamines and pseudosynthetic polymers the man-made poly (amino acids) poly(L-lysine), poly(L-glutei acid), poly(malic acid), and poly(aspartamides).

Natural polymers have the disadvantage of high biodegradability, while synthetic polymers can be synthesized and modified in a controlled manner to produce constant and homogenous physical and chemical properties and stability. However, the latter are biologically inert and do not offer the therapeutic advantages of natural polymers. Ongoing research is focused on the development of hybrid biodegradable materials for specific applications using new resorbable biomaterials and applying computational and combinatorial approaches to develop biomimetic polymer structures with unique chemistries, thus increasing diversity [4–8]. Tsao and coworkers [9] used a polyelectrolyte complex consisting of chitosan and (-poly (glutamic acid) (-PGA) as a wound dressing material. This complex combination offered good mechanical properties, suitable moisture content, and favorable removal without the damage of regenerated tissue. To form a suitable wound dressing Kim and coworkers [10] used chitosan with poloxamer. A semi-interpenetrating polymer network provided enhanced compatibility and mechanical strength. Novel wound dressing applications have been devised by novel bioprocesses and advances in organic chemistry, thus enabling the development of enhanced smart polymers as candidates for specialized wound dressings that elicit favorable biological, physical, and chemical responses. Hydrocolloid dressings, the most widely used dressings, are obtained from colloidal materials that are gel-forming agents combined with other materials such as gelatin, elastomers, carboxymethylcellulose, pectin and adhesives. These agents can be bonded together to produce a thin film, sheet, or foam with the properties of hydrocolloids, thus forming a gel on the wound surface to promote moist wound healing Recent advances have been made in wound healing and dermal substitution. The main component of every wound is the connective tissue matrix, thus there is an overall consensus that in order to heal wounds effectively, it is necessary to ensure the effective substitution of the main component. Wound dressings are an essential part of wound management and care in order to enhance the natural wound healing process. The development of new intelligent dressings is under way, and these promise to play an active role in promoting healing of both acute and chronic wounds [11].

Other important nano-biopolymer applications are drug delivery devices and tissue engineering. Innovative delivery strategies based on nano and micro-particulate systems are currently being investigated for pulmonary delivery to improve drug transport to its target [12], and to develop a compatible small-diameter engineered tissue as a scaffold for vascular graft. This requires a compliant polymer scaffold to which endothelial cells can adhere, form an anti-thrombogenic luminal surface, exhibit vasoactive properties, and improve patency, and within which smooth muscle cells can migrate, deposit functional vascular ECM, and become contractile [13]. When choosing a material for surgical or pharmacological devices, it must meet specific requirements, and of utmost importance is the biocompatibility and biodegradability. The aim of this chapter is to provide an overview of the biological activity of innovative nanoparticles and non–woven tissue.

2. Bionanotechnology

Recent improvements have been made in the technique to produce biopolymers in the form of nanosized fibers [13], biodegradable polymeric nanospheres, nanorods, and nanotubes. Several important characteristics can be seen by reducing the diameters of polymer fiber materials from micrometers to submicron or nanometers, such as a very large surface area to volume ratio; the ratio of nanofiber in comparison with that of a microfiber may be as much as 10^3 greater. This also provides greater mechanical support in the functionalities, such as tensile strength and stiffness, and improved flexibility when compared to any other known form of the same material [14,15].

Electrospinning, a spinning technique, is a widely used technology for electrostatic fiber formation, which utilizes electrical forces to produce polymer fibers with diameters ranging from 2 nm to several micrometers using polymer solutions of both natural and synthetic polymers. This is a unique approach to produce fine fibers from polymer solutions or melts, which have a larger surface area and a thinner diameter (from nanometer to micrometer) than those obtained from conventional spinning processes. There are two standard electrospinning setups, vertical and horizontal [14,16]. With the expansion of this technology, more sophisticated systems that can fabricate more complex nanofibrous structures in a more controlled and efficient manner have been developed. Electrospinning is conducted at room temperature under atmospheric conditions. An electrospinning system has three components: a high voltage power supply, a spinneret (e.g., a pipette tip), and a collecting plate (usually a metal screen, plate, or rotating mandrel). It utilizes a high voltage source to inject a charge of a certain polarity into a polymer solution or melt, which is then accelerated towards a collector of opposite polarity. Most polymers are dissolved in solvents before electrospinning, and when they completely dissolve, they form a polymer solution. The polymer fluid is then inserted into the capillary tube for electrospinning. In the electrospinning process, a polymer solution held by its surface tension at the end of a capillary tube is subjected to an electric field, thus inducing an electric charge on the liquid surface. When the electric field applied reaches a critical value, the repulsive electrical forces overcome the surface tension forces. A charged jet of the solution is then ejected from the tip of the Taylor cone and an unstable and rapid whipping of the jet occurs in the space between the capillary tip and collector, which leads to the evaporation of the solvent, leaving a polymer behind. The jet is only stable at the tip of the spinneret and after that instability starts [17–19] (Figures 3 and 4). A wide range of polymers that are able to form fine nanofibers within the submicron range suitable for varied applications are used in electrospinning. Electrospun nanofibers have been produced from natural polymers, synthetic polymers, or a blend of both, including nucleic acids, proteins, and even polysaccharides. More than 200 polymers have been electrospun successfully from natural polymers and characterized according to their applications. When used in biomedical applications, naturally occurring polymers normally exhibit good biocompatibility and low immunogenicity when compared to synthetic polymers. An advantage of natural polymers for electrospinning is their capacity to bind to cells since they carry specific protein sequences, such as RGD, arginine/glycine/aspartic acid sequences [17]. Nanoparticles can be defined as particulate matter having at least one dimension that is less than 100nm and an exceptionally high surface to volume ratio, which contributes to their unusual properties and behavior. Furthermore, because of their high surface area, the surface structure also differs from that of the core. Specific and suitable functional groups can be attached to the surface of a nanoparticle to allow it to reach its target and interact with the biological system [11].

Polymeric nanoparticles (NP) are a promising resource for drug delivery. These carriers must be designed for a specific purpose since the material used affects the distribution inside of the body and the uptake into certain cells. The NP size affects the ability to penetrate barriers in the body. It is important to select an appropriate approach for the specific drug (hydrophobic vs. hydrophilic) and delivery route. Drug incorporation by adsorption or entrapment can alter the size and physicochemical properties that determine the NP interactions [20,21].

Different methods are used for gelatin NP preparations: the desolvation method and the water-in-oil emulsification method. The protocols for chitosan NPs include the ionotropic gelation method and another method that includes complex coacervation, emulsion evaporation, nanoprecipitation, and radical polymerization [22]. For synthetic polymer preparations, such as PLGA nanoparticles, which are made from a copolymer of polylactic acid and polyglycolic acid, the salting out method or nanoprecipitation, also known as solvent diffusion or solvent displacement, is used. The emulsification methods include emulsion diffusion, a top-down technique that starts by dissolving the pre-formed polymer in an organic solvent that is partially miscible with water (e.g., ethylacetate, dichloromethane, or acetone/methanol) and emulsion evaporation, another top-down method used for the formation of PLGA NPs. The emulsion dispersion polymerization method is used to prepare polyalkyl(cyano)acrylate (PCA) nanoparticles [23].

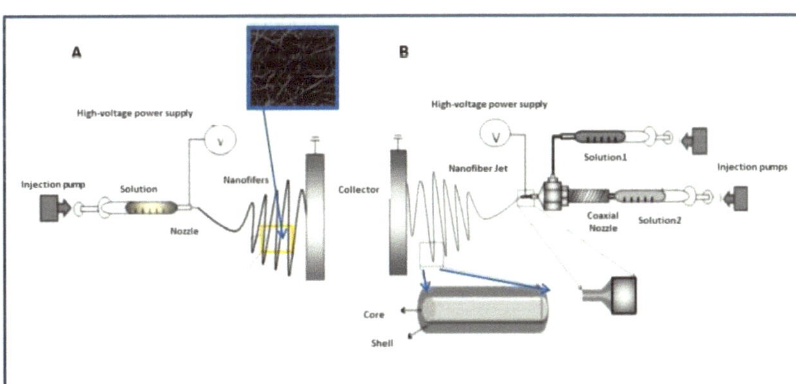

Figure 3. Schematic diagram of electrospinning setup consisting of injection pump/s, power supply, nozzle and conducting collector. A: One type of solution. B: Two type of solution and a core-shell nozzle design used to encapsulate drugs within the nanofiber. Adapted from Shin, S.H. et al. [24].

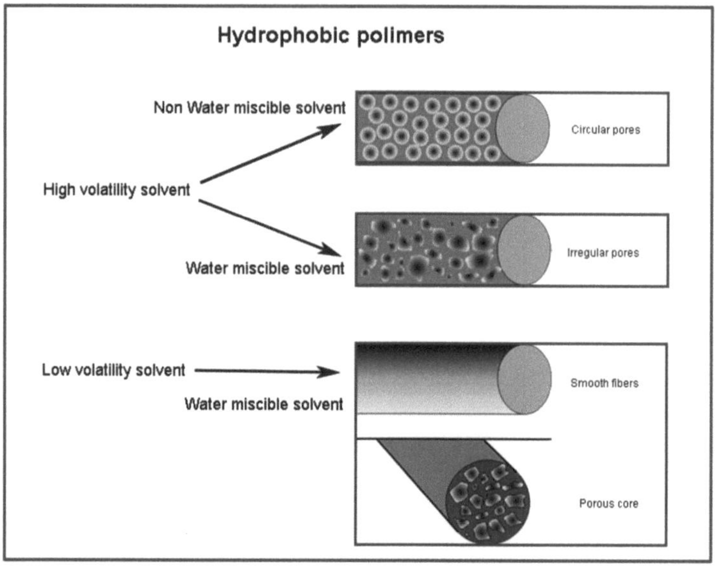

Figure 4. Fibers for hydrophobic polymers, electrospun at high percent RH, obtained with different solvent, show different morphology. Adapted from Nezarati, R.M. et al. [19].

3. Biological Activity

Like any biomaterial, a functional polymeric system aimed at serving for a limited period of time before degradation and elimination from the body must first fulfill strict criteria related to biocompatibility and biofunctionality. Therapeutic devices made from biomaterials, especially polymers, must permit in vivo: no or acceptable toxicity, no immune reaction, no carcinogenic mutation and no thrombus formation to meet the strict criteria of biocompatibility. Regarding biofunctionality, these devices must offer adequate mechanical, chemical, thermal, physical, and biological properties, and must be sterile and easy to handle [23]. Increasing attention to the environment has brought about an in-depth study of the biological activity of natural polymers or artificial polymers that respect the characteristics of biodegradability and biocompatibility, such as PLA or PLGA. Besides the application of chitin as a starting material for the synthesis of chitosan and chito-oligosaccharides, chitin itself has been a center of many therapeutic applications, and is thought to be a promising biomaterial for tissue engineering and stem-cell technologies. In 2013, Bae and coworkers [25] demonstrated that the oral administration of chitin (α and β forms) is beneficial in preventing food allergies; the oral administration of chitin was accomplished by milling it to particle size less than 20 m and mixing it with the feed. Their results showed that the α-form reduced the serum levels of peanut-specific IgE and both the forms decreased the levels of interleukin IL-5 and IL-10, and increased the levels of IL-12. Dietary supplementation of chitin was shown to exert positive immunomodulatory effects; the antibacterial activity of chitin, prepared from shrimp-shell waste, was reported by Benhabiles et al. [26]. Chitosan, a natural nontoxic biopolymer that is produced by the deacetylation of chitin, has been noted for its application as a film-forming agent in cosmetics, a dye binder for textiles, a strengthening additive in paper, and a hypolipidic material in diets. It has been used extensively as a biomaterial owing to its immune-stimulatory activities, anticoagulant properties, antibacterial and antifungal action, and its action as a promoter of wound healing in surgery. The antimicrobial activity of chitosan has been demonstrated against many bacteria, filamentous fungi, and yeasts. Chitosan has a wide spectrum of activity and high killing rate against Gram-positive and Gram-negative bacteria but a lower toxicity toward mammalian cells. Due to the presence of hydroxyl, amine, and acetylated amine groups, chitosan, low molecular weight chitosan, and chito-oligosaccharides (COS) interact with various cell receptors that trigger a cascade of interconnected reactions in organisms, which results in anti-inflammatory, anticancerogenic, antidiabetic, antimicrobial, anti-HIV-1, antioxidant, antiangiogenic, neuroprotective, and immunostimulative effects [27]. Both in the form of nanoparticles or nanofibers these biopolymers play an important role in helping the human body in its outstanding ability to self-repair.

3.1. Nanoparticles

Nanoparticles can be defined as ultradispersed solid supramolecular structures with a submicrometer size ranging from 10 to 1000 μm, used especially to produce drug delivery devices (Figure 5). The nanoparticle matrix with the drugs (dissolved, entrapped, encapsulated, or attached) acts as a reservoir for particulate systems and plays an important role as a drug delivery device, e.g., in oncology [28]. Nanoparticles fabricated from polysaccharides, proteins and biocompatible/biodegradable polymers, such as polyethylene glycol (PEG), poly(γ-benzyl l-glutamate) (PBLG), poly(D,L-lactide), poly(lactic acid), poly(D,L-glycolide), poly(lactide-co-glycolide), polycyanoacrylate, chitosan, gelatin, and sodium alginate are called PNPs [27]. The various materials used in NP preparation show different drug release kinetics, which can be achieved through desorption of surface-bound drugs, diffusion through the NP polymer wall, or erosion of the NP matrix. Drugs can be released with a burst mechanism or in a sustained manner. Synthetic polymers can be more easily engineered to produce a sustained release than natural polymers. In addition to the choice of polymer, the loading method has a strong impact on the release rate. However, the choice of the loading method depends on the drug, as certain drugs are more effective when released in a large burst, but controlled release is better for long-term effects; the size of the particle is equally important. The drug encapsulated will have an impact on the nanoparticle size, but it is difficult to predict which method of encapsulation will impact the size. This parameter is crucial to determine which organs the particle may reach and whether or not it will be opsonized by macrophages. Once NPs are in the circulation they can pass, according to their size, through fenestrations in the endothelial barrier. Table 1 shows the materials commonly used for NP preparation.

Chitosan has been the impetus for the development of safe and effective drug delivery devices. Its primary hydroxyl and amine groups that are located on the backbone permit chemical modifications to control its physical properties. The interaction of the hydrophobic moiety with a chitosan molecule determines an amphiphile that can form self-assembled nanoparticles that are able to encapsulate a quantity of drugs and deliver them to a specific site of action. Chemical attachment of the drug to the chitosan throughout the functional linker may produce useful pro-drugs, exhibiting the appropriate biological activity at the target site [29].

Synthetic polymers, when compared to natural polymers, are generally more homogenous in composition and have a higher purity, thus making the preparation of NPs more reproducible. However, it should be noted that not all of the synthetic polymers are suitable for drug delivery as they need to be biodegradable and exert low cytotoxicity. NPs made from synthetic polymers are polyesters, which include poly(lactic acid), poly(glycolic acid), their co-polymer: poly(lactide-co-glycolic acid) (PLGA), as well as polyalkyl(cyano)acrylates. These are not new and have been extensively used in clinical settings, e.g., PLA has been used for surgical sutures and implants, while PCAs have been studied for their use in sealing wounds. Knowledge of the degradation kinetics of such polymers allows for the preparation of better formulations for a more controlled release. Because of the large amount of materials used for NP preparations (Table 1) and the extensive field of application as drug delivery devices, it is not possible to provide a comprehensive overview. We have limited our intervention to some recent studies that investigated the use and biological activity of different NPs.

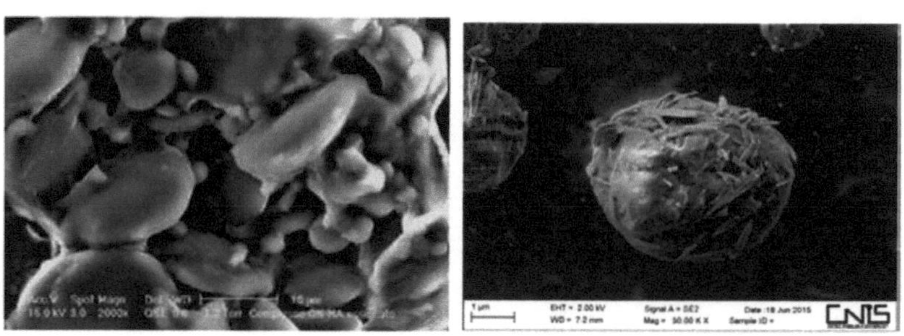

Figure 5. Micro/nanoparticles of the complex chitin nanoparticle-nanolignin. From Morganti, P. et al. [4].

Table 1. List of organic materials used to formulate nanoparticle drug carriers.

Polymeric Based Nanoparticles	Non-Polymeric Based Nanoparticles
Natural sources	
Gelatin Chitosan Alginate Nano-crystalline cellulose	Carbon-based carriers Liposomes Solid lipid nanoparticles
Synthetic sources	
Poly(lactic-co-glycolic)acid Poly-n-(cyanoacrylate) Polycaprolactone	

Lai [22], Dinavard [30], Nitta [31], and coworkers offer a good overview of the preparation and ability of polymeric nanoparticles as a drug-delivery system for gene deliver, anticancer drug delivery, and antibiotics delivery. Recently, in the treatment of pulmonary infections, the use of antibiotics for inhalation has gained increasing attention, particularly for cystic fibrosis (CF) patients. Aerosolized antibiotics offer an interesting way to deliver high drug concentrations directly to the site of infection, which reduces the toxicity and enhances the therapeutic potential of the antimicrobial agents against resistant microorganisms.

In 2009, Tahara et al. [32] demonstrated in vitro that chitosan-modified PLGA NSs (CS-PLGA NSs) are preferentially taken up by human lung adenocarcinoma cells (A549). Cellular uptake of PLGA NS was confirmed using fluorescence spectrophotometry and was visualized in A549 cells with confocal laser scanning microscopy (CLSM). The cytotoxicity of non-and CS-PLGA NS systems were compared in vitro using a 3-(4,5-dimethylthiazol-2-yl)-5-(3-carboxymethoxyphenyl)-2-(4-sulfophenyl)-2Htetrazolium (MTS) assay; CS-PLGA NSs did not show cytotoxicity to A549 cells. The cellular uptake of non- and CS-PLGA NSs is a time-, temperature-, and concentration-dependent saturable event mediated by clathrin-coated pit endocytic pathways, and that of PLGA NSs is related to a particle diameter. CS modification through electrostatic interactions between the CS adsorbed to the NS surface and the negatively charged cell membrane improved the cellular uptake of PLGA NSs. PLGA NS and CS-PLGA NS (200-nm) were internalized by A549 cells through adsorptive endocytosis started by nonspecific interactions between NS and cell membranes, triggered partially by a clathrin-mediated process. NSs have been evaluated for the delivery of different types of therapeutic agents, such as DNA, peptides, and proteins. Moreover, NSs protect the encapsulated drugs from enzymatic degradation and therefore provide a sustained release. In conclusion,

CS-PLGA NSs are better drug carriers because of a high cellular uptake due to their strong interaction with the cells and a low cytotoxicity.

Ungaro and coworkers [33] designed and developed a pulmonary delivery system for antibiotics based on spray-dried lactose/PLGA nano-embedded microparticles (NEM), engineered at the nanosize and microsize levels. To test the use of NEM, Tobramycin (Tb), the first choice antibiotic in lung infections for CF treatment, has been chosen. The results showed that to improve the size and modulate the surface properties of Tb-loaded PLGA NPs, poly(vinyl alcohol) (PVA), and chitosan are essential, and the use of alginate (Alg) allows an efficient Tb entrapment within NPs and its release up to one month. The optimized formulations of NP spray-drying with lactose offered NEM with particular flow and aerosolization properties without altering the NP features. The Tb-loaded PLGA NPs showed good in vitro antimicrobial activity against *P. aeruginosa* planktonic cells. In addition, the bio-distribution studies showed that PVA-modified Alg/PLGA NPs reached the deep lung, while CS-modified NPs remained in the upper airways lining lung epithelial surfaces. In conclusion, the composition of PLGA NP plays a crucial role in determining the technological features of NPs, and when NPs are processed in the form of NEMs, their in vitro/in-vivo deposition pattern is also modified

Recently Piras et al. [34] evaluated as an antimicrobial protein drug model, chitosan in a new formulation of nanoparticles loaded with lysozyme (LZ). LZ-loaded nanoparticles (LZ-NPs) of 150 nm diameter were prepared by inotropic gelation. They demonstrated that, these nanoparticles preserved the antibacterial activity of the loaded enzyme, which was slowly released over three weeks in vitro and remained active against *Staphylococcus epidermidis*, up to five days of incubation. So, beyond the intrinsic antibacterial activity of CS and LZ, the LZ-NPs showed a sustained antibacterial activity that resulted in about a 2-log decrease in the number of viable *S. epidermidis* compared to plain CS nanoparticles, and showed a full in vitro cytocompatibility towards murine fibroblasts.

It is well known that the development of small interfering RNA (siRNA)-controlled-release NPs may improve the therapeutic efficacy of RNA interference (RNAi) by prolonging their release to allow long-term gene silencing. In 2014 Shi et al. [35], proposed an NP platform with sustained siRNA-release properties, which can be self-assembled using biodegradable and biocompatible polymers and lipids, with an excellent silencing efficacy. The temporal release of siRNA from the NPs continued for over one month. When tested in vitro on luciferase-expressed HeLa cells and A549 cells after short-term transfection, the siRNA NPs showed greater sustained silencing activity than lipofectamine 2000-siRNA complexes. More importantly, the NP-mediated sustained silencing of prohibitin 1 (PHB1) generates more effective tumor cell growth inhibition in-vitro and in vivo than the lipofectamine complexes.

Morganti and coworkers [36] are working on the formulation of complex chitin–hyaluronan nanoparticles as a multifunctional carrier to deliver anti-aging active ingredients through the skin. They are evaluating in vitro its antioxidant capacity, anti-collagenase activity, and metalloproteinase, and anti-inflammatory mediator release. These copolymeric nanoparticles are able to entrap different kinds of active ingredients and release them at different times, depending on the productive process adopted and fine size of the micro/macro particles designed [37,38]. It is interesting to underscore how these nanoparticles, based on the use of chitin nanofibrils and obtained from crustacean waste, support the industrial sustainability [39] and respect the indications of the in-progress green economy [40].

3.2. Non-Woven Tissue

The current direction in the research area of non-woven tissue is to create a scaffold that mimics the structure and function of the native extracellular matrix (ECM) (Figure 6). The best scaffold for clinical use is one that has both structural integrity and allows for normal cellular function and interaction [41]. There is growing evidence that nanofibers amplify certain biological responses, such as cellular contact guidance and differentiation [42]. According to the medical application, i.e., wound dressing, small-diameter vascular graft, several parameters must be analyzed, such as: material selection, scaffold design, porosity, mechanical properties, fiber morphology, and cytocompatibility. Chitosan (CS) is a natural chitin-derived polysaccharide that is extensively used as a biomaterial in different engineering applications due to its low cost, large-scale availability, antimicrobial activity, biodegradability, and biocompatibility [43–46]. Besides nanoparticles, many authors have focused their studies to evaluate in vitro the biocompatibility of new co-polymer based on chitosan.

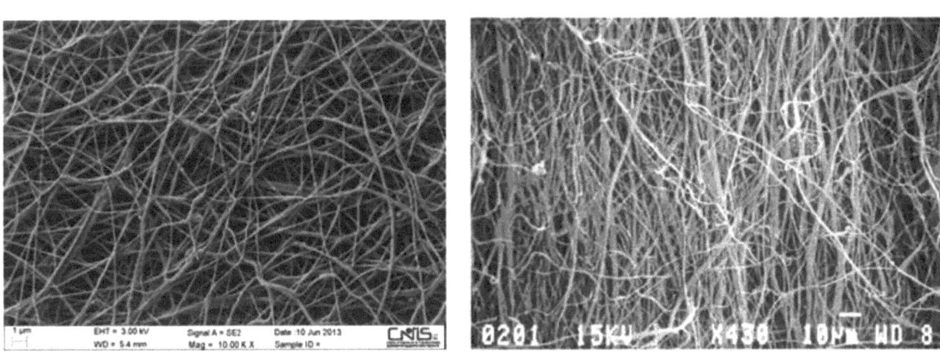

Figure 6. Similarity between the structure of Chitin Nanofibril scaffold (left) and skin ECM (right). From Morganti, P. et al. [4].

In 2013 Wang et al. [13], evaluated the biocompatibility of an electrospun chitosan/collagen complex as a scaffold in vascular tissue engineering to support the cell adhesion, proliferation, and phenotypic expression of endothelial cell markers by PIECs in-vitro. The chitosan/collagen scaffold showed endothelial cell adhesion, and did not adversely affect cellular function. In addition, the chitosan/collagen scaffold (*w*/*w*; 50/50, 20/80), in particular, showed the highest potential for vascular tissue engineering.

Naseri and coworkers [47], developed electrospun chitosan/polyethylene oxide-based randomly oriented fiber mats, reinforced with 50 wt % chitin nanocrystals (ChNC) for wound dressing. The results showed that the electrospun porous random mats comprising ChNC were free from any defects because of a homogeneous dispersion of ChNC in the chitosan matrix, indicating good chemical compatibility between the matrix and the chitin. The addition of chitin nano-crystals improved the moisture stability of these mats and facilitated water-mediated cross-linking processes. The cross-linked nanocomposite fiber mats with 50 wt % chitin nano-crystals had a high surface area (35 m^2 g^{-1}), a high tensile strength of 64.9 MPa and modulus of 10.2 GPa, and were at the same time flexible, and, therefore, were considered as beneficial for wound healing. The water vapor transmission rate of these mats was between 1290 and 1548 g m^{-2} day^{-1}, and was in the range for injured skin or wounds. The compatibility towards adipose-derived stem cells of the electrospun fiber mats confirmed their potential use as wound dressing materials [32]. Many other associations have been tested in the last few years. Enrione et al. [48] designed a gelatin/chitosan/hyaluronic acid biopolymer using a thermophysical approach for use in tissue engineering. The gelatin/chitosan/hyaluronic acid biopolymeric scaffold was made by applying a modification of the method described by Liu et al. [49]. Its thermal characterization was performed using differential scanning calorimetry (DSC), and its physical characterization by gas pycnometry and scanning electron microscopy. The effects of the gelatin (Ge) content and cross-linking on the thermophysical properties were evaluated by means of a factorial experiment design (central composite face centered). The Ge content was the main factor that affected the thermophysical properties (microstructure and thermal transitions) of the scaffold for tissue engineering, which were studied by seeding skin cells on the biopolymers. Different amounts of Ge did not affect cell attachment, while the cell growth rate increased linearly with the decrease in the Ge content. This relationship, together with the thermophysical characterization, can be used to design scaffolds for tissue engineering. Finally, the authors concluded that a Ge stock solution of 0.8% was adequate to formulate Ge/Ch/Ha-scaffolds to seed fibroblasts.

Different types of chitin- and chitosan-based wound dressing materials are commercially available. Chitin and chitosan in the form of composites, gels, nanofibers, films, non-wovens, and scaffolds have been used to regenerate wounded tissues. Nowadays, commercial products are available for topical application in wound repair based on chitin nanofiber polymers. Our research group [50] tested in vitro a novel combination of chitin nanofiber/lignin in different ratios for their anti-inflammatory and wound repair activity in experimental models of human keratinocytes. For the evaluation of the anti-inflammatory effect of chitin /lignin nanofibers, the IL-8, IL-1α and TNF-α expressions were analyzed on human keratinocytes treated with lipopolysaccharide of *P. aeruginosa* (LPS). Also evaluated were the expression of the beta-defensin 2 (hBD-2) and metalloproteinases 2 and 9 (MMP-2 and -9), which are well known to be involved in the mechanisms of tissue regeneration. The polymer tested significantly reduced the pro-inflammatory cytokines that were induced by LPS in human keratinocytes and modulated the expression of MMPs and hBD-2. This suggests that the association may improve the ability of chitosan/chitin polymers in their scaffold function.

Bacterial cellulose also seems to be a promising material for the construction of polymers in a nano, micro, and macro scale to use in wound dressing. It is composed of a pure cellulose nanofiber mesh that is spun by bacteria. Its high water content contributes to its biocompatibility [51]. Recently, Harkins et al. [52], tested a novel composite containing chitosan and cellulose for its antimicrobial activity, absorption of anticoagulated whole blood, and anti-inflammatory activity through the reduction of tumor necrosis TNF-α and IL-6, and the biocompatibility with human fibroblasts. The composites tested inhibited the growth of both Gram-positive and -negative microorganisms, such as *Escherichia coli* (ATCC 8739), methicillin-resistant *Staphylococcus* (ATCC 33591), and vancomycin-resistant *Enterococcus faecalis* (ATCC 51299) by 78, 36, and 64%, respectively. In addition, they showed no toxicity *vs.* fibroblasts responsible for the formation of the connective tissue matrix. The composites proved to be a good absorbent for anticoagulated whole blood and were able to maintain a moisture balance for wound healing. For successful tissue repair, several factors, including blood clotting and cellular survival are necessary. Therefore, the dressing material should possess anti-inflammatory activity, since proinflammatory cytokines (TNF-α and IL-6) contribute to the inflammation in chronic wounds, which stalls and prevents them from proceeding to the proliferative phase of tissue regeneration. The significant reduction in TNF-α and IL-6 by stimulated macrophages obtained with the Cel + CS composites clearly indicates their biodegradability, biocompatibility, and non-toxicity.

Advances in medicine have led to a significant increase in life expectancy but significant advances have also been made in cosmetics. The increasing proportion of women and men interested in skin rejuvenation has created a rapidly growing demand for anti-aging remedies to rejuvenate photo-damaged skin. Wrinkling, slackening, and irregular pigmentation, and symptoms of age-associated skin damage, are in fact influenced by environmental factors, particularly lifetime sun exposure. In the last few years, different techniques for rejuvenation, such as injections with fillers and bio stimulating agents for wrinkle treatment, correction of scars, and soft-tissue augmentation, have been proposed. To obtain beauty and wellness inside and outside, there has been an increased demand from plastic surgeons for new effective medical devices and procedures, and an increased use of cosmeceuticals and nutricosmetics. Morganti and coworkers [53] developed and studied a new medical device to treat facial lines and body contours by balancing the skin-cell turnover and metabolism. To this purpose, block-polymer nanoparticles (BPN) of linoleic acid–rich phosphatidylcholine nanocomplexed with hyaluronan and chitin nanofibrils (PHHYCN) were formulated by encapsulating into them cholesterol, creatine, caffeine, melatonin, vitamins E and C, and the amino acids glycine and arginine. The BPN quickly re-establish the skin-barrier function thanks to their high content in linoleic acid and phosphatidylcholine. The phosphatidylcholine-fatty acids contained inside the BPN contribute to balancing the disturbed composition and organization of the lipids at the level of the epidermal keratinocytes, and consequently of corneocyte lamellae, while the high content of linoleic acid contributes to reintegrating the reduced level of ceramide 1, a structural and stabilizing component of the stratum corneum. These BPN seem to be useful in improving the activity of permanent fillers, making them useful as an anti-aging remedy in plastic surgery. This innovative biostimulating medical device may be used for wrinkle treatment and skin rejuvenation, as well as an adjuvant in soft-tissue augmentation and stretch-mark corrections.

4. Conclusions

Developments in the field of nanotechnology and the increasing interest for the environment have promoted the formulation of new nano-polymeric materials based on natural raw materials, and/or biodegradable and biocompatible synthetic polymers to use as innovative strategies in the fields of medicine, pharmacology, agriculture, and cosmetology [54–56]. Thus, the increased use of natural polysaccharides, such as chitin derivatives and lignocellulosic polymers, can reduce greenhouse emissions and improve our way of living.

Author Contributions: Giovanna Donnarumma and Maria Antonietta Tufano wrote the article; Brunella Perfetto, Adone Baroni and Iole Paoletti were responsible for the preparation of the corresponding figures and supervising the writing of the manuscript; Paola Del Ciotto and Pierfrancesco Morganti provided the description of chitin nanofibrills properties.

Conflicts of Interest: The authors declare no conflict of interest.

References

1. Sajid, M.; Ilya, M.; Basheer, C.; Tariq, M.; Daud, M.; Baig, N.; Shehzad, F. Impact of nanoparticles on human and environment: Review of toxicity factors, exposures, control strategies, and future prospects. *Environ. Sci. Pollut. Res.* **2015**, *22*, 4122–4143. [CrossRef] [PubMed]
2. Baiguera, S.; Urbani, L.; Del Gaudio, C. Tissue engineered scaffolds for an effective healing and regeneration: Reviewing orthotopic studies. *BioMed. Res. Int.* **2014**, *2014*, 398069. [CrossRef] [PubMed]
3. Naderi, H.; Matin, M.M.; Bahrami, A.R. Review paper: Critical in tissue engineering: Biomaterials, cell sources, angiogenesis, and drug delivery system. *J. Biomater. Appl.* **2011**, *26*, 383–417. [CrossRef] [PubMed]
4. Morganti, P.; Febo, F.; Cardillo, M.; Donnarumma, G.; Baroni, A. Chitin nanofibril and nanolignin: Natural polymers of biomedical interest. *J. Clin. Cosmet. Dermatol.* **2017**, *1*, 1–7.
5. Klapiszewski, L.; Wysokowski, M.; Majchrzac, I.; Szatkowski, T.; Nowacka, M.; Siwinca-Stefanka, K.; Szwarc-Rezepca, K.; Bartczak, P.; Ehrlich, H.; Jesionowski, T. Preparation and characterization of multifunctional chitin/lignin materials. *J. Nanomater.* **2013**, *2013*, 425726. [CrossRef]
6. Wysokowski, M.; Petrenko, I.; Stelling, A.L.; Stawski, D.; Jesionowski, T.; Ehrlich, H. Poriferan Chitin as a Versatile Template for Extreme Biomimetics. *Polymers* **2015**, *7*, 235–265. [CrossRef]
7. Wolf, M.T.; Dearth, C.L.; Sonnenberg, S.B.; Loboa, E.G.; Badylak, S.F. Naturally derived and synthetic scaffolds for skeletal muscle reconstruction. *Adv. Drug Deliv. Rev.* **2015**, *84*, 208–221. [CrossRef] [PubMed]
8. Dragojevic, S.; Su Ryu, J.; Raucher, D. Polymer-Based Prodrugs: Improving tumor targeting and the solubility of small molecule drugs in cancer therapy. *Molecules* **2015**, *20*, 21750–21769. [CrossRef] [PubMed]
9. Tsao, C.T.; Chang, C.H.; Lin, Y.Y.; Wu, M.F.; Wang, J.L.; Han, J.L.; Hsieh, K.H. Antibacterial activity and biocompatibility of a chitosan-gamma-poly(glutamic acid) polyelectrolyte complex hydrogel. *Carbohydr. Res.* **2010**, *345*, 1774–1780. [CrossRef] [PubMed]
10. Kim, I.Y.; Yoo, M.K.; Seo, J.K.; Park, S.S.; Na, H.S.; Lee, H.C.; Kim, S.K.; Cho, C.S. Evaluation of semi-interpenetrating polymer networks composed of chitosan and poloxamer for wound dressing applications. *Int. J. Pharm.* **2007**, *341*, 35–43. [CrossRef] [PubMed]
11. Mayet, N.; Choonara, Y.E.; Kumar, P.; Tomar, L.K.; Tyagi, C.; Du Toit, L.C.; Pillay, V. Comprehensive review of advanced biopolymeric wound healing systems. *J. Pharm. Sci.* **2014**, *103*, 2211–2230. [CrossRef] [PubMed]
12. D'Angelo, I.; Conte, C.; La Rotonda, M.I.; Miro, A.; Quaglia, F.; Ungaro, F. Improving the efficacy of inhaled drugs in cystic fibrosis: Challenges and emerging drug delivery strategies. *Adv. Drug Deliv. Rev.* **2014**, *75*, 92–111. [CrossRef] [PubMed]
13. Wang, P.; Liu, J.; Zhang, T. In Vitro Biocompatibility of Electrospun Chitosan/Collagen Scaffold. *J. Nanomater.* **2013**, *2013*, 958172. [CrossRef]

14. Colantoni, A.; Boubaker, K. Electro-spun organic nanofibers elaboration process investigations using comparative analytical solutions. *Carbohydr. Polym.* **2014**, *101*, 307–312. [CrossRef] [PubMed]
15. Kriegel, C.; Arrechi, A.; Kit, K.; Mcclements, D.J.; Weiss, J. Fabrication, functionalization, and application of electrospun biopolymer nanofibers. *Crit. Rev. Food Sci. Nutr.* **2008**, *48*, 775–797. [CrossRef] [PubMed]
16. Agarwal, S.; Wendorff, J.H.; Greiner, A. Progress in the field of electrospinning for tissue engineering applications. *Adv. Mater.* **2009**, *21*, 3343–3351. [CrossRef] [PubMed]
17. Bhardwaj, N.; Kundu, S.C. Electrospinning: A fascinating fiber fabrication technique. *Biotechnol. Adv.* **2010**, *28*, 325–347. [CrossRef] [PubMed]
18. Shin, S.H.; Purevdorj, O.; Castano, O.; Planell, J.A.; Kim, H.W. A short review: Recent advances in electrospinning for bone tissue regeneration. *J. Tissue Eng.* **2012**, *3*, 1–11. [CrossRef] [PubMed]
19. Nezarati, R.M.; Eifert, M.B.; Cosgriff-Hernandez, E. Effects of humidity and solution viscosity on electrospun fiber morphology. *Tissue Eng. Part. C* **2013**, *19*, 810–817. [CrossRef] [PubMed]
20. Salatin, S.; Maleki Dizaj, S.; Yari Khosroushahi, A. Effect of the surfacemodification, size, and shape on cellular uptake of nanoparticles. *Cell Biol.* **2015**, *39*, 881–890.
21. Wu, J.; Kamaly, N.; Shi, J.; Zhao, L.; Xiao, Z.; Hollett, G.; John, R.; Ray, S.; Xu, X.; Zhang, X.; et al. Development of multinuclear polymeric nanoparticles as robust protein nanocarriers. *Angew. Chem. Int. Ed. Engl.* **2014**, *53*, 8975–8979. [CrossRef] [PubMed]
22. Lai, P.; Daear, W.; Löbenberg, R.; Prenner, E.J. Overview of the preparation of organic polymeric nanoparticles for drug delivery based on gelatine, chitosan, poly(D,L-lactide-co-glycolic acid) and polyalkylcyanoacrylate. *Colloids Surf. B Biointerfaces* **2014**, *118*, 154–163. [CrossRef] [PubMed]
23. Vert, M. Not any new functional polymer can be for medicine: What about artificial biopolymers? *Macromol. Biosci.* **2011**, *11*, 1653–1661. [CrossRef] [PubMed]
24. Shin, S.H.; Purevdorj, O.; Castano, O.; Planell, J.A.; Kim, H.W. A short review: Recent advances in electrospinning for bone tissue. *J. Tissue Eng.* **2012**, *3*, 1–11. [CrossRef] [PubMed]
25. Bae, M.J.; Shin, H.S.; Kim, E.K.; Kim, J.; Shon, D.H. Oral administration of chitin and chitosan prevents peanut- induced anaphylaxis in a murine food allergy model. *Int. J. Biol. Macromol.* **2013**, *61*, 164–168. [CrossRef] [PubMed]
26. Benhabiles, M.S.; Salah, R.; Lounici, H.; Drouiche, N.; Goosen, M.F.A.; Mameri, N. Antibacterial activity of chitin, chitosan and its oligomers prepared from shrimp shell waste. *Food Hydrocoll.* **2012**, *29*, 48–56. [CrossRef]
27. Lodhi, G.; Kim, Y.S.; Hwang, J.W.; Kim, S.K.; Jeon, Y.J.; Je, J.Y.; Ahn, C.B.; Moon, S.H.; Jeon, B.T.; Park, P.J. Chitooligosaccharide and Its Derivatives: Preparation and Biological Applications. *Biomed. Res. Int.* **2014**, *2014*, 654913. [CrossRef] [PubMed]
28. Dobrovolskaia, M.A.; Aggarwal, P.; Hall, J.B.; McNeil, E.S. Preclinical studies to understand nanoparticle interaction with the immune system and its potential effects on nanoparticle biodistribution. *Mol. Pharm.* **2008**, *5*, 487–495. [CrossRef] [PubMed]

29. Park, J.H.; Saravanakumar, G.; Kim, K.; Kwon, I.C. Targeted delivery of low molecular drugs using chitosan and its derivatives. *Adv. Drug Deliv. Rev.* **2010**, *62*, 28–41. [CrossRef] [PubMed]
30. Dinarvand, R.; Sepehri, N.; Manoochehri, S.; Rouhani, H.; Atyabi, F. Polylactide-co-glycolide nanoparticles for controlled delivery of anticancer agents. *Int. J. Nanomed.* **2011**, *6*, 877–895. [CrossRef] [PubMed]
31. Nitta, S.K.; Numata, K. Biopolymer-based nanoparticles for drug/gene delivery and tissue engineering. *Int. J. Mol. Sci.* **2013**, *14*, 1629–1654. [CrossRef] [PubMed]
32. Tahara, K.; Sakai, T.; Yamamoto, H.; Takeuchi, H.; Hirashima, N.; Kawashima, Y. Improved cellular uptake of chitosan-modified PLGA nanospheres by A549 cells. *Int. J. Pharm.* **2009**, *382*, 198–204. [CrossRef] [PubMed]
33. Ungaro, F.; D'Angelo, I.; Coletta, C.; d'Emmanuele di Villa Bianca, R.; Sorrentino, R.; Perfetto, B.; Tufano, M.A.; Miro, A.; La Rotonda, M.I.; Quaglia, F. Dry powders based on PLGA nanoparticles for pulmonary delivery of antibiotics: Modulation of encapsulation efficiency, release rate and lung deposition pattern by hydrophilic polymers. *J. Control. Release* **2012**, *157*, 149–159. [CrossRef] [PubMed]
34. Piras, A.M.; Maisetta, G.; Sandreschi, S.; Esin, S.; Gazzarri, M.; Batoni, G.; Chiellini, F. Preparation, physical-chemical and biological characterization of chitosan nanoparticles loaded with lysozyme. *Int. J. Biol. Macromol.* **2014**, *67*, 124–131. [CrossRef] [PubMed]
35. Shi, J.; Xu, Y.; Xu, X.; Zhu, X.; Pridgen, E.; Wu, J.; Votruba, A.R.; Swami, A.; Zetter, B.R.; Farokhzad, O.C. Hybrid lipid-polymer nanoparticles for sustained siRNA delivery and gene silencing. *Nanomedicine* **2014**, *10*, 897–900. [CrossRef] [PubMed]
36. Morganti, P.; Palombo, M.; Tishchenko, G.; Yudin, V.E.; Guarneri, F.; Cardillo, A.; Del Ciotto, P.; Carezzi, F.; Morganti, G.; Fabrizi, G. Chitin-Hyaluronan Nanoparticles to Deliver Anti aging Ingredients through the skin. *Cosmetics* **2014**, *1*, 140–158. [CrossRef]
37. Morganti, P.; Chen, H.D.; Gao, X.H. Nanoparticles of Chitin Nanofibrils-Hyaluronan Block Polymers Entrapping Lutein as UVA Protective Compound. In *Carotenoids: Food Sources Production and Health Benefits*; Yamaguchi, M., Ed.; Nova Science Publishers: New York, NY, USA, 2013; pp. 237–259.
38. Morganti, P.; Del Ciotto, P.; Fabrizi, G.; Guarneri, F.; Cardillo, A.; Palombo, M.; Morganti, G. Chitin Nanofibrils-Hyaluronic Nanoparticles Entrapping Lutein. Note I "Nanoparticles Characterization and Bioavailability". *SOFW-J.* **2013**, *139*, 12–23.
39. Morganti, P. Innovation, nanotechnology and Industrial Sustainability by the use of natural underutilized by-products: The EU support to SMEs. *J. Mol. Biochem.* **2013**, *2*, 137–142.
40. Morganti, P.; Morganti, G.; Morganti, A. Nanobiotecnologia e Bioeconomia. *ICF* **2014**, *5*, 26–30.
41. Ayres, C.E.; Jha, B.S.; Sell, S.A.; Bowling, G.L.; Simpson, D.G. Nanotechnology in the design of soft tissue scaffolds: Innovation in structure and function. *Wiley Interdiscip. Rev. Nanomed. Nanobiotechnol.* **2010**, *2*, 20–34. [CrossRef] [PubMed]

42. Nisbet, D.R.; Forsythe, J.S.; Shen, W.; Finkelstein, D.I.; Horne, M.K. Review paper: A review of the cellular response on electrospun nanofibers for tissue engineering. *J. Biomater. Appl.* **2009**, *24*, 7–29. [CrossRef] [PubMed]
43. Gomes, S.; Leonor, I.B.; Mano, J.F.; Reis, R.L.; Kaplan, D. Natural and genetically engineered proteins for tissue engineering. *Prog. Polym. Sci.* **2012**, *37*, 1–17. [CrossRef] [PubMed]
44. Pashneh-Tala, S.; MacNeil, S.; Claeyssens, F. The tissue-engineered vascular graft—Past, present, and future. *Tissue Eng. Part B.* **2016**, *22*, 68–76. [CrossRef] [PubMed]
45. Fu, W.; Liu, Z.; Feng, B.; Hu, R.; He, X.; Wang, H.; Yin, M.; Huang, H.; Zhang, H.; Wang, W. Electrospun gelatin/PCL and collagen/PLCL scaffolds for vascular tissue engineering. *Int. J. Nanomed.* **2014**, *9*, 2335–2344. [CrossRef] [PubMed]
46. Sridhar, R.; Sundarrajan, S.; Venugopal, J.R.; Ravichandran, R.; Ramakrishna, S. Electrospun inorganic and polymer composite nanofibers for biomedical application. *J. Biomater. Sci. Polym. Ed.* **2013**, *24*, 365–385. [CrossRef] [PubMed]
47. Naseri, N.; Algan, C.; Jacobs, V.; John, M.; Oksman, K.; Mathew, A.P. Electrospun chitosan-based nanocomposite mats reinforced with chitin nanocrystals for wound dressing. *Carbohydr. Polym.* **2014**, *109*, 7–15. [CrossRef] [PubMed]
48. Enrione, J.; Díaz-Calderón, P.; Weinstein-Oppenheimer, C.R.; Sánchez, E.; Fuentes, M.A.; Brown, D.I.; Herrera, H.; Acevedo, C.A. Designing a gelatin/chitosan/hyaluronic acid biopolymer using a thermophysical approach for use in tissue engineering. *Bioprocess. Biosyst. Eng.* **2013**, *36*, 1947–1956. [CrossRef] [PubMed]
49. Liu, H.; Mao, J.; Yao, K.; Yang, G.; Cui, L.; Cao, Y. A study on a chitosan-gelatin-hyaluronic acid scaffold as artificial skin in vitro and its tissue engineering applications. *J. Biomater. Sci. Polym. Ed.* **2004**, *15*, 25–40. [CrossRef] [PubMed]
50. Morganti, P.; Fusco, A.; Paoletti, I.; Perfetto, B.; Del Ciotto, P.; Palombo, M.; Chianese, A.; Baroni, A.; Donnarumma, G. Anti-inflammatory, immunomodulatory and tissue repair activity on human keratinocytes by green innovative nanocomposites. *Materials* **2017**, *10*, 843. [CrossRef] [PubMed]
51. Peters, N.; Gatenholm, P. Bacterial cellulose-based materials and medical devices: Current state and perspectives. *Appl. Microbiol. Biotechnol.* **2011**, *9*, 1277–1286. [CrossRef] [PubMed]
52. Harkins, A.L.; Duri, S.; Kloth, L.C.; Tran, C.D. Chitosan-cellulose composite for wound dressing material. Part 2. Antimicrobial activity, blood absorption ability, and biocompatibility. *J. Biomed. Mater. Res. B Appl. Biomater.* **2014**, *102*, 1199–1206. [CrossRef] [PubMed]
53. Morganti, P.; Palombo, P.; Palombo, M.; Fabrizi, G.; Cardillo, A.; Svolacchia, F.; Guevara, L.; Mezzana, P. A phosphatidylcholine hyaluronic acid chitin–nanofibrils complex for a fast skin remodeling and a rejuvenating look. *Clin. Cosmet. Investig. Dermatol.* **2012**, *5*, 213–220. [CrossRef] [PubMed]
54. Kim, H.N.; Jiao, A.; Hwang, N.S. Nanotopography-guided tissue engineering and regenerative medicine. *Adv. Drug Deliv. Rev.* **2013**, *65*, 536–558. [CrossRef] [PubMed]

55. Eslicka, J.C.; Yeb, Q.; Parkb, J.; Toppc, E.M.; Spencerd, P.; Camardaa, K.V. A computational molecular design framework for crosslinked polymer networks. *Comput. Chem. Eng.* **2009**, *33*, 954–963. [CrossRef] [PubMed]
56. Khorshidi, S.; Solouk, A.; Mirzadeh, H.; Mazinani, S.; Lagaron, J.M.; Sharifi, S.; Ramakrishna, S. A review of key challenges of electrospun scaffolds for tissue-engineering applications. *J. Tissue Eng. Regen. Med.* **2016**, *10*, 715–738. [CrossRef] [PubMed]

Clinical Activity of Innovative Non-Woven Tissues

Tommaso Anniboletti, Marco Palombo, Simone Moroni, Agostino Bruno, Paolo Palombo and Pierfrancesco Morganti

Abstract: Tissue engineering and the study of engineered materials are ever-expanding disciplines for the development of technological applications, focused on the reconstruction of mammalian tissue. In this study, we explained the surgical problems connected with the wound healing of burned skin, reporting the solutions recovered by the use on non-woven tissues (that we called MAVI dressing) made prevalently by chitin nanofibril (CN)-binding Ag^+ ions. Chitin-based materials and their derivatives, in fact, are receiving increased attention in tissue engineering because of their unique and appealing biological properties, useful to support the skin anatomical structure and its physiological functions. To this purpose, the in vivo results obtained on 30 patients will be shown by photographic images. The biocompatibility and the histological immune responses are shown by these advanced medications connected with the modulating effectiveness of the cytokine cascade. The obtained results recovered in vitro on a culture of keratinocytes and fibroblasts, were confirmed by the quick regenerative activity shown on people affected by burns of the first and second grade. In conclusion, these medications have shown that, while CN seems to be a natural polymer of choice for rapidly regenerating the burned skin, Ag^+ ions, bonded to its fiber structure in very low concentration, possess a sufficient antibacterial effectiveness to control the skin microbiological growth without showing side effects. Therefore, this hybrid biomaterial (chitin nanofibril/Ag nanoparticle composite) can be used for manufacturing advanced medications, solving both the problem of increasing its effectiveness for wound healing, and to slow down the bacterial growth connected with the wounded and burned skin. Moreover, the use of these advanced medications reduces the cost/h of plastic surgeons and the ancillary costs, shortening the time of the wound healing process.

Regenerative medicine has developed extremely rapidly during the last few years, so that the ideas, aspirations, and expectations of cell biologists, material scientists, engineering chemists, biochemists, and of course dermatologists and plastic surgeons have flourished [1–3]. Thus, naturally occurring nanostructures and biomaterials useful for the life science sectors have been a source of inspiration for new nanotechnological designs and to make innovative building blocks and produce products. The biological use of raw biomaterials for wound dressings, such as the natural chitin/chitosan polymers and nanocomposites from fishery waste and the lignocellulosic compounds such as lignin from plant biomass [4–6], has shown that these natural ingredients possess interesting cicatrizing and reparative activity on wounded skin and organs [7–9]. In addition these compounds, totally biodegraded from environmental chitinases and human chitotriosidases [10], have been shown to be non-toxic, skin-friendly and environmentally-friendly, being catabolized to molecules normally used from all the living cells. Among the chitin derivatives, chitin nanofibril (CN) represents the patented purest crystalline form of chitin [11]. This crystalline polymer has been used to make block copolymeric nanoparticles which, embedded into nanoemulsions and non-woven tissues, may release the entrapped ingredients at different skin layers and different times [12–14]. These nanoparticles may be regarded as interesting carrier for gels and emulsions, and important potential filler materials for the enhancement of the physical and mechanical properties of the polymer matrices used to produce the non-woven-tissue that we called Mavi dressing as the name of the factory that furnished the free sample-test for this study [15].

On the other hand, they are useful to entrap active ingredients and Ag-ions that, when embedded into CN electrospun fibers, assume structural similarities compared to the native extracellular matrix (ECM). The obtained Ag-non-woven tissue made of chitin nanofibers (CN) and chitosan (CS) has shown the capacity to reduce the burden bacteria of infected skin (Table 1), and to enhance the reparative capacity of the skin (Figure 1), showing interesting antibacterial activity together with wound healing effectiveness [16,17]. The consensus within wound therapy recommends, in fact, that modern wound dressings should preserve the skin humid environment, while creating a barrier against mechanical stress and secondary infections [18]. Moreover, on the one hand, advanced dressings have to absorb wound exudate and reduce potential microorganism growth, while on the other hand they have to be safe and a non-irritant. Other important properties of these non-woven tissues are their acceptability to the patient and cost per unit [19]. It must be remembered that wound infections are the most serious complications related to burn injuries that affect up 1% of the worldwide population each year. This is the reason that wound therapy represents one of the challenging areas in drug product development; in the USA more than 6.5 million patients are involved annually, with an estimated cost of treatment per year of US$25 billion [20].

While the use of biomaterials for the treatment of wounded and burned skin has revealed to a huge potential for skin repair [21–24], natural fibers, such as CN, form an interesting option for most widely-applied polymers in medical technology, because of their capacity to form scaffolds, mimicking the structure and organization of the extra-cellular matrix (ECM) [25,26].

Table 1. Antibacterial activity of CN-Ag non-woven tissue [27].

Sample	Bacterial Growth (CFU/g)
Agar + culture of bacteria from bioburden skin tissue (CB)	10^7
Agar + CB+ CS – CN nanocomposite film	10^5
Agar + CB + CS - CN – AG nanocomposite film	10^3

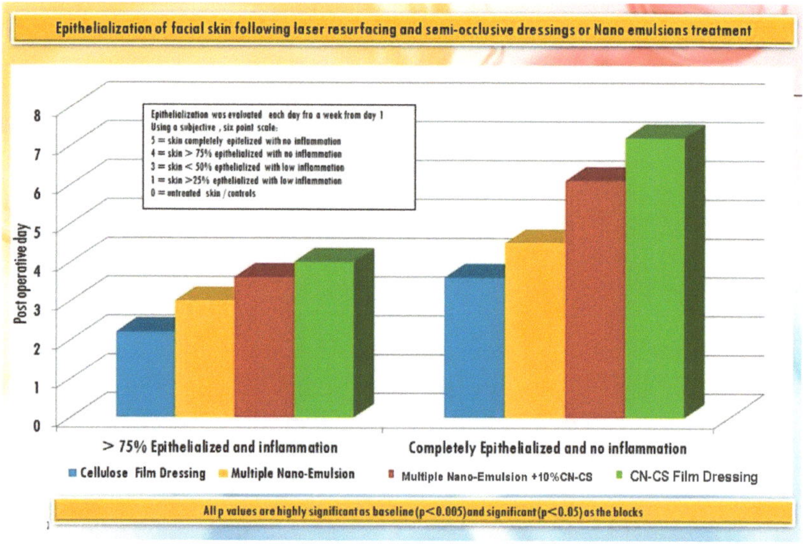

Figure 1. Effect of CN-CS dressing type on the epithelialization of facial skin following laser resurfacing. The epithelialization was evaluated daily during a week beginning from the 1st post-operative day. The appearance of the facial skin was as follows: >25% of the epithelialized skin with low inflammation (1st post-operative day), <50% of the epithelialized skin with low inflammation (3rd), 75% of the epithelialized skin with no inflammation (4th), completely epithelialized skin with no inflammation (5th) [27].

Assesment of antibacterial activity of Naonocomposite CN-CS and Cn-Cs-Ag films after 18 h incubation at 36 °C on agar Containing the Culture of Bacteria taken from Bioburden Skin Tissue. However, only recently the nature of the cellular environment, required for optimal tissue repair and regeneration, has been understood. A major consideration in tissue engineering is, in fact, the first pursuit of scaffolds that provide an architecture on which seeded cells are directed to proliferate and differentiate to form new tissues [27]. This tissue-engineering approach allows the production of new extra-cellular matrices which, resembling the native ECM, replace or regenerate the injured tissues. The creation of this engineered matrix that mimics the ECM requires a scaffold that, serving as a cell carrier, provides structural support until native tissue is again formed in vivo [28–30]. The pore size of a scaffolding system is, in fact, one of the more important structural design parameters, as it influences cellular infiltration, spreading, intercellular communication, and the transport of nutrients and metabolites [28–32]. On the other hand, the stiffness and mechanical properties of the used substrate, acting as a physical support for the transport of cell signals, influence cellular behavior such as adhesion, spreading, motility, survival, and differentiation.

Together with the physicochemical characteristics of these matrices, the concept of the slow-release of antiseptic agents has been accepted, which, delivering antimicrobial ingredients without being detrimental to the healing process, impedes the micro-organisms proliferation. Wound infections, in fact, continue to be an important complication of chronic wounds in terms of patient morbidity and medical resources. Based on these findings, bioactive, antiseptic ingredients and materials for in situ tissue regeneration play an important role in the control of the skin microenvironment. Over time, many products have been manufactured, particularly in the burns field. In superficial partial-thickness burns, some authors suggest the use of paraffin gauze because there is a low risk of infection. This medication tends to dry out with burn exudate, causing pain in the dressing change and impairment in movement. In deep partial-thickness burns, the incidence of infection is higher than in superficial partial-thickness burns, thus the use of medications that can prevent infection is advocated. The use of silver-containing dressings is therefore recommended. Chitin is a natural, high-molecular-weight linear polymer of β-(1,4) linked *N*-acetylglucosamine (*N*-acetyl-2-amino-2-deoxy-D-glucopyranose) units. Chitosan, a copolymer of glucosamine and *N*-acetyl glucosamine units linked by 1–4 glucosidic bonds, is a cationic polysaccharide obtained by a partial (~60%) alkaline deacetylation of chitin, Industrial chitin nanofibrils (CNs) [4–9] have been shown to be a 1:1 copolymer of N-acetyl glucosamine and glucosamine. However, the role of chitin and chitosan are amazing, attracting increasingly more attention due to their biological and physicochemical characteristics, so that different scientific papers and patents have been published [33,34].

These natural polymers are biocompatible, biodegradable, and nontoxic, showing anti-inflammatory, anti-microbial and hydrating actives, and, therefore, have good biocompatibility and positive effects on wound healing. Previous studies have shown that chitin-based dressings can accelerate the repair of different tissues, facilitating wound contraction and regulating the secretion of inflammatory mediators and innate immunity, depending on their dimensions [35]. Thus, mean-sized chitin has shown pro-inflammatory activity, while small-sized chitin has shown an anti-inflammatory function activating both TNF and IL-10 in macrophages. For this purpose, chitin nanofibrils have a mean dimension of 240 × 7 × 5 nm [36].

Chitosan and chitin provide a non-protein matrix for 3D tissue growth and activate macrophages stimulating cell proliferation and a specific hierarchical tissue organization [33,34]. Moreover, they also have a hemostatic activity, which helps in natural blood clotting and blocks nerve endings, hence reducing pain. Both chitosan and chitin gradually depolymerize to release *N*-acetyl-β-D-glucosamine, which, initiating fibroblast proliferation, helps ordered collagen deposition. Finally, the stimulating synthesis and increased production of natural hyaluronic acid at the wound site helps with faster wound healing and scar prevention. This the probably reason why the different CN-Ag-Lignin/PEO non-woven tissues made by electrospinning or casting technologies, have been shown to have interesting anti-inflammatory, anti-microbial [27,36] and cicatrizing [37] activities, without revealing toxic side-effects, being skin-friendly and environmentally-friendly [35–38].

Chapter 1

Burns are a complex skin lesion that are recognized in different depth degrees involving the skin extensive body surface with underlying tissue destruction. In consideration of the histological depth of the loss of tissue, three burn degrees are reported [39].

1st degree burn: only the epidermis is involved with raising of the stratum corneum, dermal edema and vascular dilatation. It heals spontaneously in seven days (Figure 2).

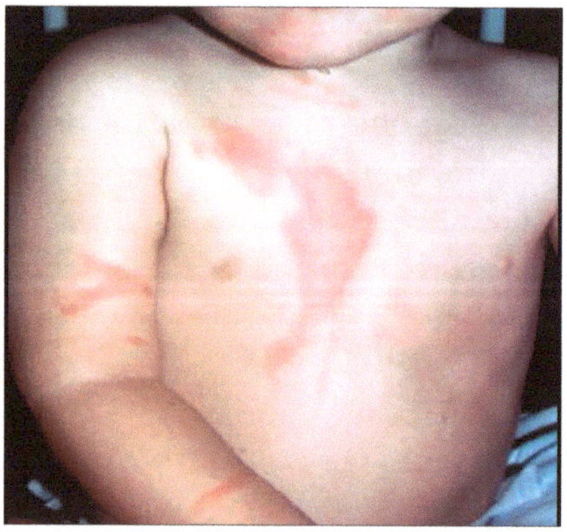

Figure 2. Clinical case of a child with first degree burn on the chest and the right arm.

Superficial 2nd degree burn: in this case the destruction of the epidermis and the superficial layer of the dermis is observed with the development of blisters and a detachment of the epidermis due to the pressure of the transudate fluids. This type of burn is characterized by erythema, edema, blisters and intense pain. Skin is pink, warm, and painful to touch. Bleeds easily and clears the acupressure. The hairs are present. It heals spontaneously in 15 days (Figure 3).

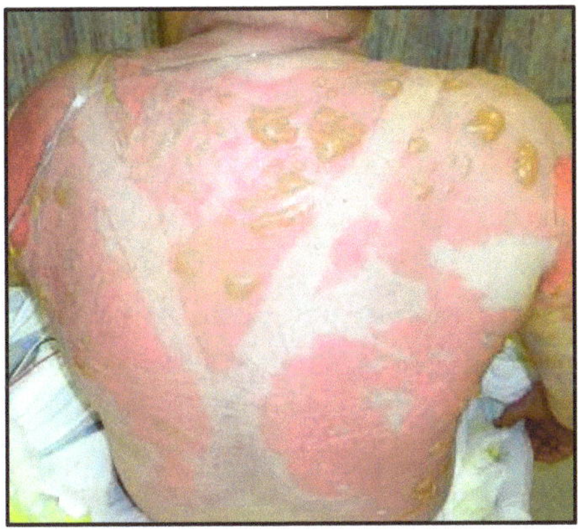

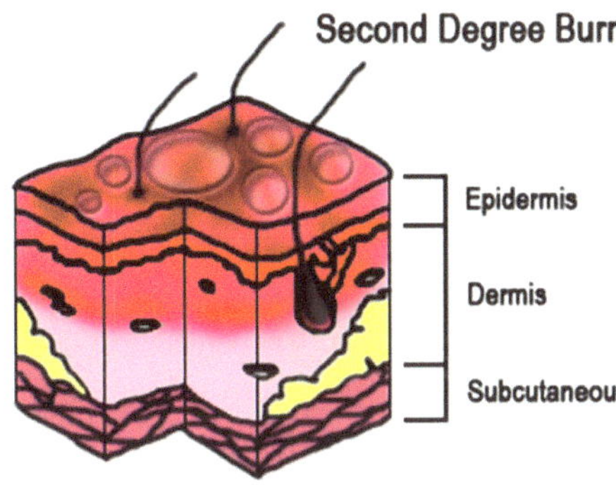

Figure 3. (**Left**): schematic skin section of superficial second-degree burn; (**Right**): clinical case of a patient with superficial second-degree burn on the back due to scald.

Deep 2nd degree burn: Damage that involves the epidermis, dermis, and many superficial neural structures, causes pain and burning. The de-epithelialized areas appear whitish, with eschar expression of more serious damage not uncommon. Under the blisters the skin is pale or bright red (dermal edema and vasodilatation) and it is not painful to touch. Heals in 3–4 weeks with frequent scars and often requires surgical treatment (Figure 4).

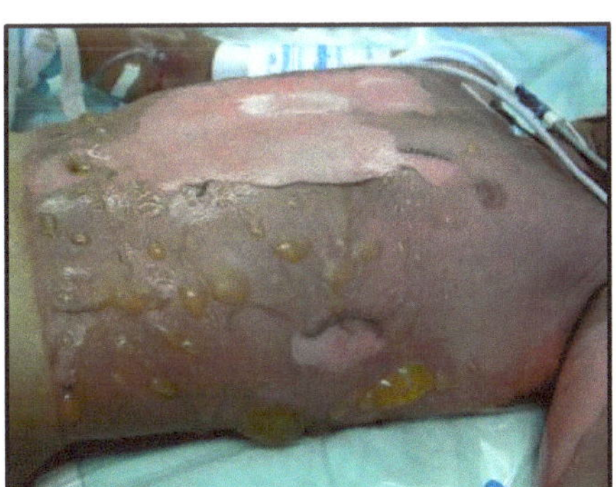

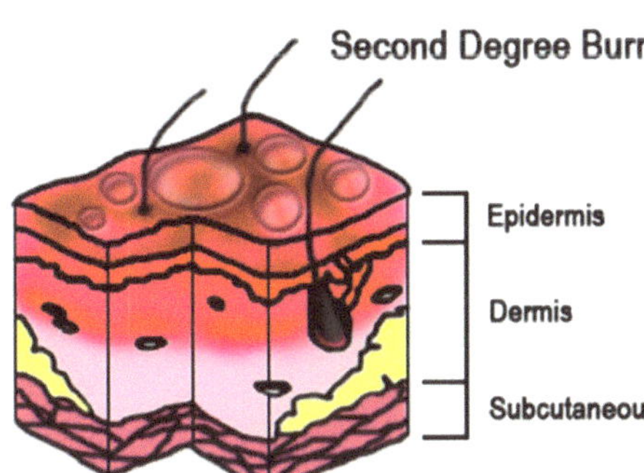

Figure 4. (**Left**): Schematic skin section of a deep second-degree burn; (**Right**): clinical case of a patient with deep second-degreeburn on the trunk.

Third 3rd degree burn: complete destruction of epidermis and dermis is observed. In more severe cases, the exposure of muscles, tendons and bones is possible. If the burning agent is fire or a hot body, the area of necrotic eschar looks blackish or brownish and is dry, cold, hard, painless and non-bleeding. If the agent is a scald, the necrotic area appears whitish or gray in color and of soft consistency. The presence of thrombosed vessels in waxy skin is a pathognomonic sign of this injury. Pain is absent because of the destruction of nerve endings, while only sensitivity to deep pressure is preserved. The hair, even when present, can easily be removed. The treatment is surgical and it heals with scarring (Figures 5 and 6). The goal of the surgical treatment in burns is to prevent invasive local infection and sepsis, to avoid that the burn damage becomes deeper, to get the best skin coverage as quickly as possible, to avoid hypertrophic scars and scar contractures, and to obtain complete healing in the shortest time and with the least number of operations. The operations consist in the removal of all necrotic tissue, which is extended up to the achievement of a vital plane.

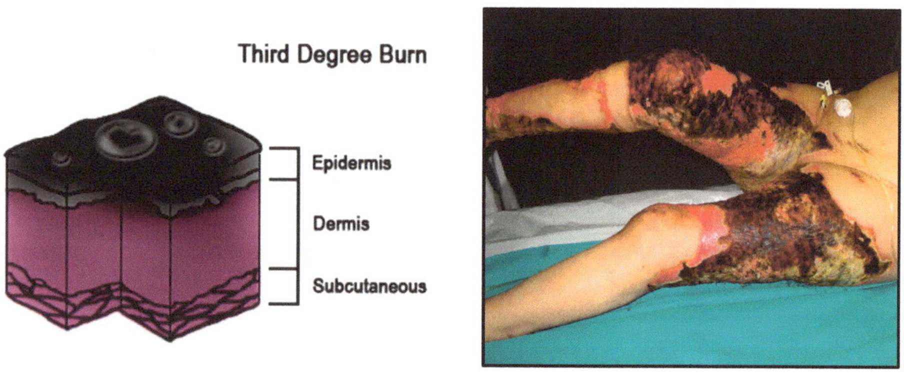

Figure 5. (**Left**): schematic skin section of third-degree burn; (**Right**): clinical case of a patient with third degree burn on the legs due to fire.

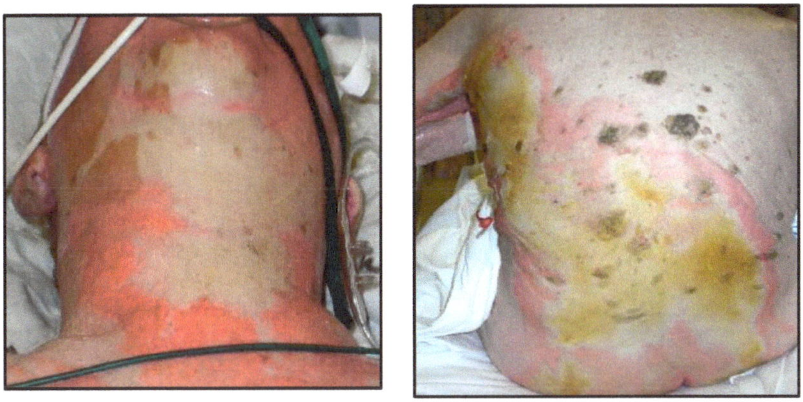

Figure 6. (**Left**): third-degree burn of the back due to scald; (**Right**): third-degree burn of the neck due to scald.

Based on the burn depth, a superficial tangential, a deep tangential and a fascial escharotomy—depending on the depth of the of necrotic tissue removal—can be distinguished. The superficial tangential escharotomy consists of the removal of the superficial dermis, until the dermal capillary plexus is reached; dermal plexus bleeding indicates the achievement of the vital plane on which the skin graft can be performed. The deep tangential escharoctomy is carried out at different levels of depth, sacrificing the entire dermis and hypodermis; the unviable dermis (hard and gray) also does not provide a substrate for skin grafting, but healthy fat looks shiny and yellowish and contains blood vessels that allow a high possibility that the graft will take. In the fascial escharotomy, the excision of necrotic tissue is done to the muscolaris fascia; the subcutaneous tissue is brown, necrotic and hemorrhagic, while the fascial smooth surface is a well-vascularized recipient site (perforating arteries) and allows skin grafts to take easily. Obviously, the main disadvantages of this procedure are the permanent cosmetic deformity (because the fat does not regenerate), risk of nerve and superficial tendon damage (skin denervation and permanent loss of sensation), and bone and tendon exposure. The autologous skin graft is the only definitive method to cover a loss of substance. The donor sites are, in order of preference: the upper and anterior-lateral thigh, the anterolateral area of the legs, upper arms, abdomen, back, chest and buttocks (Figure 7).

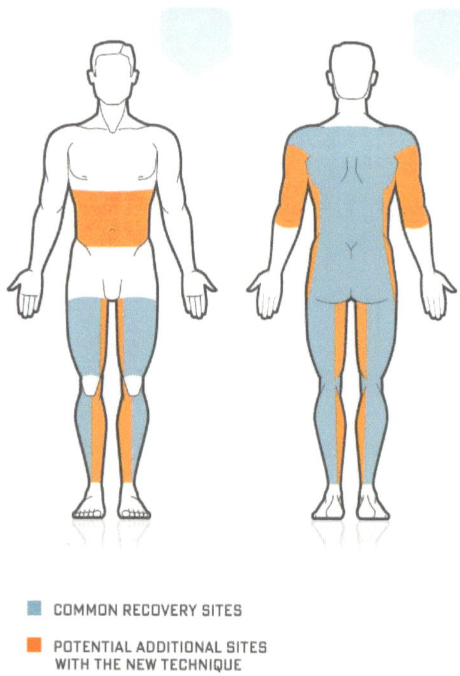

Figure 7. Schematic representation of full and split-thickness skin graft donor sites (Source: by the courtesy of World Press Media).

To take the dermal-epidermal area, skin grafts are used: Brown dermatome (electric or air) electric Padgett or Zimmer dermatome, and a range of freehand knives. Autografts are taken with a thickness of 0.010 to 0.012 cm and applied in a laminar or meshed structure. The mesh graft in the skin produces a series of perforations in a geometric pattern, to allow the expansion of the skin by increasing the initial size. In this way a limited donor area can cover a wider receiver area. The dermal-epidermal skin graft is prepared with mesh graft in ratio of 1 to $1\frac{1}{2}$ or 1 to 3, until even, in some cases, 1 to 6 or 1 to 9.

The skin grafts, both laminar and meshed, lie on the vital area (after adequate and careful hemostasis), are regularized and adapted to the loss of substance, and anchored using mechanical staplers (small grafts and those made in the face are fixed with thin sutures and moulage) (Figures 8 and 9). The meshed skin grafts are used in patients with large burns and limited donor sites. The presence of the "breaks" in the structure of the graft allows excellent drainage of the serous and blood collections that can be formed under laminar skin grafts.

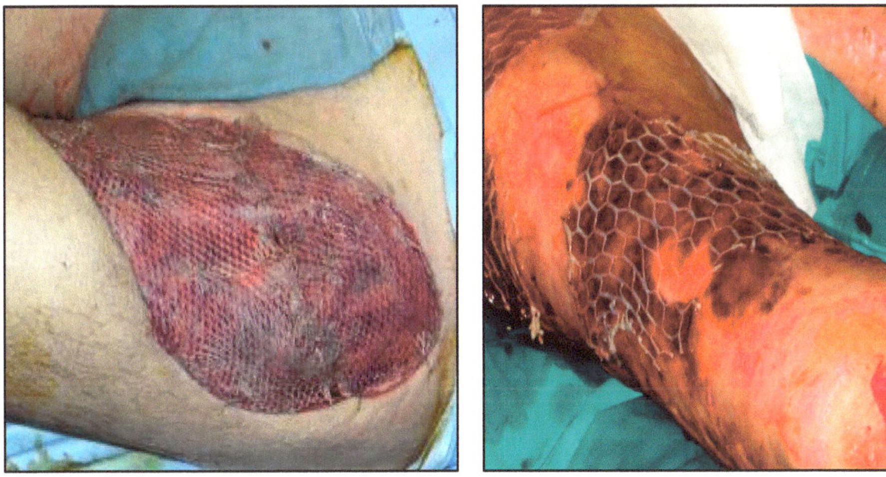

Figure 8. Skin grafts prepared with mesh graft in ratio of 1 to 1:1.5 (**Left**) or 1:6 (**Right**).

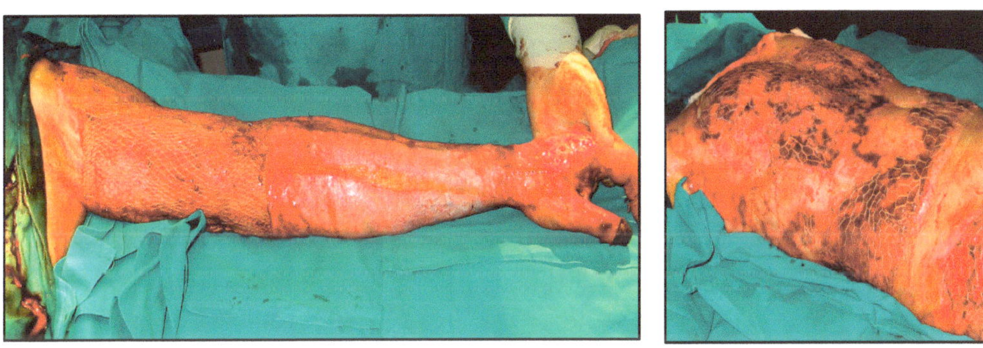

Figure 9. Clinical cases of patients with large burns and limited donor sites treated with meshed split-thickness skin grafts.

Ideal cutaneous wound repair should involve skin regeneration, bringing all the anatomical and physiological capabilities back to normal, without scarring. Unfortunately, wound healing in adult mammals too often results in a fibrotic normal or abnormal scar formation. Abnormal scarring that leads to hypertrophic scar or keloid formation that also invades the healthy tissue seems to be due to a persistent skin-barrier perturbation [29,30]. Thus, the prevention of unsatisfactory scarring begins before the treatment by the selection of both the dressings to be used and the post-operative care. A key part of the preoperative discussion is informed consent, which requires adequate information for the patient so that she/he can to make a decision regarding the procedure, indicating the procedure selected with the relative risks and benefits. Naturally, with regard to scarring, the patient should be aware of potential scarring that may occur considering which areas of the body can be prone to this unusual wound contraction, as well as she/he has to know that smoking, alcohol and certain medications such as retinoids, may have negative impact on the final scar [31]. However, scarring, which cannot be prevented but controlled, is the inevitable final stage of wound healing. To what degree the resultant scar affects functional and cosmetic outcomes is dependent on early and consistent treatment. This is the reason why our group dedicates great attention to the control of the location and depth of injury, together with the pretreatment and cleansing of the burned/wounded skin, and the relative selection of the non-woven tissues to be used.

Chapter 2

Since natural polymers are an interesting option for most widely-applied fibers in composite non-woven technology, this study aimed to use an innovative non-woven tissue made prevalently by CN, functionalized by the use of marine collagen peptides, lignin and Ag^+ ions.

This innovative engineered composite, proposed as the most promising polymer matrix in wound-dressing development, has been shown in vitro to possess interesting antibacterial properties (Figure 10) connected with good cell adhesion and proliferation capabilities to guide cell differentiation [32].

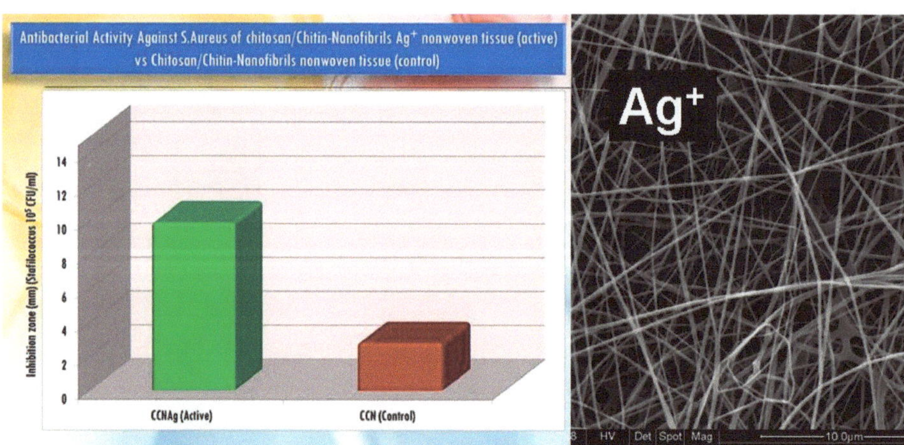

Figure 10. Antibacterial activity of chitosan/chitin fibers Ag^+ treated.

It is interesting to underline that this medical dressing is composed of two different layers. The outside layer (blue in color and composed of polypropylene of pharmaceutical grade) slows down the transcutaneous perspiration, increasing the cicatrizing activity of CN and the bactericidal power of the Ag^+ ions cross-linked to the chains of the CN-polymer. The inside very thin layer, in direct contact with the wound and composed of natural dispersible materials, mimics the ECM architecture, influencing the cell life and reproduction. The methodology to crosslink Ag^+ to crystal Chitin, before obtaining the non-woven tissue by electrospinning, gave the possibility not only to reduce the toxic effects of the metallic ions (nano-concentrated) at the level of the skin cells, but also to enhance the fibers' anti-inflammatory and bactericidal activity [36–38].

The aim of the study was to use these innovative, biodegradable, and skin-friendly non-woven tissues to facilitate skin repair, serving as either temporary or permanent replacements for burned tissue.

In the treatment of superficial second-degree burns and split-thickness skin graft donor sites, the goal is to achieve complete re-epithelialization. Thus, the effectiveness of MAVI dressing in the healing of split-thickness skin graft donor sites was evaluated on superficial second-degreeburns.

After local institutional ethical committee approval, we collected data from ambulatory patients of our burn center from June 2014 to August 2014.

Exclusion criteria were admitted patients, full-thickness burns, and operated patients that came to the ambulatory facility for postoperative follow up.

Out-patient treatment was selected according to the classification of the American Burn Association: a second-degreeburn with TBSA (total burn surface area) less that 15% in adults (10% in children), burns not involving eyes, ears, face, hands, feet or perineum; burns not derived from electrical injuries, not associated with inhalation, not in poor-risk patients.

The burn depth was clinically assessed. No systemic antibiotics were used. Patient's burns were diagnosed as completely healed up when re-epithelisation was complete in all affected areas. For each patient, we collected the age, sex, cause of burn (scald or flame), type of dressing used and the days required for complete healing (healing time). We used three different preparations of MAVI dressings.

MAVI n.1 was utilized in 10 patients, treating the coverage of split-thickness skin graft donor sites in four patients and superficial second-degreeburns in six patients.

(MAVI n.1 is made of CN-Ag-lignin-PEO and collagen peptides made by electrospinning).

MAVI n.2 was utilized in 10 patients, treating the coverage of split-thickness skin graft donor sites in seven patients and superficial second-degreeburns in three patients.

(MAVI n.2 is made of CN-Ag-lignin-PEO made by electrospinning).

MAVI n.3 was utilized in 10 patients, treating the coverage of split-thickness skin graft donor sites in two patients and superficial second-degreeburns in eight patients.

(MAVI n.3 is made of CN-Ag-lignin-chitosan made by casting technology).

A prospective randomized clinical study was performed on 30 burn patients to investigate the effectiveness, safety and tolerability of MAVI dressings.

The MAVI dressing was applied on superficial second-degree burns and split-thickness skin graft donor sites in patients aged between seven and 71 years old, the mean average age was 43.6. The cause of the burn was scalding in 83% of cases and fire in the remaining 17%. The donor sites and the superficial second-degreeburn wounds were assessed on postoperative days one, five, 14, 21, and long-term for infection, hyperemia, pruritus, pain, exudate level, and adherence to the wound bed (Table 2).

Table 2. Different preparations and applications of MAVI dressings.

	Characteristics	Patients	Donor Sites/Second-Degree Burns
MAVI I	CN-Ag-lignin-PEO and collagen peptides	10	4/6
MAVI II	CN-Ag-lignin-PEO	10	2/8
MAVI III	CN-Ag-lignin-chitosan	10	7/3

At the follow-up visits, the donor sites and the superficial second-degree burn wounds were assessed again for pruritus and pain, patient comfort and convenience for the doctor. Touch-pressure, thermal and pain sensibility tests were performed preoperatively and on postoperative follow-up together with the assessment of color and texture of the re-epithelialized areas.

In all patients, re-epithelialization was completed between five and 13 days (mean eight days) after the application of the MAVI dressing.

Six patients out of 30 required pain killers over the first three days after burns with no significant differences between the type of dressing used.

There were neither significant differences between donor sites and the superficial second-degree burn wounds regarding pain, hyperemia, pruritus, exudate, and final appearance (color and texture).

The areas dressed with MAVI completely healed within 5–13 days in a significantly higher proportion than the traditional dressings, showing during the whole study less incidence of exudates and peri-lesional erythema.

The aesthetic outcome of the treated lesions after healing was significantly better for the MAVI dressing and it required very few renewals of the medication during the first week of treatment. The high interval time between dressing changes reduced the amount of medication, patient suffering, overall costs and human resources, according to the reported following cases.

After the application of MAVI n.1 we observed the complete re-epithelialization of the split-thickness skin graft donor sites in three patients and a healing delay due to exudate in one patient; and complete superficial second-degree burn healing in four patients (Figure 11).

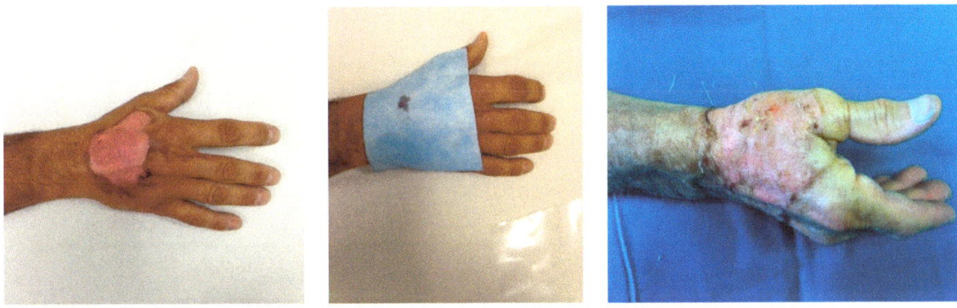

(a) (b) (c)

Figure 11. Clinical case n°1: (**a**) superficial second-degree burn on the dorsum of the right hand; (**b**) Application of MAVI dressing number 1; (**c**) Complete re-epithelialization after five days post-burn.

After the application of MAVI n.2, we observed the complete re-epithelialization of the split-thickness skin graft donor sites in five patients and a healing delay due to exudate in two patients; complete superficial second-degree burns healing in two patients and the failure of the procedure in one patient due to infection. After the application of MAVI n.3 (Figure 12), we observed the complete re-epithelialization of the split-thickness skin graft donor sites in all patients, and complete superficial second-degree burns healing in seven patients.

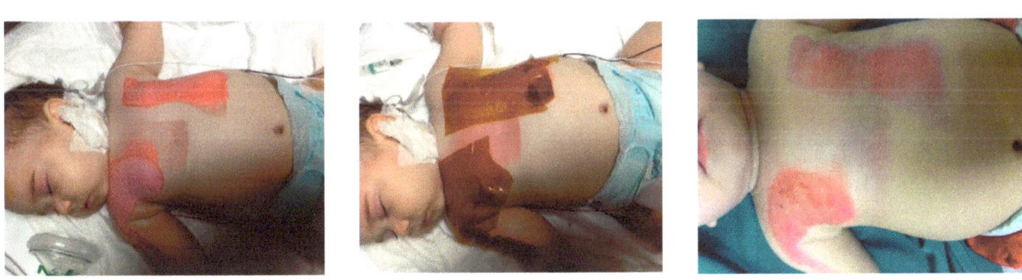

(a) (b) (c)

Figure 12. Clinical case n°2: (**a**) Second-degree burn of the chest and right shoulder; (**b**) Application of Mavi dressing n°3; (**c**) Complete skin re-epithelialization after six days post burn.

Chapter 3

The self-renewing property of the epidermis plays a major role during wound healing. Within moments of wounding, keratinocytes not only inform each other that the barrier has been broken, but also communicate with the dermis and the local immune system about the urgent need to repair the gap, maintaining a constant flow of information.

The purpose of such complex signaling cascades is to regulate gene targets during wound healing, thus controlling the keratinocyte activation cycle.

The most common initiator of keratinocyte activation is Interleukin-8 (IL-8), while defensine-2 (Figures 13 and 14 [36,37]) was found to be closely associated with the progression of wound healing.

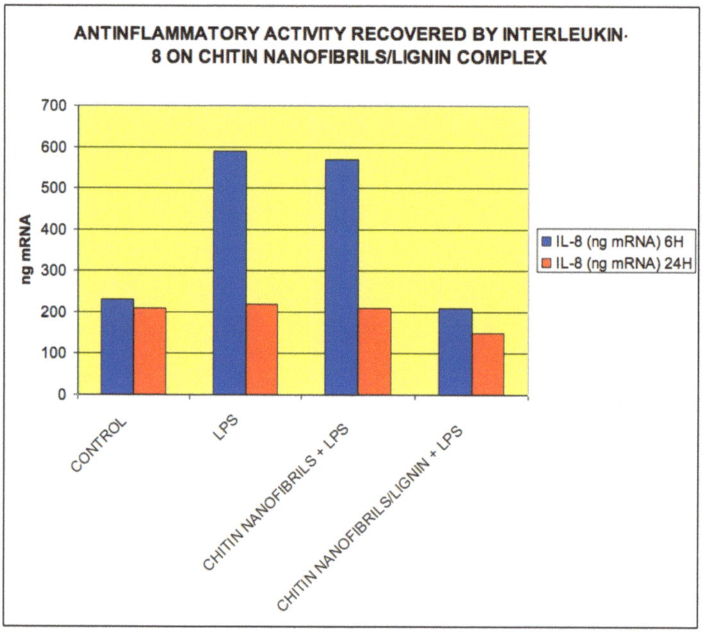

Figure 13. The figure illustrates gene expression modulation in human cell lines of pro-inflammatory cytokines in HaCaT keratinocytes. The cell lines were treated with: Control = untreated cells; LPS (Lipopolysaccharide of *P. Aeruginosa*) = pro-inflammatory substance; LPS + the Complex Chitin Nanofibrils/Lignin; LPS + Chitin Nanofibrils [37].

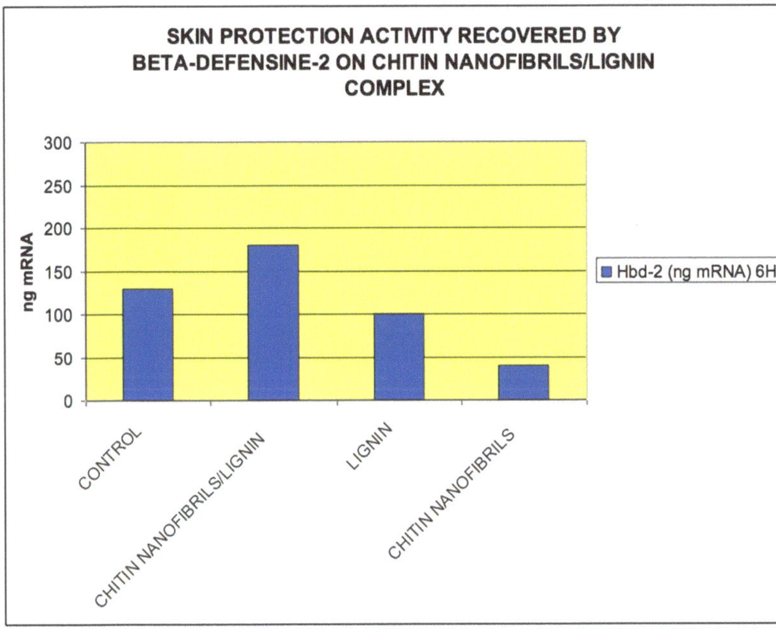

Figure 14. The figure illustrates the modulation of the gene expression of beta-defensin in human keratinocyte HaCaT cell lines. The cell lines were treated with: Control = untreated cells; Lignin; The Complex Chitin Nanofibrils/Lignin; Chitin; An over-expression in the values of these markers is an indication of tissue regeneration activity [37].

Therefore, IL-8 and defensine 2 appear to be accurate prognostic markers of wound repair.

The in vitro results of our studies on keratinocytes cultures were in agreement with the in vivo studies. An interesting modulating activity was observed on both IL-8 and defensine-2 on the non-woven tissue made by the natural fibers embedded in keratinocyte cultures.

On the other hand, the in vivo double-blind study made on the burned and wounded skin of a patient-group recovered in hospital notably accelerated the skin repair and cicatrizing activities of the same non-woven tissues. Moreover, due to the bactericidal and anti-fungal activity of the Ag-metal-ions, capable of regulating microorganisms invasion, it was possible to keep the dressing in place for one week, sensibly reducing the general cost of the in-hospital treatment.

It is interesting to underline how the obtained nano-fibers, made prevalent by the polymer chitin nanofibril bound to lignin and peptides, establishing each with other a stable ionic bonds, show an interesting skin-modulating activity on the synthesis of defensines and interleukines released from keratinocytes, also modulating the metalloproteinase activity [37]. Moreover, Ag^+ ions, strictly bound to the Chitin chains, gave an interesting antibacterial activity to the medical dressing.

By this approach it seems desirable to use these innovative, chitin-based, non-woven tissues for medical purposes, due to their particular effectiveness in mimicking the size and arrangement of native collagen, accelerating the normal repairing activity of the skin, altered for burn or wound healing. Additionally, the electrospun scaffolds, obtained by a layer-by-layer method for their high surface-to-volume ratio and interconnecting pores, seem capable of facilitating cell adhesion and the formation of cell–cell junctions, accelerating the cicatrizing process [35–39]. These nano-fibrous scaffolds, in fact, seem to induce favorable cell—ECM interactions, increasing the cell proliferation rate, maintaining the cell phenotype, supporting cell differentiation, and promoting in vivo-like three-dimensional matrix adhesion [36–38]. Moreover, they probably activate cell-signaling pathways, by providing the chemical and physical stimuli to cells necessary for faster skin-repairing activity. The sequential, multi-layering electrospinning of the different natural polymers employed in combination with natural peptides, surely enhanced the mechanical integrity and dimensional stability of the electrospun meshes, while the nanostructured Ag^+ silver probably maintained the skin microbiota in equilibrium, impeding the excessive growth of the opportunistic microorganisms [36,37].

These innovative medical dressings seem to represent new, cost effective, non-woven tissues and procedures to be used for skin regeneration and tissue transplantation. Their activity, compared to the traditional non-woven tissues normally used in our hospital, was shown to be more effective in a shorter time to regenerate the skin affected by burns of the first and second degree. It is interesting to underline the capacity that these non-woven tissues have to prevent microbiota growth for the right period of time, extending the medication period. In conclusion, the MAVI dressing seems to be a safe and effective dressing for the re-epithelialization of skin graft donor sites and superficial second-degree burn wounds, showing higher activity than traditional dressings.

The goal of such a treatment strategy based on the use of innovative, bioengineered, non-woven tissues, which are totally biodegradable and environmentally-friendly, is, in fact, to act as smart band-aids, to replace altered or senescent resident cells and reestablish the anatomy and physiology of burned/wounded skin [40,41]. Well-designed randomized clinical trials will involve our group in the next phase in order to scrutinize the true potential of these natural polymers for regenerative medicine.

Acknowledgments: We thank MAVI SUD Srl, Aprilia, Italy, for the sample-test given for this study.

Author Contributions: T.A. and P.M. verified the analytical methods. T.A., A.B., S.M., M.P., P.P. investigated and carried out the clinical part experiments. T.A. and P.M. supervised the findings of this work. P.M. had the idea and realized the tissue engineering dressing. All authors discussed the results and contributed to the final manuscript.

Conflicts of Interest: We declare hat Paolo Palombo, Marco Palombo, Simone Moroni, Agostino Bruno and Tommaso Anniboletti have no conflicts of interest. Pierfrancesco Morganti works as the Head of the R&D Centre of Nanoscience, MAVI SUD, Srl, Italy.

Abbreviations

CS	Chitosan
CN	Chitin nanofibrils
S. aureus	Staphylococcus aureus
CFU	colony forming unit
CB	bio-burden skin bacteria
PEO	polyethylene oxide
Ag	Silver

References

1. Chapekar, M.S. Tissue engineering: Challenges and opportunities. *J. Biomed. Mater. Res.* **2000**, *53*, 617–620. [CrossRef]
2. Chan, G.; Money, D.J. New Materials for tissue engineering: Towards greater control over the biological response. *Trends Biotechnol.* **2008**, *26*, 382–392. [CrossRef] [PubMed]
3. Nieponice, A. Clinical Translation of Tissue Engineering and Regenerative Medicine Technologies. In *Biomaterials and Stem Cells in Regenerative Medicine*; Ramalingam, M., Ramakrishna, S., Best, S., Eds.; CRC Press: New York, NY, USA, 2012; pp. 521–532.
4. Morganti, P.; Del Ciotto, P.; Carezzi, F.; Morganti, G.; Chen, H.-D. From Waste Material a New and Safe Anti Aging Compound: A Chitin Nanofiber Complex. *SOFW J.* **2012**, *138*, 128–138.
5. Morganti, P.; Li, Y.-H. Healthy products from waste materials. *Eurocosmet. China Spec. Issue* **2012**, *20*, 60–63.
6. Morganti, P.; Morganti, G.; Morganti, A. Transforming nanostructured chitin from crustacean waste into beneficial health products: A must for our society. *Nanotechnol. Sci. Appl.* **2011**, *4*, 123–129. [CrossRef] [PubMed]
7. Morganti, P.; Tiscenko, G.; Palombo, M.; Kelnar, L.; Brozova, L.; Spirkova, M.; Pavlova, E.; Kobera, L.; Carezzi, F. Chitin nanofibrils for biomimetic products: Nanoparticles and nanocomposite chitosan films in health-care. In *Marine Biomaterials: Isolation, Characterization and Application*; Kim, S.-K., Ed.; CRC Press: New York, NY, USA, 2013; pp. 681–715.

8. Morganti, P.; Chen, H.D.; Gao, X.H.; Del Ciotto, P.; Carezzi, F.; Morganti, G. Nanoparticles of Chitin Nanofibril-Hyaluronan block polymer entrapping lutein as UVA protective compound. In *Carotenoids: Food Source, Production and Health Benefits*; Nova Science Publishers, Inc.: New York, NY, USA, 2013; pp. 237–259.
9. Morganti, P.; Carezzi, F.; Del Ciotto, P.; Tishchenko, G.; Chianese, A. A Green Multifunctional Polymer from Discarded Material: Chitin Nanofibril. *Br. J. Appl. Sci. Technol.* **2014**, *4*, 4175–4190. [CrossRef]
10. Eide, K.B.; Norberg, A.L.; Heggset, E.B.; Lindbom, A.R.; Varum, K.M.; Eijsink, V.G.H.; Soile, M. Human Chitotriosdase-Catalyzed Hydrolisis of Chitosan. *Biochemistry* **2012**, *51*, 487–495. [CrossRef] [PubMed]
11. Morganti, P.; Muzzarelli, C. Spray-Dried Chitin Nanofibrils, Method for Production and Uses Thereof. U.S. Patent 12/094912, 23 November 2006.
12. Morganti, P.; Fabrizi, G.; Guarneri, F.; Palombo, M.; Palombo, P.; Cardillo, A.; Ruocco, E.; Del Ciotto, P.; Morganti, G. Repair Activity of Skin Barrier by Chitin-Nanofibrils Complexes. *SOFW J.* **2011**, *137*, 10–26.
13. Morganti, P.; Del Ciotto, P.; Morganti, G.; Fabien-Soulé, V. Application of Chitin Nanofibrils and Collagen of Marine Origin as Bioactive Ingredients. In *Cosmeceuticals: Trends and Prospects*; Kim, S.-K., Ed.; CRC Press: New York, NY, USA, 2012; pp. 267–289.
14. Morganti, P.; Del Ciotto, P.; Fabrizi, G.; Guarneri, F.; Cardillo, A.; Palombo, M.; Morganti, G. Safety And Tolerability of Chitin Nanofibrils-Hyaluronic Acid Nanoparticles Entrapping Lutein. Note 1: Nanoparticles Characterization and Bioavailability. *SOFW J.* **2013**, *139*, 12–23.
15. Dobrovol'skaya, I.P.; Katatkin, I.A.; Yudin, V.E.; Ivan'kova, E.M.; Elokhovskii, V.Y. Supramolecular structure of chitin Nanofibrils. *Polym. Sci. Ser. A* **2015**, *57*, 52–57. [CrossRef]
16. Morganti, P. Chitin Nanofibrils and Their Derivatives as Cosmeceuticals. In *Chitin, Chitosan, Oligosaccharides and Their Derivatives. Biological Activities and Application*; Kim, S.-K., Ed.; CRC Press: New York, NY, USA, 2010; pp. 531–542.
17. Morganti, P.; Del Ciotto, P.; Gao, X.-H. Skin Delivery and Controlled Release of Active Ingredients Nanoencapsulated by Chitin Nanofibrils: A New Approach. *Cosmet. Sci. Technol.* **2010**, *20*, 136–142.
18. Hurler, J.; Nataša, Š.-B. Potentials of Chitosan-Based Delivery Systems in Wound Therapy: Bioadhesion Study. *J. Funct. Biomater.* **2012**, *3*, 37–48. [CrossRef] [PubMed]
19. Glossens, A.; Cleenewerck, M.B. New Wound dressings: Classification and Tolerance. *Eur. J. Dermatol.* **2010**, *20*, 24–26.
20. Sen, C.K.; Gordillo, G.M.; Roy, S.; Kirsner, R.; Lambert, L.; Huntb, T.K.; Gottrup, F.; Gusrantir, G.C.; Longsker, M.T. Human skin wound S: A major snowballing threat to Public Health and Economy. *Wound Repair Regen.* **2009**, *17*, 763–771. [CrossRef] [PubMed]
21. Murphy, K.D.; Lee, J.O.; Herndon, D.N. Current Pharmacotherapy for the treatment of severe burns. *Expert Opin. Pharmacother.* **2003**, *4*, 369–384. [CrossRef] [PubMed]

22. Eisner, J.J.; Euozi, D.; Ullman, Y.; Berdicevsky, I.; Shefy-Pleg, A.; Ziberman, M. Novel biodegradable composite wound dressings with controlled release of antibiotics: Results in a guinea pig burn model. *Burns* **2011**, *37*, 896–904. [CrossRef] [PubMed]
23. Kabelitz, D.; Shroder, J.M. *Mechanisms of Epithelial Defence*; Karger Medical and Scientific Publishers: Basel, Switzerland, 2005.
24. Falanga, V. *Cutaneous Wound Healing*; Martin ad Unix Ed.: London, UK, 2001.
25. Rover, D.T.; Maibach, H.I. *The Epidermis in Wound Healing*; CRC Press: New York, NY, USA, 2004.
26. Yudin, V.E.; Dobrovolskaya, I.P.; Neelov, I.M.; Dresvyanina, E.N.; Popryadukhin, P.V.; Ivan'kova, E.M.; Elokhovskii, V.Y.; Kasatkin, I.A.; Okrugin, B.M.; Morganti, P. Wet spinning of fibers made of chitosan and chitin nanofibrils. *Carbohydr. Polym.* **2014**, *108*, 176–182. [CrossRef] [PubMed]
27. Morganti, P.; Chen, H-D.; Gao, X-H. Chitin Nanofibril: A Natural Eco-Friendly and Immunoadjuvant Actuve Carrier for Medical Use. *J. Appl. Cosmetol.* **2016**, *34*, 141–154.
28. Kim, J.; Joo, S.; Ko, I.; Atala, A.; Yoo, J.; Lee, S. Smart Biomaterial Scaffold for In Situ Tissue Regeneration. In *Biomaterials and Stem Cells in Regenerative Medicine*; CRC Press: New York, NY, USA, 2012. [CrossRef]
29. Seidi, A.; Ostrovidov, S.; Ramalingam, M. Fabrication and Application of Gradient Hydrogels in Cell and Tissue Engineering. In *Biomaterials and Stem Cells in Regenerative Medicine*; CRC Press: New York, NY, USA, 2012. [CrossRef]
30. Khan, A.J.; Cook, B. Wound healing and scar formation. In *Surgical Technique for Cutaneous Scar Revision*; Harahap, M., Ed.; Marcel Dekker Inc.: New York, NY, USA, 2000; pp. 1–14.
31. Elias, P.M.; Wood, L.C.; Feingold, K.R. The skin and its barrier function: Implications for wound healing. In *Cutaneous Wound Healing*; Falanga, V., Ed.; Martin Dunitz: London, UK, 2001; pp. 5–14.
32. McDonald, M.; Stasko, T. Prevention of unsatisfactory scarring. In *Surgical Technique for Cutaneous Scar Revision*; Harahap, M., Ed.; Marcel Dekker Inc.: New York, NY, USA, 2000; pp. 53–80.
33. Morganti, P.; Carezzi, F.; Del Ciotto, P.; Morganti, G.; Nunziata, M.L.; Gao, X.-H.; Chen, H.-D.; Tischenko, G.; Yudin, E.V. Chitin Nanofibrils: A Natural Multifunctional Polymer. Physicochemical characteristics, effectiveness and safeness. *Nanobiotechnology* **2014**, 1–31.
34. Jayakumar, R.; Prabaharan, M.; Kumar, P.S.; Nair, S.V.; Furuike, T.; Tamura, H. Novel Chitin and Chitosan Materials in Wound Dressing. *Biomed. Eng. Trends Mater. Sci.* **2011**. [CrossRef]
35. Morganti, P.; Febo, P.; Cardillo, M.; Donnarumma, G.; Baroni, A. Chitin Nanofibril and Nanolignin: Natural Polymers of Biomedical Interest. *J. Clin. Cosmet. Dermatol.* **2017**, *1*. [CrossRef]
36. Morganti, P.; Del Ciotto, P.; Stoller, M.; Chianese, A. Antibacterial and antiinflammatory green nanocomposites. *Chem. Eng. Trans.* **2016**, *47*, 61–66.

37. Morganti, P.; Fusco, A.; Paoletti, I.; Perfetto, B.; Del Ciotto, P.; Palombo, M.; Chianese, A.; Baroni, A.; Donnarumma, G. Anti-inflammatory, Immunomodulatory, and Tissue Repair Activity on Human Kerstinocytes by Green Innovative Nanocomposites. *Materials* **2017**, *10*, 843. [CrossRef] [PubMed]
38. Morganti, P.; Stoller, M. Chitin and Lignin: Natural Ingreduents from Waste Materials for making Innovative and Healthy Products for Humans and Plants. *Chem. Eng. Trans.* **2017**, *60*, 319–324.
39. Gibson, A.L.F.; Bennett, D.D.; Taylor, L.J. Improving the histologic characterization of burn depth. *J. Cut. Pathol.* **2017**. [CrossRef] [PubMed]
40. Bottino, M.; Vohra, Y.; Thomas, V. Spatially Designed Nanofibrous Membranes for Periodontal Tissue Regeneration. In *Biomaterials and Stem Cells in Regenerative Medicine*; CRC Press: New York, NY, USA, 2012.
41. Phipps, M.C. Mesenchymal Stem Cell Responses to Bone-Mimetic Electrospun Matrices Composed of Polycaprolactone, Collagen I and Nanoparticulate Hydroxyapatite. *PLoS ONE* **2011**, *6*, e16813. [CrossRef] [PubMed]

EU Cosmetic Regulation: Quality Enhancement of Consumer and Environment Protection, Market Development

Sonia Selletti

Abstract: The purpose of this article is to outline the main features of the recent European recast of the Cosmetics Regulation (Regulation EC No. 1223/2009), focusing on the core aims of the legislator: harmonization throughout member states (hence the choice of a Regulation as the appropriate legal instrument) and the reduction of administrative burden and ambiguities, to enhance the protection of human health and the environment, thus fostering quality in the market to the benefit of consumers, who may rely on strengthened in-market controls. Specific attention is also paid to the justification of claims, in order to lead consumers to make informed choices based on clear, transparent, and honest claims, counting on a set of 'common criteria' laid down in the specific Regulation (EC) No. 655/2013. This framework makes room for innovation in cosmetic research, since it regulates the use of nanomaterials in cosmetic products, respecting the environment.

---

1. Introduction

A revision of regulations governing cosmetics has recently taken place in Europe thanks to the (EU) Regulation no. 1223, adopted on 30 November 2009, which came into force for all member states on 11 July 2013 [1].

The new regulation is a re-casting of the rules on cosmetics adopted with directive 76/768—which was the object of as many as 65 amendments—and is a fundamental step in the enhancement of the quality of cosmetics, both in terms of consumer safety and market development. Although consumer safety has taken a place of pride in the legislator's production, the latter has made clear that quality and safety are the main pro-competitive elements, holding equal prominence, which may be beneficial to market development [2].

2. Main Features of the Regulation

Although the Regulation appears at times to be a harmonized coding of the previous regulations, it does in fact sum up almost 40 years' experience in the application of the rules in question, taking this experience to heart and thus setting the stage for an even more effective protection. Thus, the issue of safety is a lynchpin concept to such an extent that Article 3 prescribes that "cosmetic products made available on the market shall be safe for human health when used under normal or reasonably foreseeable conditions of use...". From this initial indication, one may well understand the kind of responsibility that companies are required to shoulder if they intend to place cosmetic products on the market.

The safety of cosmetics is pursued using a number of different "tools":

- the Regulations are accompanied by Annexes that list the substances that are subject to prohibitions or restrictions to safeguard health (these lists may be modified over time based on scientific advancements),
- for the purpose of imputing cosmetics on to the market, they must be subjected to a safety assessment (the EU Commission on 25 November 2013 issued the specific guidelines for the performance of an appropriate safety evaluation) [3].
- cosmetic production must take place while complying with good manufacturing practices,
- to ensure supervision and guarantee compliance with the obligations indicated in the Regulations, the figure of the 'responsible person' has been introduced in Article 4 (for example, the importer is, according to the law, a responsible person),
- a system has been introduced that enables the identification of the supply chain in order to guarantee the traceability of the cosmetics, supported by the institution of a centralized notification procedure through an EU portal managed by the Commission (so called CPNP—Cosmetic Products Notification Portal) and the identification of the subjects that operate in various capacities in the production process (manufacturer; importer; distributor) that have been assigned specific obligations,
- reinforced surveillance procedures have been outlined, to be implemented by the authorities with the aim of curtailing the counterfeiting that even afflicts the cosmetic sector and may be detrimental to the consumer's health; this system is based on the principle of cooperation between the various authorities and envisages the active participation of the responsible person and other subjects involved in the production process,
- customer protection is also achieved through a cosmetic vigilance program that involves the reporting and collection of information on undesirable effects; a specific reporting of serious undesirable effect (SUE) has been introduced and is ongoing in each member state,
- the Regulation also envisages a special protection related to the claims made by cosmetics to ensure that the consumer can make informed decisions based on objective and not deceitful elements; this has meant the introduction of "common criteria" (with EU Commission Regulation No. 655 of 10 July 2013) to "inform end users about the characteristics and quality of products" seeing

as these are "essential in order to differentiate between products" and help to "stimulate innovation and foster competition" [4,5].

- concerning the environment, special recommendations have been expressed by the Unfair Commercial Practices Directive (UCPD) Guidance Document to encourage industries to make good and accurate environmental claims, avoiding the use of "green claims" regarding the composition and process of products if not well-documented or not documented at all. Greenwashing is in fact a coined expression to underline the act of potentially misleading consumers regarding the environment practices of a company or the environmental benefits of a product or service. According to the revision of the UCPD Criteria Guidelines on Environmental Claims [6] completed on May 25, 2016, a definition for "environmental claims" has been provided (the expressions "environmental claims" or "green claims" refer to the practice of suggesting or otherwise creating the impression (in the context of a commercial communication, marketing, or advertising) that a product or a service is environmentally friendly (i.e., it has a positive impact on the environment) or is less damaging to the environment than competing goods or services. This may be due to, for example, its composition, the way it has been manufactured or produced, the way it can be disposed of, and the reduction in energy or pollution which can be expected from its use. When such claims are not true or cannot be verified, this practice can be described as "greenwashing"). Furthermore, a couple of main principles address the behavior of traders that: (i) must, above all, present their environmental claims in a specific, accurate, and unambiguous manner; and (ii) must have scientific evidence to support their claims and be ready to provide it in an understandable way in case the claim is challenged [7].

3. Innovation

The theme of innovation in a cosmetic context, often referred to by the legislator, offers scope for a brief consideration of the room that the regulations allow for innovative products, both in terms of the research and development of molecules, formulae, and technologies, as well as new kinds of products.

Thus, research is stimulated by the fact that the restrictions already imposed on the use of certain substances do not curtail the development and research of other substances, formulae, and technologies, the use of which will then have to be further investigated through the safety assessment procedures that the manufacturer is responsible for carrying out. The manufacturer is therefore fully entitled to head down innovative paths, being fully aware of the parameters that the company is required to comply with.

In relation to product types, it should be recalled that the legislator has underlined the need to hold firm on a clear demarcation between cosmetics and similar health products (medicines, medical devices, biocides, food integrators) in an attempt to avoid overlapping classifications relative to so-called borderline situations so that the correct sector regulations that apply can be identified.

Thus, by way of example, plenty of discussions and comparisons have been held on the cosmetic—or other—nature of certain products such as teeth whitening chewing gum, mascara that enhances eyelash growth, and adhesive patches used to fight unsightly body fat deposits (or cellulite) or to improve hair growth. There are many other interesting examples that have been assessed in an EU Commission manual on borderline products (Manual on the Scope of Application of the Cosmetic Regulation EC No. 1223/2009, November 2013), which has turned out to be a very useful tool not only for those seeking to interpret the regulations but also, and especially, for those engaged in product research and development [8].

The issue of innovation and development in the field of cosmetics finds its natural source in the very definition of the cosmetic product, which, as we know, is based on what one may term a binary system, which refers to the application site of the cosmetic product (external surfaces of the human body: epidermis, hair and hair follicle, nails, lips, external genital organs, teeth, mouth mucous) and the functions that are prevailingly or primarily performed (cleaning, perfuming, change of appearance, protection, maintenance, correction of body odors). Thus, the function of cosmetics, despite being established at a legal level, leaves plenty of scope for research, development and innovation in terms of functions such as "protection" and "maintenance", onto which one may graft the most advanced cosmetic qualities that are in a position to establish said products as functional cosmetics, even if they are also recommended in other specific or complementary contexts such as therapeutic contexts (while maintaining the prohibition of boasting therapeutic effects).

The easiest example, and also the most significant one, is found in the dermatological context, where the treatment for relevant pathologies is not only based on innovative medicinal products, but also on the contribution provided by functional cosmetics that enable the patient to improve their quality of life from every point of view. Thus, the cosmetic product plays an important complementary role to the therapy.

4. Nanomaterials

It is worth spending a few words on the issue of "nanomaterials", which have been specifically regulated by the European Cosmetics Regulation No. 1223/2009 as well as from a legal point of view, as an explicit acknowledgement that these ingredients are undergoing considerable development.

The Regulation has introduced several articles with implications for products containing nanomaterials, starting with the definition of a nanomaterial, for the purposes of the Cosmetics Regulation, provided under Article 2.1 (k) as "an insoluble or bio persistent and intentionally manufactured material with one or more external dimensions, or an internal structure, on the scale from 1 to 100 nm". While it is the responsibility of the manufacturer (under the aegis of the responsible person) to decide whether any ingredients they are using are nanomaterials (to this end, it is important to acquire proper information from raw material suppliers), the Regulation states (Article 16) that a high level of protection of human health should be ensured for any cosmetic products containing nanomaterials. Moreover, the Regulation requires that a specific evaluation be carried out, as part of the safety assessment, in order to determine any toxicological effects due to particle sizes, including nanomaterials (i.e., Annex 1—Cosmetic Safety Report). To help assessors evaluate nanomaterials appropriately, the Scientific Committee on Consumer Safety (SCCS) has published a report entitled Guidance on the Safety Assessment of Nanomaterials in Cosmetics [9,10].

Both the authorities and consumers shall be specifically informed about the presence of nanomaterials in cosmetic products: the former by means of a centralized notification procedure (cosmetic products that contain nanomaterials will need to be notified six months before the product is placed on the market), and the latter by means of the product label (all ingredients present in the form of nanomaterials shall be clearly indicated in the list of ingredients followed by the word 'nano' in brackets, so as to enable consumers to make informed decisions).

There are some exemptions for nanomaterials intended to be used as colorants, preservatives, and UV filters, which should be listed in Annexes IV to VI in order to be permitted for such uses: these are never subject to the nano-notification requirements, irrespective of the size of the ingredients, since the positive listing in the Annexes supersedes the need for nano-notification. Products containing ingredients listed in Annex III (list of substances which cosmetic products must not contain except subject to the restrictions laid down) in the form of a nanomaterial need not be notified.

Conflicts of Interest: The author declares no conflict of interest.

References

1. The European Parliament and the Council of the European Union. Regulation (EC) No. 1223/2009 of 30 November 2009 on cosmetic products. *Off. J. Eur. Union.* **2009**, *L342*.
2. The Commission of the European Communities. Directive (EC) No. 768 of 27 July 1976 on the approximation of the laws of the member states relating to cosmetic products. *Off. J. Eur. Union.* **1976**, *L262*.
3. The Commission of the European Communities. European Commission implementing decision of 25 November 2013 on Guidelines on Annex I to Regulation (EC) No 1223/2009 of the European Parliament and of the Council on cosmetic products. *Off. J. Eur. Union.* **2013**, *L315*.

4. The Commission of the European Communities. Regulation (EC) No. 655/2013 of 10 July 2013 laying down common criteria for the justification of claims used in relation to cosmetic products. *Off. J. Eur. Union.* **2013**, *L190.*
5. The Commission of the European Communities. Guidelines to Commission Regulation (EC) No.: 655/2009, Laying down common criteria for the justification of claims used in relation to cosmetic products. *Off. J. Eur. Union.* **2009**, *L190.*
6. Multi-Stakeholder Dialogue on Environmental Claims. Compliance Criteria on Environmental Claims 25 May 2016. Available online: http://ec.europa.eu/consumers/consumer_rights/unfair-trade/unfair-practices/files/mdec_compliance_criteria_en.pdf (accessed on 11 October 2017).
7. Van Laer, J. *Consumer market study on environmental claims for non-food products*; European Commission: Brussel, Belgium, 2014.
8. Manual on the Scope of Application of the Cosmetic Regulation EC; No. 1223/2009, Version 1.0. (November 2013). Available online: https://ec.europa.eu/docsroom/documents/13033?locale=en (accessed on 11 October 2017).
9. European Commission, Health & Consumers. Scientific Committee on Consumer Safety (SCCS)—Guidance on the Safety Assessment of Nanomaterials. Available online: http://ec:europa.eu/health/scientific_committees/consumer_safety/docs/sccs_s_005.pdf (accessed on 11 October 2017).
10. EU Commission. *Report from the Commission to the European Parliament*; The Council and the European Economic and Social Committee, European Parliament and Council: Brussel, Belgium, 2015.

The Patent Protection of Nanotechnological Inventions: The European View

Claudio Germinario

Abstract: The patent protection of nanotechnological innovations represents an exciting challenge for all those in the specialized circle. The scope of the present work is to highlight how the protection of nanotechnological inventions implies a new interpretation and application of the general requirements of patentability. By relying on the Case Law of the Board of Appeal of the European Patent Office, the author will illustrate with examples how the requirements of novelty, inventive step, industrial applicability, clarity, and sufficiency of disclosure apply in the nanotechnological platform.

1. Introduction

What is a patent?

The patent is a "contract" involving three parties: inventor, authority, and public. The inventor makes the innovation and applies for a patent; the authority grants the patent, which confers exclusive rights to the inventor (i.e., a monopoly for the commercial exploitation of the invention); and the public is allowed, and must be enabled, to use and exploit the invention after expiry of the patent terms, normally 20 years. Therefore, the very concept of "patent" represents the compromise of two opposite interests: the inventor's interest and the public's interest.

2. The Patent Application

It is fundamental to diffuse the culture and knowledge of these innovative, environmentally friendly nanomaterials and nanotechnologies by scholastic teaching, together with the use of newspapers, television, and any other mean of mass media communication.

The patent application is a written document normally consisting of two sections: (i) the description of the invention and (ii) the claims.

(i) The description discloses the invention in all its technical aspects. The description may contain specific examples of concrete invention embodiments and make reference to drawings. In order that the public's interest (i.e., the possibility of exploiting the invention after expiration of the patent) is safeguarded, the patent application shall disclose the invention in a manner sufficiently clear and complete for it to be carried out by a person skilled in the field, i.e., the public.

(ii) The "claims" are the legal section determining the extent of the protection conferred by the patent or patent application.

2.1. The Basic Principles of Patentability

In order for an invention to be patentable, it must be novel, involve an inventive step and be susceptible to industrial application.

Additionally, as seen above, the invention must be clearly and sufficiently disclosed.

An invention is considered to be new if it does not already form part of the state of the art; in other words, if it was not made available to the public by means of a written or oral description, by use, or by any other way before the date of filing the patent application.

An invention is considered to involve an inventive step if it is not obviously derivable from the state of the art and it is considered to have industrial application; if it can be made or used in any kind of industry, including agriculture.

3. Patentability in the Nanotech Field

This section evaluates how the usual requirements of novelty, inventive step, industrial applicability, clarity, and sufficiency of disclosure apply to inventions in the nanotechnology area [1].

First, it is necessary to identify an accepted and shared meaning of "nanotechnological invention". It seems there is a consensus in the scientific and patent literature, on the point that "nanotechnology" is not simply the science of the infinitesimally small (i.e., at least one dimension of less than 100 nm), but rather, the science of the infinitesimally small accompanied by at least one new property (physical, chemical, pharmaceutical, etc.) directly derived or caused by the nano-size.

Under this definition, a patentable nanotech invention may be reasonably considered as a patentable "selection-invention". A selection invention is an invention based on the recognition of a sub-range extracted from a previously known larger range, with said sub-range causing a technical effect not recognized or attributed to the whole larger range.

3.1. Novelty

If, prior to the filing date of the patent application seeking protection for an invention, the same invention had already been described in a written publication, oral presentation, or public prior use, then the invention no longer fulfils the requirement of novelty necessary for patentability. For example, if a natural product, such as chitin, is generically described in a scientific publication, with its chemical structure and some characterizing parameters, then this product will be excluded from subsequent patent protection since it is lacking novelty.

However the question is whether the same principle also applies when we move from the macro scale to the nano scale. For instance, if the invention consists of "nanocrystals" or "nanofibrils" of chitin, can this "nano-form" of chitin be considered novel over the chitin described in the prior art?

An indirect answer can be found in the case law of the Board of Appeal of the EPO, which laid down general principles in assessing the novelty of polymorphisms, but that also find application in the field of nanocrystalline substances.

For instance, the Board of Appeal recognized the novelty of a substance either in crystalline or in amorphous form, although the very same substance was known the prior art in a different physical form (either amorphous or crystalline, respectively) [2–4]. The Board of Appeal also recognized the novelty of Aspartame type IIa, whcih is a specific polymorphic form of aspartame, over Aspartame III, which is another polymorphic form of the same aspartame dipeptide [5].

By translating the same principles, a substance in nano-form, such as nanocrystals or nanoparticles, may be considered novel, i.e., a different entity, over the same substance in macro-size form. This is indeed the case of the chitin nanofibrils (or chitin-whiskers), over the normal amorphous chitin or chitin long fibers obtained by electro-spinning. The same principles have been applied in assessing the novelty of composite materials, comprising a nano-sized particulate phase of a given chemical compound.

In a first decision [6], the Board acknowledged the novelty of a composite material comprising a layer of nanocrystalline nickel obtained by electrodeposition, on the basis of the size of the nanocrystals (less than 11 nm). Prior art documents disclosed essentially identical materials, yet comprising a layer of crystalline nickel having micro/macro-size crystals (>100 nm).

In a second decision [7], the Board of Appeal was faced with the problem of evaluating the novelty of a cigarette filter, consisting of a cellulose acetate tissue comprising nanoparticles of TiO_2 of size less than 100 nm as a photodegrating agent. A prior document disclosed essentially identical cigarette filters, but comprising TiO_2 particles of a much larger size, about 500 nm (pigment grade).

4. The Board Recognized the Novelty of the Invention

4.1. Inherency

The situation may prove even more intriguing when the down-scaling to nano-size confers to the "known" substance some novel properties, either chemical, physical, or biological.

In the macro world, it is a generally accepted rule that the discovery of a novel property of a known substance is unable to restore the novelty of this substance. It is in fact considered that the novel property was inherent to the substance itself.

In other words, if the novel property had been searched for in the substance, it would have been immediately revealed.

This does not necessarily happen to nanomaterial, since the novel properties characterizing the material in the form e.g., of a nanoparticle or a quantum-dot may directly result from the infinitesimal size of the material and, accordingly, could not be detected in the same material at the macro scale, even if it was purposely looked for. For this reason, the inherency approach, and its impact on the novelty of macro materials, no longer applies when we move from the macro to the nano scale.

4.2. Inventive Activity

The claimed subject matter is considered to involve an inventive activity if, for the skilled person, it is not obviously derivable from the state of the art.

This requirement, applied to nanoscience, raises a number of questions, specifically in consideration of the interdisciplinary nature of the nanotech platform. For example, which should be the reference technical field for inventions overlapping different technical fields, such as biotechnology, electronics, and semiconductors? Also, who should be the skilled person?

The case law of the EPO in the biotechnological field already offers some answers, which can be extended to the nanotech field.

The Board has recognized in many cases in the biotech field that the skilled person is indeed represented by a team of persons, each of whom is an expert on a different aspect of the same technical field. Furthermore, in the nanotech platform it is likely that this fictitious person will be embodied by a team of persons, each of whom is an expert on a different technical field: for instance, gene technology and semiconductors. Yet, this multidisciplinary approach raises the bar in assessing the requirement of inventiveness, making this requirement more difficult to meet. In fact, what seems to be inventive to one single "expert", may prove obvious when evaluated in the light of the combined knowledge of two or more experts.

A further question is whether the miniaturization of an object can in itself endow said object with an inventive step. In the macro domain, the answer is in the negative. The Board of Appeal has already set forth [8] that the easy miniaturization of devices or equipment is suggested by the well-known industrial need of optimizing space and materials.

However, moving into nano-size can be regarded as an obvious step only if the necessary tools and technology are available to the skilled person. It is well known that the top-down method has technical limits. Overtaking these limits by an alternative top-down method or by novel bottom-up methods could never a priori be regarded as obvious.

Moreover, as already seen above, miniaturization to the nano scale is very often accompanied by a novel and unpredictable effect. Under these circumstances, the result of miniaturization should always be considered inventive, regardless of the apparent obviousness of the means used to achieve the result.

5. The New Technical Effect Directly Caused by the Nano-Size

There are in the patent literature many examples of "novel properties" or "novel effects", caused by miniaturization, that justified the patentability of nano-substances or nanomaterials already known in macro-form.

For instance, chitin nanofibrils show, in addition to optimal mechanical properties, a surprising resistance to hydrolytic enzymes as compared to natural amorphous chitin. See Table 1 (kindly provided by Prof. Pierfrancesco Morganti—Mavi Sud S.r.l.).

Table 1

	CHITIN NANOFIBRILS		Weight variation		CHITIN NATURAL		Weight variation	
	INITIAL	FINAL	g.	%	INITIAL	FINAL	g.	%
CELLULASE	0.10028	0.10021	- 0.00007	0.07%	0.10060	0.10004	- 0.00056	0.56%
LYSOZYME	0.09999	0.09991	- 0.00008	0.08%	0.10010	0.09900	- 0.00110	1.09%
PECTINASE	0.10110	0.10109	- 0.00001	0.0099%	0.10234	0.10120	- 0.00114	1.11%
AMYLASE	0.10031	0.10029	- 0.00002	0.02%	0.10905	0.10305	- 0.00600	5.5%
COLLAGENASE	0.02563	0.02560	- 0.00003	0.12%	0.10958	0.10008	- 0.00950	8.67%

This unexpected property makes chitin nanofibrils a very promising material in dermatology and cosmetic surgery, for protecting wounds or damaged skin or as filler in the treatment of wrinkles, showing a high resistance to hydrolytic endogenous enzymes.

Thank to this novel effect, chitin nanofibrils have been claimed for use in dermo-cosmetic surgery and in therapy [9]. They have been claimed in complexes with negatively charged polymers for use as carriers for medicament [10], for the preparation of films comprising anti-bacteria metals, as well as for many cosmetic and therapeutic applications and food supplementation or preservation [11].

Other examples include nano-constructs with pharmacological activity comprising an inorganic metallic nanoparticle with an amphiphilic polymeric coating and an active antiviral peptide bound thereto [12]. The nanoparticles used as a carrier of the medicament enable an effective transportation of the active molecule through the cellular membrane and directly into the cell cytosol, thus escaping the endosome pathway and accordingly the lysosomal degradation, i.e., an unpredictable behavior caused by the nano-structure.

Yet another example is the "antireflective coating for use in photolithography" of Advanced Micro Devices Inc. [13,14]. An antireflective Titanium Nitride (TiN) layer with a thickness between 25 and 40 nm minimized the percent of reflected radiation to less than 5%. Thickness values below 25 or above 40 nm, as disclosed in the prior art, resulted in a significant increase of the reflected radiation up to 30%, as illustrated in Figure 1.

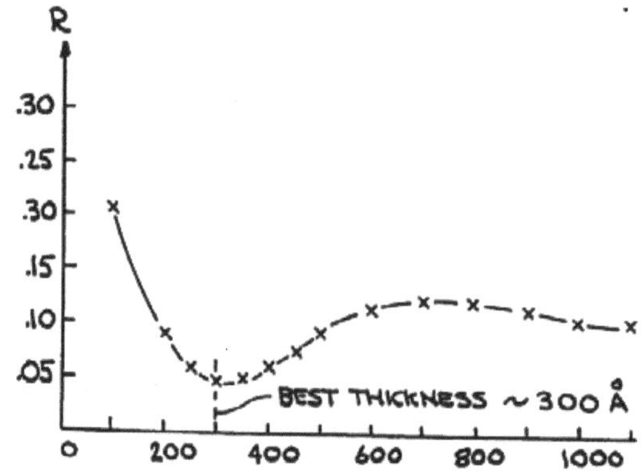

Figure 1. Reflected radiation of about 5% for 25 to 40 nm (i.e., 250–400 Å).

An additional case [15,16] (Rohm and Haas Company) related to a coating composition comprising a dispersion of copolymer particles having a size range of 20 to 70 nm. The miniaturization of the particles to 70 nm or less, although achieved by "normal" methods, was able to endow the claimed coating with an unpredictable optimal effect, namely with excellent transparency and clarity, where the effect was lost for larger particle-sizes. The closed prior art document described comparable coating compositions comprising polymer particles spread along a much broader range (10–1000) and preferably along the range of 100–500 nm. Accordingly, the Board of Appeal recognized the novelty and the inventive step involved in the selection of the range of 20–70 nm over the coating compositions described in said prior document.

6. The Reasonable Expectation of Success

Another important factor playing a role in evaluating the existence of an inventive step is the "reasonable expectation of success". This principle, which was developed and frequently applied in the biotechnology field, may give a strong support to inventiveness in any pioneer field such as nanotechnology. In fact, the prejudicial effect, for the patentability of an invention, represented by an earlier suggestion of the invention in the prior art depends on the circumstances. In a traditional technical field, all procedures and tools needed for realizing a "suggestion" are normally known and available to the skilled person and thus the results easily predictable. Under these circumstances, putting to practice the suggested teaching of a prior art document is always accompanied by a high expectation of success, and does not involve any inventive merit.

In reverse, in an emerging field in which the technology is not necessarily available and the procedures are not always standardized or reproducible, the result of an action is hardly predictable. In this situation, the realization of a "suggestion" is not always accompanied by a "reasonable expectation of success". Thus, achieving the desired result may prove inventive even though it is theoretically suggested.

For example, the above discussed decision [6] related to a composite material comprising a layer of nanocrystalline nickel of less than 11 nm, obtained by electrodeposition. As many as eight prior documents apparently suggested the same process of the invention, but in the context of microcrystalline structures (i.e., in the order of 100- to 1000-fold larger). Under these circumstances, the Board found that "...the skilled person had no obvious reason to foresee that the prior teaching could still be successfully extrapolated to structures smaller by at least two orders of magnitude, if not with the benefit of hindsight". In other words, the skilled person could not have any expectation of success when putting to practice the prior art teaching on nano-size level.

7. Sufficiency of Disclosure

The patent application describes the invention in a manner sufficiently clear and complete for it to be carried out by the skilled person. This requirement means that the skilled person should be able to repeat the invention, essentially over the whole ambit of the claim, based on the technical teaching given by the application (corroborated by the common general knowledge), and the realization of the invention, with the achievement of the declared effects of the invention, may not depend on chance. These conditions are essential in the patent protection of nanotech inventions.

In fact, when the technical effect achieved by, and characterizing, the invention strongly depends e.g., on the (nano) size of the claimed material or on the delicate conditions necessary to obtain said material and to realize the invention, the requirement of sufficiency of disclosure plays a decisive role and leaves no room for practices broadly shared in other technical fields, such as arbitrary generalization of the real invention or the extension of the protection to non-investigated equivalents.

Coming back to the antireflective coating for use in photolithography of decision T 453/97 (Note [14] and Figure 1) the effect characterizing the invention was the percentage of reflected versus incident radiation of less than 5%. This effect was quickly lost outside the very limited thickness range of 25 to 40 nm. Therefore, any arbitrary generalization of this range would make the realization of the invention and the achievement of the characterizing effect impossible.

The same applies to the concept of "equivalents", and strongly suggests that in the nanotech field there appears to be no room for any form of speculation or arbitrary extension of the scope of protection to cover embodiments theoretically "equivalent" to the real invention, but not yet investigated. This means that the need for a sufficient number of examples and experimental results substantiating all described embodiments of the invention is felt in the nanotech field more than in any other technical field.

For instance, in a case relating to an electro-magnetic transducer [17], one element of the invention was a semiconductor layer produced by molecular beam epitaxy using, as a doping element, atomic (N°) nitrogen radicals.

A similar method was described in a prior art document which illustrated with all experimental data an embodiment making use of molecular (N2°) nitrogen radicals, but which also cited as an alternative the use of atomic (N°) radicals, though without any support of examples or experimental results. Lacking any experimental confirmation, the Board considered that the use of atomic (N°) radicals in the prior art document was a merely a speculative attempt to disclose not yet investigated subject matter and accordingly disregarded this prior art teaching.

8. Metrology

A specific aspect of the requirement of "sufficiency of disclosure" is metrology. For the scientist, the definition of metrology is "the science of weights and measures" [18], thus including procedures, techniques, instruments, sensors, or any other tool for measuring.

On the contrary, for the patent attorney, "metrology" refers to an important aspect of patentability, namely the requirement of "sufficiency of disclosure" which enables the "repeatability" of the invention by the skilled person. Thus, metrology is indeed a condition for patentability.

To meet this requirement, the expert, supported by the technical teaching in the patent specification, optionally integrated by common general knowledge, should be able to put to practice the invention, in addition to achieving the technical effects the invention is said to produce. It goes without saying that the skilled person must be aware of all necessary instructions for monitoring the parameters characterizing the invention and for measuring the results obtained, in order to understand whether or not he/she has actually achieved the declared effect of the invention.

In a traditional technical field all analytical procedures, tools, and techniques are usually a part of the common general knowledge, and thus they need not be exhaustively described in the patent specification. However, this is not always the case in a pioneer field, such as nanotechnology.

In fact, nanotechnological inventions, like all inventions in any emerging fields, are often characterized by non-conventional features and effects: by unusual parameters, e.g., nano-size, monoatomic layers; by unusual processes, e.g., epytaxis, spattering, mono-atomic deposition; by unusual effects, e.g., detection of a single analyte molecule or detection of fluctuation of a specific index. Under these circumstances, the patentability of an invention strongly depends on the detailed description in the patent text of all technical aspects falling within the concept of "metrology".

This requirement has been clarified in many decisions of the Board of Appeal of the EPO. In a very recent case [19], the patent claimed: (i) "A silica glass for EUV lithography containing TiO, characterized in that the fluctuation of the refractive index (Δn) is at most 2 × 10-4 within an area of 30 mm × 30 mm in each of two orthogonal planes".

The validity of the patent was contested in an opposition procedure based on the following opponent's arguments: " ... that the patent did not provide sufficient information on the measurement of the parameters appearing in the claims: i.e., 'the fluctuation of the refractive index (Δn) ... '". Therefore, the skilled person repeating the invention did not know for sure whether the obtained glass fell within the scope of protection of the claims.

Another decision [20] related to a patent protecting "A silica gel, characterized, inter alia, by the feature represented by the silanol-group density of 6–20 ... " in combination with other parameters.

Yet, the patent specification taught an incomplete and erroneous method for determining and measuring the silanol-group density cited in the claim. Nor could the skilled person find any more reliable teaching in the prior art documents. For this reason, the Board revoked the patent, affirming that: " ... an erroneous method of measurement or determination of an essential parameter of the claimed product unavoidably results in a lack of sufficiency of disclosure of the claimed product ... ".

The same principles were recently confirmed by still another decision [21] relating to an adsorbent article, whose absorbent capability was defined in the patent claims by totally inconsistent values produced by an arbitrary test method.

The Board revoked the patent, arguing that: " ... the purpose of a parameter contained in a claim is to define an essential technical feature of the invention. ...The method specified for determining that parameter should therefore be such as to produce consistent values, so that the skilled person will know, when he carries out the invention, whether what he produces will fall within the scope of the invention".

9. Clarity

This requirement ensures that the claims define the object of the protection and are clear, concise, and based on the description.

A simple question arises here: which is the most suitable claim format, for a nanotech invention, that is able to meet the requirement of clarity without unduly reducing the scope of protection?

In fact, there exists no claim form specific to nano-invention, but there are claim forms which are certainly more suitable than others, as illustrated below.

10. Product-by-Process Claim

At the nano scale, it may prove difficult to properly define an invention, product, substance, or material by way of its structural features. It may be easier and safer to identify the invention by way of its preparation process, using wording such as: "product obtainable by the following preparation process comprising . . . ". For this reason, the product-by-process claim is extensively used to define inventions in the nanotech platform.

For example, the "nanocrystal metallic material" object of decision T0915/00 [6] was defined as being " . . . obtainable by a process comprising the electrodeposition of the material on a substrate in a solution containing ions of the metal characterized . . . by passing a D.C. current . . . at pulse intervals . . . ".

The meaning recognized by the Board of Appeal of this type of claim is quite different when the invention is on the macro scale or in the nano field. At the macro scale, there is a generally recognized principle according to which a new process for preparing a known product is unable to restore the novelty of the obtained product.

Yet, at the nano scale, it is difficult to prove that even minor modifications in the process parameters are neutral as regards the structural features of the obtained product. This has been the position taken by the Board of Appeal in more than one decision.

In the aforementioned decision [6], Nickel nanocrystals with the size of 11 nm were obtained by electrodeposition by passing in an ion solution a D.C. current at pulse intervals. These nanocrystals have been considered different from Nickel nanocrystals of nearly the same size obtained by electrodeposition by passing in the ion solution a D.C. current having the same characteristics, but applied continuously. The Board accepted the appellant's argument that the different process resulted in a different crystalline structure, though having the same size.

By applying the same criteria, the Board acknowledged in a subsequent decision [22] the novelty of a nanocrystalline product obtained by sputtering versus the same nanocrystalline product obtained by electrodeposition.

In an earlier case [23], a device having a specific surface topography was defined by means of its preparation process involving epitaxial growth. The Board recognized the novelty of the device, accepting that the process gave rise to peculiar characteristics.

11. Functional Definitions

Another frequently used claim type for the characterization of nanotech inventions is the functional claim, which defines the claimed compound by way of its functional properties. As seen above, the miniaturization to the nano scale is frequently accompanied by a novel specific property. Accordingly, it is sometime easier and clearer to characterize the miniaturized material by way of its novel property caused by the nano-size rather than to characterize the material by way of its structural features, which eventually produce the novel property. For this reason, the functional definition of the invention is extensively used in this field.

12. Terminology

A final aspect of clarity is the clarity of the terminology used to describe the invention. In a pioneer multidisciplinary area, the technical meanings of the terms are not necessarily standardized and equally recognized by the experts of the different overlapping fields. This circumstance may render the interpretation of the claim, and accordingly the interpretation of the scope of protection, subjective and unpredictable. Therefore, all nanotech patent applications should contain a self-standing "glossary" clearly explaining the meaning, for the purpose of the invention, of all technical terms used in the claims. This glossary should be intended not only to facilitate the examination of the application, but, with a view to the future, to enable the national courts to make reliable decisions on the question of patent infringement.

Conflicts of Interest: The author declares no conflict of interest.

References and Notes

1. Germinario, C. Learning The Lesson: Can Nanotech Patenting Learn From The Past Biotechnology Experience? *Patent Word* **2006**, *181*, pp. 26–29.
2. E.I. du Pont de Nemours and Company. Mineral filled copolyamide compositions. Decision case number T 0037/01, 18 June 2003.
3. Enichem Synthesis S.p.A. Product in amorphous form. Decision case number T 0020/94, 04 November 1998.
4. Warner-Lambert Company LLC. Atorvastatine new polymorphic form. Decision case number T 0777/08, 24 May 2011.
5. Ajinomoto Co. Inc. Dipeptide crystals. Decision case number T 0509/92, 22 July 1997.
6. Integran Technologies Inc. Nanocrystalline metals. Decision case number T 0915/00, 19 June 2002.
7. Celanese Acetate LLC. Photodegradable cellulose ester tow. Decision case number T 0006/02, 03 February 2005.
8. Trustees of the University of Pennsylvania. Fluid handling in microfabricated analytical devices. Decision case number T 0070/99, 23 January 2003.

9. Mavi Sud. Preparation of Chitin and Derivatives Thereof for Cosmetic and Therapeutic Use. International Patent Application No. PCT/IB2005/053576; Publication No. WO/2006/048829, 17 October 2006.
10. Mavi Sud. Method of Preparation of Chitin and Active Principles Complexes and the so Obtained Complexes. International Patent Application No. PCT/IB2012/051961; Publication No. WO/2012/143875, 26 October 2012.
11. Mavi Sud. Suspensions and Materials Comprising Complexes of Chitin Nanofibrils with Metals. International Patent Application No. PCT/IB2014/065015; Publication No. WO/2015/049656, 9 April 2015.
12. Università degli Studi di Milano Bicocca; Università degli Studi Statale di Milano. Nanoconstructs with Pharmacological Activity. International Patent Application No. PCT/IB2013/055943; Publication No. WO/2014/013473, 23 January 2014.
13. Advanced Micro Devices Inc. Patent Application No. EP0289174, 2 November 1988.
14. Advanced Micro Devices Inc. Antireflection coatings for use in photolithography. Decision case number T 0453/97, 22 July 1998.
15. Rohm and Haas Company. Method of treating or coating a substrate with an aqueous composition and the use of such compositions. Patent Application No. EP0429207, 28 February 1996.
16. Rohm and Haas Company. Method of treating or coating a substrate with an aqueous composition and the use of such compositions. Decision case number T 0952/01, 15 January 2004.
17. Minnesota Mining and Manufacturing Company, et al. Electro-magnetic transducers. Decision case number T 288/02, 22 April 2004.
18. *Collins English Dictionary*, 12th ed.; Harper Collins Publ.: UK, 2014.
19. Asahi Glass Company. Silica Glass Containing TIO2 and Optical Material for EUV Lithography. Decision case number T 0023/11, 13 February 2014.
20. Degussa AG. Fällungskieselsäuren. Decision case number T 1250/01, 14 Mai 2003.
21. The Procter & Gamble Company. Absorbent articles comprising a material having high flux capabilities. Decision case number T 0815/07, 15 July 2008.
22. Dana Corporation. Bearings. Decision case number T 0753/02, 04 March 2004.
23. British Telecommunications. Optoelectronic device. Decision case number T 0950/97, 26 March 2002.

Subject Index

Glossary and Abbreviations

Acetic acid: a two-carbon, carboxylic acid, colorless and with a pungent odor, which is the most essential component of vinegar.

Acetyl Glucosamine: molecule composed of glucose with a chain of acetyl groups linked to an amino group.

Advanced medications: non-woven tissues or specialized tools for medical purposes, produced by innovative molecules and methodologies.

Agglomerate: collection of weak, la-bound particles or an aggregated mixture of the two in which the resulting external surface area is similar to the sum of the surface areas of the individual components.

Aggregate: particle comprising strongly bonded or fused particles in which the resulting external surface area may be significantly smaller than the sum of calculated surface areas of the individual components.

Alginates: water-solubile and anionic, linear hetero-polysaccharide composed of two different monomers (beta-D-mannuronic acid and alpha-L-glucuronic acid, (1--4)-beta-linked).

Allograft: transplantation of cells, tissues, or organs from a nonidentical donor of the same species into a recipient.

Angstrom: one-tenth of a nanometer.

Antimicribial agent: substance that inhibits the growth of microorganisms such as bacteria, fungi, or protozoans.

Antioxidant: molecule capable of inhibiting the oxidation of other molecules.

Autograft: transplantation of cells, tissues, or organs from one part of the body to another in the same individual.

Allotrope: forms of an element with different arrangements of atoms.

Amylase: enzyme that is present in human saliva for beginning the chemicals process of digestion and catalyzing the breakdown of starch into sugars.

Angiogenesis: growing new blood vessels by sprouting existing ones.

Anti-inflammatory: activity of a specialized network of compounds that slow downs and regulates the inflammatory cascade of cytokines.

Antimicrobial: activity of specialized compounds to modulate the growth of pathogens and microorganisms.

Antioxidant activity: protective activity performed by ingredients such as A, E, and C vitamins and enzymes to protect the skin and human body from environmental aggressions.

Anthocyanin: large group of red-blue plant pigments(flavonoids) with antioxidant properties, occurring mostly in flowers and fruits.

Apoptosis: also known as programmed cell death, is the process that deletes cells from a population in a deliberate manner without eliciting an immune response.

Arginine: polar, semi-essential amino acid involved in many metabolic processes. It is one of the constituents of NMF, natural moisturizing factor of the skin.

Artificial polymer: polymer man-made by synthesis.

Attract: move closer together, opposite charges (positive and negative) and opposite magnetic poles(north and south) are attracted to each other.

Bacterium: unicellular prokaryote microorganism, classified as a lower form of life.

Bioavailability: the presence of active molecules or products where they are needed in the body to be most effective.

Bio-based: material derived from plant, living matter, or renewable biological resources.

Biodegradable plastic: plastic compound that undergoes biodegradation via naturally-occurring microorganisms, such as bacteria, fungi, and algae.

Biodegradation: chemical dissolution of materials by bacteria or other biological means.

Bio-economy: economy based on the use of agricultural and industrial by-products with low consumption of water and energy, which safeguards social necessities.

Bioethylene: obtained from plant biomass, via bio-ethanol, represents a chemically identical alternative of ethylene from petrochemical feedstocks. Its synthesis reduces the production of greenhouse gas emissions (GHG), retaining a biodegradable compound.

Biomass: natural source of raw materials remaining from the agricultural and industrial processes used in the production of bio chemicals regarded nowadays as an alternative to petrochemical technologies.

Bio-mineralization: the process by which living organisms produce minerals to harden or stiffen existing tissues.

BIOBOARD: EU project that deals with the development of sustainable, protein-based paper and paperboard coating systems to increase the recyclability of food and beverage packaging materials.

Biocompatibility: properties of materials that, being biologically compatible, do not elicit any local or systemic responses from a living system.

Bio-composite: biopolymers or synthetic polymers reinforced with natural fibers.

Biodegradability: the ability of organic substances and materials to be broken down in carbon dioxide, methane, and water through the enzymatic action of microorganisms into a specific period of time and environment.

Biodegradable: material capable of disintegration by biological means.

Biodegradable compostable packaging: specific type of biodegradable and compostable package that may be intended for the composting of organic waste. The resulting compounds may be used as fertilizer.

Biodegradable (non-compostable) **packaging:** packaging that can be degraded in the presence of microorganisms, releasing CO2, water and methane into the atmosphere under certain time, heat, humidity, and oxygen conditions.

Biodiversity: represents the variety of different types of life found on earth, measuring all the organisms found in its ecosystems.

Bioinformatics: interdisciplinary field that addresses biological problems through the use of computational techniques.

Biological environment: complex of physical, chemical, and biotic factors in which life forms can exist, positively acting upon an organism or an ecological community.

Biomass waste: is represented by all the by-products that remain in the environment after agricultural and industrial production.

Biomaterials: defined as a material or combination of materials, synthetic or natural in origin, that can be used to flair, replace, or model tissues and organs *in vitro* and *in vivo.*

Biomaterials science: multidisciplinary field involving materials, mechanical, and biomedical engineering; molecular cell biology clinical medicine; histology; bioethics; and all the disciplines connected with the life.

Biomimetic production: green methodology that, looking to the productive methods adopted from animals and plants, makes ingredients and products with the use of enzymes without using chemicals.

Biomimetic product: product made by green methodologies with low consumption of water and energy, and use of waste products as raw materials.

BIO-MIMETIC: EU project based on the use of lignin and chitin nanofibrils, obtained from biomass, to produce innovative household products and cosmetics through the use of enzymatic technologies with low consumption of water and energy.

Biomimicry: imitation of some systems in nature used to solve human problems.

Bionanomolecules: naturally occurring building blocks, recognition modulae, and machines, such as ribosomes.

Bionanotechnology: use of biological principles for nanotechnological applications; an overlap between biology and nanotechnology with the prevalent use of biomolecules of natural origin obtained from agricultural and industrial waste, as an alternative to chemicals that are petrol-derived.

Bioplastic: whole family of materials that are bio-based, biodegradable, or both, derived from renewable biomass sources, such a fishery's, plant waste, or microbiota.

Biopolyester: as *cutin*, this is a compound composed of hydroxy and epoxy acids, functioning as a barrier between the aerial parts of higher plants and their environment.

Biopolyethylene: polymer made through the use of renewable feedstocks that emits less greenhouse bases when compared to fossil-based polyethylene. Its production helps reduce greenhouse gas emissions, but the polymer is not biodegradable like polyethylene.

Bioprocess: any process that uses complete living cells or their components to obtain desired products.

Biorefinery: facility that integrates biomass conversion processes and equipment to produce fuels, power, and chemicals.

Biosynthesis: production of chemicals that are enzyme-catalyzed by living organisms.

Biotechnology: any technological processes and applications that use biological systems or living organisms to make products.

Blow Molding Process: most advanced method for processing plastic bottles. Transparent bottles are made of PET, while opaque ones are made of HDPE or PL.

Bottom-up: process based on the formation of complex structures through the coordinated assembly of simple building blocks.

Bulk material: material ordered, stored, issued, and sold by weight.

Burn: severe skin damage, caused by heat, electricity, chemical, or irradiation, where many of the cells affected die. It is categorized by severity as first, second, and third degree, when the skin turns white and loses sensation.

Carbohydrates: polymers of simple sugar molecules that are bonded together.

Carrageenan: family of high molecular weight-sulfated polysaccharides obtained by extraction from some categories of the algae class.

Casting Technology: low- or high-pressure process that allows one to develop, under controlled condition, thin films by using natural/man-made fibers or other special materials.

CBC: productive process characterized by three main bio transformations of lignocellulosic materials: production of saccharolytic enzymes, hydrolysis of cellulose and hemicellulose to sugar, and fermentation of next se and pentose sugars.

Cell cycle: series of events that takes place in a cell, leading to its division and duplication.

Cell culture: process in which cells grow under controlled conditions, outside their natural environment.

Cell differentiation: process in which a precursor cell acquires its final specialized character.

Cell membrane: biological membrane that separates the interior of all cells from the outside environment.

Cell signaling: process in which cells communicate with each other or are stimulated or inhibited by extracellular signals.

Cell viability: method to control the state of cells, to determine their possibility of living or dying , after a specific treatment made *in vitro* by cellular cultures. Cell viability tests might calculate the effectiveness and safeness of many active ingredients.

Cellular uptake: cellular mechanisms regulating the endosomial trafficking of materials into the cells to determine their delivery effectiveness.

Cellulase: enzyme that catalyzes the hydrolysis of the glucose 1,4 bond to form polysaccharides.

Cellulose: structural linear polysaccharide consisting of repeating cellobiose units made by two glucose units connected via befa-1,4 glucosidic bond. This natural polymer provides tensile strength in the cell walls of plants.

Catabolism: set of pathways that break down molecules into smaller units, releasing energy.

Catechin: photochemical compound with antioxidant property found in high concentration in green tea and other vegetables, such as grapes, cocoa, etc.

Chemical structure: molecular geometry reporting the arrangement of the chemical bonds of the chemical compound.

Chemicals: terminology used to indicate chemical products man-made by petrochemical processes.

Chemoattractant: chemical substance that induces a cell or organisms to migrate towards it.

Chitinase: digestive enzyme that breaks down glycosidic bonds in chitin.

Chitosan: partially de-acetylated and water-soluble form of chitin.

Chitin: a natural polymer obtained from crustacean waste, formed by many molecules of glucosamine and acetyl glucosamine linked by covalent bonds.

Chitin dressing: non-woven tissue made prevalently by chitin micro/nano fibers.

Chitin nanofibrils: pure crystal-like fibrils of chitin having a mean dimension of 240x7x5 nm obtained from a patented methodology.

Cicatrization: physiological process of wound healing producing a scar tissue.

CN: Chitin Nanofibrils. The purest Crystallin portion of the raw material chitin.

CNC: Crystallin cellulose nanofibers. It is the purest part of the cellulose polymer.

Co-extrusion: process that involves the extrusion of two or more materials through a single fiber due to plastic fiber production.

Collagen: protein rich in glycine and proline that represents the major component of the ECM of animals (about 30% of all the body proteins) and confers tensile strength to the skin tissue.

Collagenase: enzyme that breaks collagen peptide bonds, often destroying their structures.

Colloidal quantum dots: semiconductor nanocrystals between ~1 and 10 nm, which are transitioning between bulk solids and molecules. Their different emission colors are size-dependent.

Competitiveness: ability and performance of a company to supply goods and services in a given market, in relation to and in competition with other companies.

Composite: building material consisting of two or more constituent materials with different physicochemical characteristics that are different from the individual components used.

Compost: result of disintegration and aerobic biodegradation of a product that can be used as a fertilizer.

Compostability: capacity of an organic material to be transformed into compost through the composting process, following the EU directive 94/62/EC on Packaging & Packaging waste.

Compostable: material that meets D6400 composting standards set by American Society of Testing and Materials (ASTM). It has to (1) break down to carbon dioxide, water, and inorganic compounds and biomass at a rate similar to paper; (2) disintegrate into small pieces within 90 days; and (3) leave no toxic residue.

Compostable polymer: plastic material that is capable of undergoing biological decomposition in a compost site as part of an available program. However, biodegradation and disintegration rate are the points that constitute the difference between biodegradable and compostable.

Copolymer: chemical compound of high molecular weight produced by polymerization of two or more monomers together.

Cosmeceuticals: terminology used to try to define cosmetic products made and controlled by more strict pharmaceutical methodologies.

Cosmesis: preservation or restoring of body beauty.

Cosmetic regulation: rules under which cosmetics are produced, claimed, and sold in different countries.

Covalent bond: chemical bond characterized by the sharing of one or more electrons between two atoms.

Cross section: shape obtained cutting transversally an object.

Crustacean waste: represents ~45% of the discarded by-material from the crustaceans' food. It is made prevalently by chitin embedded in calcium and magnesium carbonate.

Cyclodextrines: crystalline, non-hygroscopic, and cyclic oligosaccharides derived from starch, used as solubilizing and stabilizing agents.

Cytocompatibility: or cytological compatibility refers to the cytological effects produced by the temporary interaction between cells and the material of various polymers and surface topographies.

Cytokines: large group of proteins, peptides, or glycoproteins secreted by specific cells as signaling molecules to mediate and regulate the immune system, the inflammation cascade, and hemopoiesis.

Cytology: medical and scientific study of cells.

Cytotoxicity: degree to which an agent has specific toxic effects on the cells.

Kraft process: it consists of a chemical pretreatment of wood at 155/175 C by the use of a solution of Na2S/NaOH that allows the delignification of the raw material.

Dalton: unit used for indicating mass on an atomic or molecular scale. One Dalton is approximately equal to the mass of one proton or one neutron.

Defensin: family of antimicrobial peptides found in animals and plants capable of killing or inactivating Gram-positive and Gram-negative bacteria, fungi (including yeasts), parasites, and even enveloped fit uses such as HIV.

Degradable: general term used to describe all polymeric materials that disintegrate through a range of physical and chemical processes.

Detergent: surfactant or mixture of surfactants that have *cleaning properties in dilute solutions.* According to the ingredients used, the products refer to laundry or dish detergents or human cleaning agents.

Disaccharide: carbohydrate formed when two monosaccharides undergo a condensation reaction.

Disintegration: breakdown of a material into small and separate fragmentation.

Drug delivery: method or process of administering a pharmaceutical compound to achieve a therapeutic effect in humans or animals.

EC: European Economic Commission.

ECM: Extra cellular matrix. The gel composition of cellular components comprising the Hyaluronan-water network containing collagen and elastic fibers.

Ecological economics: economy embedded and supported by natural systems so that goods and processes have to be made and organized by low consumption of water and energy and by the use of agricultural and industrial waste as raw materials.

Ecosystem: systems of plants and animals that, living all in a perfect equilibrium in our planet, permit the maintenance of its biodiversity.

EDTA: ethylene-diamine tetra-acetic acid.

Effectiveness: capacity of the ingredient/product to show its efficacy.

Elagitannins: any group of tannins, related to ellagic acid, occurring in various tanning extracts.

Elastic fibers: bundles of proteins found in an extracellular matrix of connective tissue, characterized by great elasticity.

Elastin: small tropoelastic polymers used to form composite materials.

Electromagnet: metal core made into a magnet by passing a current through a wire wrapped around it.

Electron: negatively charged subatomic particle that moves around the nucleus of an atom; it acts as a carrier of electricity.

Electrospinning: technology that uses electrical charges to draw micro/nano fibers from a liquid. Using this method, it is possible to create nanofibrous scaffolds for tissue engineering.

Electrospun: non-woven tissue resulting from the electrospinning of fibers.

ELISA: enzymatic biochemical technique used for diagnostic tools, mainly in immunology, to detect the presence of an antibody or an antigen in a sample.

Environment: air, water, plant and animals that live around us.

Environmentally-friendly: ingredient/product that does not produce noxious waste.

Enzymes: macromolecules able to accelerate or catalyze bio-chemical reactions.

Epidermis: skin outer covering, composed of keratinocyte cells, that creates a waterproof barrier that is self-repairing and continually renewed.

Epigalochatechin: major Cathechin content with an antioxidant and cancer preventive capacity, accounting for 59% of the total catechins in green tea.

EPO: European Patent Office.

Epidermal growth factor (EGF): mediator implicated in keratinocyte migration, fibroblast proliferation and differentiation, and granulation tissue formation.

EU: European Union. It represents the Union of 27 different countries.

EuroPaBio: European Association for Bioindustries.

European Patent Office: EU office for applying and searching issues, grants, rules, and methodology for patents.

EVA: poly(ethylene-co-vinyl acetate). Polymer used to produce bi-layer sheets to enhance the barrier properties of food packaging.

Extrusion coating/lamination: coating of a molten web of synthetic resin onto a substrate material to form thin hot films.

Dermis: relatively thick layer of connective tissue beneath the epidermis that is composed of an Extracellular Matrix (ECM) gel rich of hyaluronic acid and water, into which collagen and elastic fibers are embedded.

FABS: Food, Agriculture, Fisheries, and Biotechnology. A flagship initiative for EU-China Cooperation on Research and Innovation on Food, Agriculture, and Biotechnologies.

FDA: Food and Drug Administration. The Health Ministry of the United States of America.

Feed stock: any bulk raw material or fuel required for an industrial process.

Fiber: structure of polymer that closely can mimic the extracellular matrix and may be used for tissue engineering and other medical applications, particularly when in nanoscale and of natural origin.

Fiber processing: methodology to produce natural or man-made fibers.

Fiber strength: any bulk raw material or fuel required for an industrial process.

Fiber structure: architectural organization of a fiber.

Fibroblast: cells that, when present in the dermis, synthesize the ECM framework.

Filler: reinforcing material for fiber nanocomposites, employed to improve the desired properties or simply reduce the cost of a nanocomposite.

Financial capital: plays an important role in our economy by reflecting the productive power of the other types of capital.

Flat die extrusion: common process and arrangement for cast film, extrusion coating, and extrusion laminating.

Flavonoids: large family of polyphenols with a common chemical structure and antioxidant properties, synthesized by plants.

Flexible packaging: package or container made of flexible or easily yielding materials that, when filled or closed, can be readily changed in shape.

Food packaging: package to provide protection, tampering resistance, and special physical, chemical, or biological needs. It plays a fundamental role in the maintenance of food benefits prior to consumption.

Force: push or pull that tends to charge the motion of an object.

FP7: Seventh Framework Program. The EU program of research finished at the end of the year 2013.

FTIR: Fourier Transform Spectroscopy analysis. It offers quantitative and qualitative measurement S for organic and inorganic samples, identifying chemical bonds and functional groups of molecules.

Furfural: colorless, oily liquid with an aromatic odor, obtained from cellulosic waste.

Gallotannins: class of hydrolisable molecules obtained by the condensation of the carboxylic group of Gallic acid. It acts as a cytoprotective agent in oxidatively stressed cells.

GDP: Gross domestic product. It represents all manufacturing products machines and services made by a country.

Gene Regulatory Network (GRN): combination of different regulators that communicate with each other and with other particles with and in the cell that control the level of gene expression in mRNA and in proteins.

Gene therapy: use of a gene to obtain a therapeutic, prophylactic, or diagnostic effect.

GHGs: greenhouse gas emissions. It is the gas that traps heat in the atmosphere.

Glycine: the simplest non-polar amino acid of human body. It can fit into hydrophilic or hydrophobic environments because of its minimal chain compose of only one hydrogen bond.

Glycan: generic term for any sugar, in free form or attached to another molecule, that is used interchangeably with carbohydrate.

Glucosamine: natural occurring amino sugar compound found in the human body, especially in the fluid around joints, where it appears as acetyl glucosamine.

Glycolic acid: smaller alpha hydroxy acid found in some sugar-crops, used in cosmetic dermatology as peeling and skin moisturizing agent.

Glycobiology: study of the structure, function, and biology of carbohydrates.

Glicomics: systemic study of all glycan structures in a biological system.

Glycopeptide: peptide having one or more covalently attached glycans.

Glycoproteins: protein with one or more covalently attached glycans.

Glycosaminoglycans: polysaccharide side chains of proteoglycans of free complex polysaccharides composed of linear disaccaride repeating units, each composed of a hexosamine and a hexose or hexuronic acid.

Glycosilation: enzyme-catalyzed covalent attachment of a carbohydrate to a polypeptid, lipid, polynucleotide, or another carbohydrate.

Gram-negative bacteria: bacteria, such as gonococci and meningococci, that cannot retain the violet stain after the decolonization step. They are especially found in the gastro-intestinal tract, are responsible for many diseases, and are resistant to many antibiotics.

Gram-positive bacteria: bacteria, such as staphylococci and streptococci, produce a positive result in the gram stain test (crystal violet), appearing purple-colored under a microscope.

Graphene: man-made material representing a new class of nanoparticles, known as nanotubes. Made by an atomic structure of carbonic atoms, nanotubes are about 200 times stronger than steel by weight.

Gravity: force of attraction between all objects; the amount of attraction depends in part on the mass of the objects and the distance between the objects.

Green economy: economical way of producing goods based on the low consumption of energy and water through the use of agricultural and industrial by-products as raw materials, while respecting the environment and society.

Green washing: *green sheen,* in which green marketing is deceptively used to promote the perception that an organization's products, aims, or policies are environmentally-friendly.

Guar gum: hydrocolloidal galactomannan structurally comprising long and straight chains of (1--4)-alpha-D-mannopyranosyl units linked together by (1--4)-beta-D-galactopyranosil units by (1--6) linkages.

HDPE: high density poly(ethylene).

Hexosamine: hexose with an amino group in place of the hydroxyl group at the C-2 position. Common examples are the N-acetylated sugars, such as N-acetyl glucosamine.

Hexose: six-carbon monosaccharide topically with an aldehyde at the C-1 position and hydroxyl groups at the all other positions.

Homogenization: methodology to reduce the size, for example, of an emulsion by reducing the size of its particles, distributing them equally throughout the suspension to render it uniform.

HPCS: hydroxy propyl chitosan. Chitosan enriched by propulsion chains.

Human capital: people's health, skills, and motivation, which are required for productive work.

Human physiology: study of how normal human tissues and organs function mechanically, physically, and in terms of biochemical functions.

Hyaluronan: mammals' natural polymer composed of repetitive molecules of glucuronic acid and acetyl glucosamine, which are capable of attracting and bonding hundred molecules of water.

Hydrogels: when derived from natural polymers, they are used biologically in stem cell engineering applications, as they are components resembling the ECM structure or exhibit properties similar to the matrix components in native tissues.

Hydrolisis: chemical process involving the breaking of a bond in a molecule using water.

Hydrolitic enzymes: complex of catalytic proteins that uses water to break down substrates in their simplest units.

Hydrophilic polymers: substrates containing polar or charged functional groups, which dissolve or become swollen in water.

Hydrophobic polymers: substrates which, in water, often cluster together to form micelles. Hydrophobic is often interchanged with lipophilic, i.e., ***fat-loving.***

Hydroxyapatite: naturally occurring mineral form of calcium phosphate and an essential component of normal bone and teeth.

IgE: Immune globulin E. An immune antibody protein that is produced by the immune system and acts as an antibody to activate allergic reactions.

IL-1 (interleukin-1): cytokine with pro-inflammatory activity.

IL-8 (interleukin-8): cytokine with pro-inflammatory activity.

IL-10 (interleukin-10): cytokine with antiinflammatory and immunosuppressive activity.

IL-12 (interleukin-12): involved in the differentiation of T cells.

IL-5 (interleukin-5): a T cell-derived glycoprotein that stimulates eosinophil production and activation.

IL-6 (interleukin-6): both pro-inflammatory and anti-inflammatory cytokine.
Imunostimulants: substances that stimulate the immune system by inducing activation or increasing activity of immune cells.
Interleukins (IL): small signaling proteins produced by a variety of cell types that participate in the body's defense system regulating many aspects of inflammation and immune and autoimmune responses.
Injection molding: manufacturing process for producing gods or parts of goods by injecting material into a mold.
Inventiveness: ability to think and/or realize new ideas and methods.
Ionic strength: measure for quantifying the concentration of ions or dissolved chemicals in a solution.
Iridescence: lustrous color caused by differential refraction of light waves, which change with the angle of view.
ISO: International Standardization Organization.
Keloid: irregular and shaped scar of fibrous tissue formed over the wound during the repairing process.
Keratinocyte: cell with the capacity to synthesize the keratin filament proteins, which give the epidermis its toughness. They are embedded in the skin and covered by lipid lamellae.
Kinetic: movement energy, i.e., the energy that possess an object because of its motion.
Kraft process: process for convention of wood into wood pulp, consisting of cellulose fibers. It consists of the treatment of wood chips with a mixture of sodium hydroxide and sodium sulfite that breaks the bonds that link lignin to the cellulose.
Lectin: protein that specifically recognizes and binds to Glycans without catalyzing a modification of the glycan.
Levulinic acid: organic compound classified as a ketoacid and produced from plant biomass, has pharmaceutical, food, cosmetic, and agricultural applications.
Lightness: perceived quality of color by which the object appears to reflect or transit the incident light.
Lignins: one of the main classes of materials that , with a structure rich in polyphenols, provides structural support for the plant, giving strength, rigidity, and resistance to environmental stresses.
Lignin derivatives: unlike lignins, they may be used in the pharmaceutical and cosmetic sectors because of their water solubility.
Lignosulfonates: compounds extracted from waste liquid of plant biomass at the end of sulfite-pulping process.
Lignocellulosic Feedstocks: inedible plant material known as biomass with polymeric structure composed principally of natural biopolymers built up by three main constituents: cellulose, hemicellulose, and lignin.
Lipectomy: surgical excision of subcutaneous fatty tissue.
Liquefaction: process by which a gas is concentrated to a liquid.
Liquid nitrogen: nitrogen in a liquid state at an extremely low temperature. Zit is produced industrially by fractional distillation of liquid air.
Lyophilization: or freeze-drying procedure: dehydration process used to preserve a perishable material.

Lysozyme: enzyme that, as part of the innate immune system, is characterized by the ability to break down the bacteria cell wall.
Lotus effect: occurring in the lotus plant, water is repelled by the surface of the leaves, rolling off and collecting dirt in the process.
Macro- : large scale.
Macromolecules: very large molecules, with a diameter ranging from ~100 to ~10,000 angstroms, commonly created by polymerization of smaller subunits (monomers).
Macrophage: specialized defensive cells that phagocyte the investing pathogens and help activate lymphocytes to mediate the sophisticated adaptive immune system responses.
Manufactured capital: comprises material goods or fixed assets, such as tools, machines, buildings, and other forms of infrastructure that contribute to the production process but do not become embodied in its output.
Matrix: composition of one or more polymers to be reinforced by the use of inorganic or organic materials called fillers.
MEM medium: Minimal Essential Media is one of the most commonly used cell culture media for *in vitro* studies.
MEMS: microelectromechanical systems.
Metrology: science of measurement that includes all the theorical and practical aspects of measurement.
Micro-: prefix meaning one-millionth (0.000.001).
Micron: one-millionth of a meter.
Microarray: also known as a biochip. Collection of microscopic DNA spots attached to a solid surface.
Microbe: microscopic living organism, such as bacterium, fungus, protozoan, or virus.
Microbiome: all of the genetic material of a microbial community sequenced together.
Microcapsule: mall sphere with a uniform wall around it.
Microspheres: spherical micro particles with diameters in micrometers ranging from 1 to 1000 millimicrons.
Mitogen: chemical substance that stimulates cell division.
Monosaccharide: basic unit of the carbohydrates.
Montmorillonite : very soft phyllosilicate group of minerals that forms a clay swelling in water, possessing a high cation-exchange capacity.
Nano-: prefix meaning one-billionth (0.000 000 001).
Nanobiomaterial: any material, surface, or construct that interacts with biological systems, having at least one dimension between 1 and 100 nm in size.
Nanobiotechnology: application of nanotechnological principles and tools to biology.
Nanocellulose: highly crystalline cellulose nanofibers, present in natural plant bodies that are useful as reinforcing agent in nanocomposites.
Nanochitin: known also as chitin whisker, chitin nanofibril, or chitin crystallite.
Nanocomposites: polymer-matrix composites in which nano-sized inorganic or organic fillers are distributed homogeneously, to give them more mechanical, thermal, and chemical resistance.
Nanocrystal: any nanometer-sized crystal.
Nanofibre: nano-object, flexible or rigid, with similar external dimensions in the nanoscale, whose third dimension is significantly larger.

Nanomaterials: material with at least one external dimension of approximately 1–100 nanometers.
Nanomedicine: application of nanotechnology in medicine.
Nanometer (mm): unit of length in the metric system, equal to one billionth of a meter.
Nani-object: material with one, two, or three external dimensions in the nanoscale.
Nanoparticle: particle of any shape with all three external dimensions between 1nm and 100 nm.
Nanoplate: Nano-object with one external dimension in the nanoscale, whose two other external dimensions are significantly larger.
Nanorod: solid nanofiber.
Nanoscale: size range from approximately 1 nm to 100 nm.
Nanoscience: study of particles on an atomic or molecular scale, whose size is measured in nanometers.
Nanotechnology: application of scientific knowledge for processing, separation, consolidation, and deformation of materials by one atom or by one molecule.
Nanotube: miniaturized cylindrical carbon-atoms structure measuring nanometers. It has composed of hexagonal graphite molecules attached at the edges.
Nanowire: electrically conducting or semi-conducting nanofiber.
Natural capital: any stock or flow of energy and matter that yields valuable goods and services.
Natural fibers: all the fibers found in nature.
Natural polymer: all polymers of natural origin.
n-CHITOPACK: EU research project to produce biodegradable and compostable food packaging.
Neuroprotective activity: the ability to exert protective activity against cytotoxic agents.
NIR: (near-infrared) simple instrument useful for direct measurements without the necessity to prepare samples.
nm: nanometer. 1 nanometer is a billionth of a meter.
NMR: powerful technique that can be used for qualitative and quantitative measurements of chemicals.
Non-woven tissue: tissue made by fibers organized randomly.
Nutricosmetics: cosmetic products made by using raw material utilized as food or food components.
OECD: Organization for Economic Co-operation and Development. An international organization helping governments tackle the economic and social governance challenges of a globalized world.
Oligomer: molecule consisting of few monomer units in contrast to a polymer that consists of an unlimited number of monomers.
Oligosaccharides: important class of polymeric carbohydrates found in all living entities.
Organosolv process: the treatment of biomass to extract lignin by using solvents/water at 180–200 C and a working pressure between 2 and 5 bars.
Osteocalcin: non-collagenous, small protein, secreted by fibroblast, found in bone and dentin.

Osteopontin: multifunctional protein molecule, part of the mineralized Matrix. It is considered a bridge between cells and minerals and as a key cytokine regulating tissue repair.

Particle: minute piece of matter with defined physical boundaries.

Packaging: wrapping material around a consumer item that serves to contain, identify, describe, display, promote, and otherwise make the product marketable and keep it clean.

Patent novelty: based on the novelty requirements necessary to protect an original invention by a Patent.

Patent protection: protection by the law of the intellectual property connected with any original invention.

PBAT: poly butyrate adipate terephthalate. A biodegradable copolymer used for plastic bags and wraps, as an alternative to polyethylene.

PBLG: poly(benzyl-L-glutamate), a biodegradable and biocompatible polymer utilized in fields of drug delivery systems, tissue engineering, and biomedical materials for safety and effectiveness.

PBS: polybutylene succinate. A thermoplastic biodegradable polymer resin of the polyester family, produced from a petrochemical-base such as nice vid and 1,4 butanediol.

PCL: Polycaprolactone. A natural nanocomposite used to replace traditional polymers for packaging applications.

PE: polyethylene.

Pectin: a structural colloidal heteropolysaccharide contained in the primary cell walls of terrestrian plants.

Peptide: short polymer composed of two or more amino acids.

PGA: poly(Glycolic acid). Common polymer used as a scaffold biomaterial.

PEG: polyethylen-glycol. It may be considered a water-soluble carrier acting as humectant, lubricant, binder, and coupling agent, depending upon its concentration and grade of polymerizations.

PEOX: polyethylene oxide. A water-soluble sinthetic polyester available in a range of molecular weights, used as carrier or component in biomedical fields.

Permeability: property or state of being permeable, i.e., how easily liquid and gas pass through something.

PET: polyethylenteraphtate. A colorless, thermoplastic, crystal-like resin used to make Food packagings.

PHA: polyhydroxy alkanoates. Biodegradable, thermoplastic polymers used for packaging productions.

PHB: semicrystalline biodegradable polymer with properties similar to polypropylene.

Pholyphenol: class of organic chemicals characterized by the presence in their molecules of large multiples of phenol structural units.

Photolithography: process that involves manipulating material through the use of light and processing geometric shapes on a mask on the surface of a silicon wafer.

Protonic crystals (PhCs): periodic optical structures that confine or control the emission and propagation of light.

PLA: poly (lactic acid) is a biodegradable and biocompatible thermoplastic polymer obtained from plant biomass.

PLGA: poly(DL-lactic-co-glycolic acid). A biocompatible and biodegradable polymer exhibiting interesting properties, used to fabricate devices for drug delivery and tissue engineering applications.

Polyacrilic acid: synthetic high molecular-weight polymers of acrylic acid repeated units, frequently used as emulsifying, stabilizing, and controlling releases for pharmaceutical and cosmetic active agents.

Polylactil esters: non-degradable polymers, transparent and elastic, that are resistant to breakage. They have a broad range of properties that make them suitable for supporting various cell cultures as the Extra Cellular Matrix (ECM).

Polyethylene: made from polymerization of ethylene. The most thermoplastic commodity used in consumer products. It possesses good elasticity and resistance to stress but it is not biodegradable. It is also a gelling agent for anhydrous hydrocarbon oils and fatty esters.

Poly(butylene adipate-co-terephtalate): biodegadable, thermoplastic polymer with excellent balanced properties useful for improving both UV and heat resistance.

Poly(butylene succinate): thermoplastic, polymer resin of the polyester family useful for Eco-Smart Food packaging.

Poly(ethylene terephthalate): thermoplastic polyester formed from ethylene glycol through a direct esterification process. It possesses Bret electrical resistance and low moisture adsorption.

Poly(gamma- benzyl l-glutamate): amphiphilic biodegradable cationic bioplymer.

Poly(lactic acid): see PLA.

Polyamide: high molecular weight polymer in which amide bounds occur among many molecules of amine. These bounds may be either natural or synthetic.

Polycaprolactone: natural nanocomposite used to replace traditional polymers in packaging applications.

Polycyanoacrylate: potential lysosomotropic, non-toxic carrier for producing materials to close wounds.

Polyester: generic name for a manufactured fiber in which the fiber-forming substance is any long chain synthetic polymer.

Polyethylene glicol: polyglycol composed of several molecules of condensate ethylen glicol.

Polymer: chemical compound made of small molecules arranged in a single repeating structure for a larger molecule.

Polymer composite: combination of a polymer and synthetic or natural inorganic filler. Fillers are employed to improve the desired properties or simply reduce the cost.

Polymerization process: any process in which relatively small molecules, called monomers, combine chemically to produce a very large, chainlike network.

Polyolefin: any class of thermoplastic polymers produced from a single olefin.

Polypropylene: thermoplastic polymer for producing films for packaging, textile, and plastic items for medical and laboratory use.

Polysaccharide: combination of 9 or more monosaccharides, linked together by glucosidic bonds.

Polystyrene: thermoplastic, synthetic resin of polymerization styrene of variable molecular weight.

Polyurethanes: class of thermoplastic polymeric materials with a wide range of properties that make them a good candidate as functional material for applications in biomedical science. They are produced by the condensation reaction of a polyisocianate and an hydroxyl-containing material, such as lignin.

POSS: polyhedral oligomeric sesquioxane. A novel chitosan nanocomposite.

Poyhydroxyalcanoate: biodegradable linear polyesters produced in nature by bacterial fermentation of sugar or lipids.

PP: polypropylene. Polymer of propylene.

PPG: polypropylene glycol.

Pro-anthocyanidin: class of polyphenols found in nature in a variety of plants, also known as condensed tannins, with very powerful antioxidant properties.

Product innovation: creation and subsequent introduction of a good or service that is either new or an improved version of a previous good or service.

Productive process: system of defined chemical or mechanical steps to create an object involving the use of raw materials, machinery, and power.

Proteins: essential components of organisms, participating in every process within cells together polysaccharides and nucleic acids.

Proteoglycan: any protein with one or more covalently attached glycosaminoglycan chains.

Proteomic: large-scale study of proteins to control their structures and functions.

PS: polystyrene

Pyrolysis: thermochemical decomposition of organic materials at elevated temperatures in the absence of oxygen.

PVP: poly(vinyl pyrrolidone), film-forming, hair fixative resin; a dispersant and adhesion promoter.

Quantum: smallest amount of physical quantity of energy that can exist independently.

Quantum dot: crystalline nanoparticle that exhibits size-dependent properties due to quantum confinement effects on the electronic states.

Raw material: any inorganic or organic material, natural or man-made, used to produce goods or energy.

REACH: actual EU Regulation on registration, evaluation, authorization, and restrictions of all the chemicals produced or imported.

Recyclability: ability of a material to be captured and separated from a waste stream for conversion or reuse.

Regenerative Medicine: study and realization of tools, non-woven tissues, and special biomaterials capable of helping natural healing processes to speed up the regrowth of damaged tissues.

Release: active ingredient in a single short to obtain an effect over a defined timespan. It is necessary to measure, for example, the effectiveness of a drug.

Renewable material: resources that have the potential to be replaced over time on a daily basis using agricultural and industrial by-products.

Resorption: process of losing substance or tissue to prove its destruction, disappearance, or dissolution by a biochemical activity.

Safeness: property of a product that defines it as safe.

Salicylic acid: ortho-hydroxybenzoic acid. It is a white powder of acrid taste and discolored by light, used in dermatology as a peeling agent.

Scaffold: 3D structure made of synthetic or natural biocompatible materials that, degraded over time, leaves only the integrated tissue in its place.
Scaffold protein: assembles the cell signaling-cascade, localizing and enhancing its efficacy.
Scar hypertrophic: cutaneous condition characterized by the deposit of an excessive amount of collagen, which produces a raised scar, which is different from a keloid scar.
Sebum: oily, lipid-containing substance secreted by the sebaceous glands of the skin.
Self-assembly: arrangement of molecules without outside guidance.
SEM: Scanning Electron Microscopy.
SHF process: Separate Hydrolysis Fermentation. A two-step process of the biomass. In the first step the lignocellulose is hydrolyzed to sugars using enzymes, while in the second step, the sugars are fermented to ethanol with the help of yeasts.
Skin-friendly: terminology used to define a product that is effective on skin and does not show negative side-effects.
Silver nanoparticles: spherical nanoparticles of silver between 1nm and 100nm.
Slurry: thin mixture of insoluble substances with a liquid, as water or oil.
Smart textile: textile possessing innovative and interesting properties.
Social capital: comprises all the different cooperative systems and organizational frameworks that people use to live and work together.
Solvolysis: chemical reaction in which the solvent, such as water or alcohol, is one of the reagents and is present in great excess so that it is required for the reaction.
Somatic cell: any cell that makes up an organism, except for a reproductive cell.
Somatic-cell therapy: use of manipulated cells or tissues to obtain biological effects.
SSCF process: simultaneous saccharification and co-fermentation. It is a process carried out by genetically engineered microbes used for Xylose-rich lignocellulosic materials.
SSF process: process in which hydrolization and fermentation of sugars take place in the same vessel, unlike SHF.
Starch: polysaccharide from grain, relatively low-cost and highly degradable, used to produce hard and flexible plastic biomaterials.
Stem cells: undifferentiated, self-renewing cells that carry on dividing to generate a mixture of different cell progeny. They are required wherever there is a recurring need to replace differentiated cells that cannot divide themselves.
Strenght: tensile strength of a polymer that quantifies how much an elongating stress will endure before its failure.
Surface: outmost or uppermost layer or area of something.
Surface tension: surface film caused by the attraction of the liquid molecules to each other.
Sustainable economy: use of raw materials obtained from agricultural and industrial by-products by processes, possibly *green,* with low consumption of water and energy.
Swelling: a time-dependent volume increase of a polymer, caused by stress changes, an increase of water content, or a combination of both.
Synthetic polymer: polymer hand-made by a process of synthesis.
T-cells: or T-lymphocytes are a type of linphocyte that plays a central role in cell-mediated immunity.
Tannic acid: natural substance widely found in natgalls and other plant parts.
Tannins: class of lower molecular weight, water-soluble polyphenols present in many plant foods, commonly used in wine circles, as well as in medicinal applications.

Thermochemical conversion: application of heat and chemical processes in the production of energy products from biomass.

Thermoforming: polymer material fed in sheets with thicknesses in the range 50–300 microns for packaging production. The method consists of heating the material in its glass transition temperature, below its melting point, to obtain a softened sheet that under vacuum is transformed to the desired shape.

Thermosetting: known also as thermoset. It describes a prepolymer material that cures irreversibly by means of heat or a chemical reaction.

Thiocidolysis: last step of the lignin degradation process, catalyzed by peroxidase enzymes.

Tissue engineering: multidisciplinary field focused on the development and application of knowledge in chemistry, physics, engineering, and life and clinical sciences to the solution of critical medical problems.

Tissue-engineered: cells or tissues that have been modified to be used for repairing, regenerating, or replacing human tissue.

TNF-alpha: tumor necrosis factor. It is a non-glycosilated protein acting as a cell signal.

Tobramycin: aminoglycoside antibiotic, derived from *Streptomices tenebrarius,* used for different bacterial infections.

Top-down: process based on the miniaturization of a complex system through the sizing-down of its components.

Toxicology: studying or detecting the detrimental effects of a substance.

Tumor necrosis factor-beta (TGF-beta): key mediator for fibroblast migration and proliferation, granulation tissue formation, increased collagen synthesis, and neovascularization.

Ultrasound: energy input used during the thermochemical conversion of lignocellulosic biomass into liquefied depolymerized products.

um: millimicron. One-thousandth micron; 1 nanometer.

UNEP: United Nations Environment Program. The voice for the environment in the United Nations systems that delivers analysis and interpretation of data and information about biodiversity.

UV: ultraviolet rays. Electromagnetic radiation with a light wave length from 400 nm to 100 nm.

Viscoelasticity: property of materials that exhibit both viscous and elastic characteristics when undergoing deformation.

Viscosity: quality of being viscous. The coefficient of viscosity is the ratio of the tangential frictional force per unit area to the velocity gradient perpendicular to the direction of flow of a liquid.

Wavelength: distance between the high points of a wave.

Washing: act of person to apply water or some other liquid for the purpose of cleaning.

Waste: any material unused and rejected as worthless , unwanted, or unusable.

Wet spinning: manufacturing process for creating polymeric fibers through continuous electrospinning.

WHEYLAYER: EU project based on the use of natural protein to make biodegradable containers.

WHO: World Health Organization. A workforce of some 8,500 people representing more than 150 nationalities with the mission to provide global leadership in public health.

Wound contraction: although an integral stage in the wound healing process, prolonged contraction results in unsightly and thickened scars.

Wound dressing: any material or methodology used to regenerate tissues and/or organs.

Wound healing: complex process whereby the skin or other body tissues repair their injuries through highly programmed phases.

Wound repair: process involving the combining activity of inflammatory, vascular, connective tissue, and epithelial cells to repair the skin injuries.

X-Ray diffraction: rapid, non-destructive analytic technique primary used for phase identification of a crystalline material.

XRD: x-Ray diffraction analytical method.

Young modulus: or elastic modulus. It is a mechanical property of linear elastic solid materials, measuring the force needed to stretch, for example, a polymer.

MDPI
St. Alban-Anlage 66
4052 Basel
Switzerland
Tel. +41 61 683 77 34
Fax +41 61 302 89 18
www.mdpi.com

MDPI Books Editorial Office
E-mail: books@mdpi.com
www.mdpi.com/books

www.ingramcontent.com/pod-product-compliance
Lightning Source LLC
Chambersburg PA
CBHW040852230926
51850CB00012BA/2930
* 9 7 8 3 0 3 8 4 2 6 9 2 9 *